Family Maps
of
Lauderdale County, Mississippi
Deluxe Edition

With Homesteads, Roads, Waterways, Towns, Cemeteries, Railroads, and More

Family Maps
of
Lauderdale County, Mississippi
Deluxe Edition

With Homesteads, Roads, Waterways, Towns, Cemeteries, Railroads, and More

by Gregory A. Boyd, J.D.

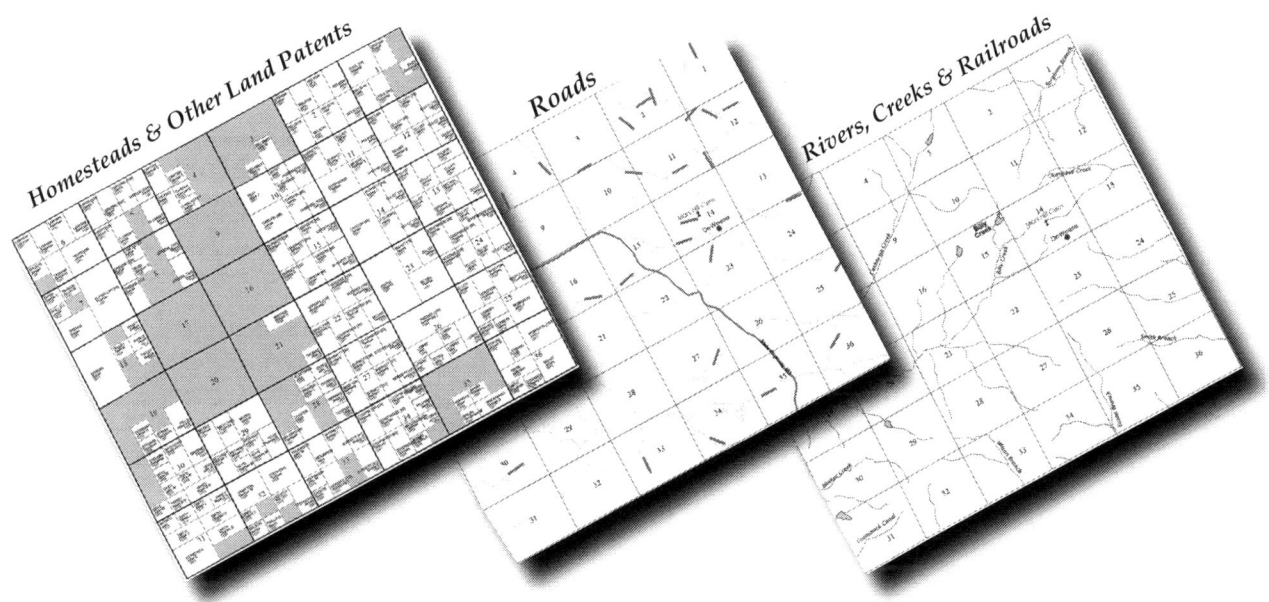

Featuring 3 *Maps Per Township...*

Arphax Publishing Co.
www.arphax.com

Family Maps of Lauderdale County, Mississippi, Deluxe Edition: With Homesteads, Roads, Waterways, Towns, Cemeteries, Railroads, and
by Gregory A. Boyd, J.D.

ISBN 1-4203-1204-9

Copyright © 2006, 2010 by Boyd IT, Inc., All rights reserved.
Printed in the United States of America

Published by Arphax Publishing Co., 2210 Research Park Blvd., Norman, Oklahoma, USA 73069
www.arphax.com

First Edition

ATTENTION HISTORICAL & GENEALOGICAL SOCIETIES, UNIVERSITIES, COLLEGES, CORPORATIONS, FAMILY REUNION COORDINATORS, AND PROFESSIONAL ORGANIZATIONS: Quantity discounts are available on bulk purchases of this book. For information, please contact Arphax Publishing Co., at the address listed above, or at (405) 366-6181, or visit our web-site at www.arphax.com and contact us through the "Bulk Sales" link.

—LEGAL—

The contents of this book rely on data published by the United States Government and its various agencies and departments, including but not limited to the General Land Office-Bureau of Land Management, the Department of the Interior, and the U.S. Census Bureau. The author has relied on said government agencies or re-sellers of its data, but makes no guarantee of the data's accuracy or of its representation herein, neither in its text nor maps. Said maps have been proportioned and scaled in a manner reflecting the author's primary goal—to make patentee names readable. This book will assist in the discovery of possible relationships between people, places, locales, rivers, streams, cemeteries, etc., but "proving" those relationships or exact geographic locations of any of the elements contained in the maps will require the use of other source material, which could include, but not be limited to: land patents, surveys, the patentees' applications, professionally drawn road-maps, etc.

Neither the author nor publisher makes any claim that the contents herein represent a complete or accurate record of the data it presents and disclaims any liability for reader's use of the book's contents. Many circumstances exist where human, computer, or data delivery errors could cause records to have been missed or to be inaccurately represented herein. Neither the author nor publisher shall assume any liability whatsoever for errors, inaccuracies, omissions or other inconsistencies herein.

No part of this book may be reproduced, stored, or transmitted by any means (electronic, mechanical, photocopying, recording, or otherwise, as applicable) without the prior written permission of the publisher.

This book is dedicated to my wonderful family:

Vicki, Jordan, & Amy Boyd

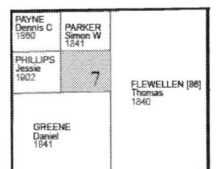

Contents

Preface 1
How to Use this Book - A Graphical Summary 2
How to Use This Book 3

- Part I -

The Big Picture

Map **A** - Where Lauderdale County, Mississippi Lies Within the State 11
Map **B** - Lauderdale County, Mississippi and Surrounding Counties 12
Map **C** - Congressional Townships of Lauderdale County, Mississippi 13
Map **D** - Cities & Towns of Lauderdale County, Mississippi 14
Map **E** - Cemeteries of Lauderdale County, Mississippi 16
Surnames in Lauderdale County, Mississippi Patents 18
Surname/Township Index 23

- Part II -

Township Map Groups

(each Map Group contains a Patent Index, Patent Map, Road Map, & Historical Map)

Map Group **1** - Township 8-N Range 14-E 52
Map Group **2** - Township 8-N Range 15-E 62
Map Group **3** - Township 8-N Range 16-E 72
Map Group **4** - Township 8-N Range 17-E 84
Map Group **5** - Township 8-N Range 18-E 98
Map Group **6** - Township 8-N Range 19-E 112
Map Group **7** - Township 7-N Range 14-E 116
Map Group **8** - Township 7-N Range 15-E 124
Map Group **9** - Township 7-N Range 16-E 134
Map Group **10** - Township 7-N Range 17-E 146
Map Group **11** - Township 7-N Range 18-E 156
Map Group **12** - Township 6-N Range 14-E 166
Map Group **13** - Township 6-N Range 15-E 174
Map Group **14** - Township 6-N Range 16-E 184
Map Group **15** - Township 6-N Range 17-E 196
Map Group **16** - Township 6-N Range 18-E 206
Map Group **17** - Township 5-N Range 14-E 216
Map Group **18** - Township 5-N Range 15-E 226
Map Group **19** - Township 5-N Range 16-E 236

Map Group **20** - Township 5-N Range 17-E..246
Map Group **21** - Township 5-N Range 18-E..254

Appendices

Appendix A - Congressional Authority for Land Patents 264
Appendix B - Section Parts (Aliquot Parts)...265
Appendix C - Multi-Patentee Groups in Lauderdale County 269

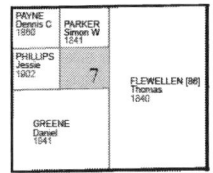

Preface

The quest for the discovery of my ancestors' origins, migrations, beliefs, and life-ways has brought me rewards that I could never have imagined. The *Family Maps* series of books is my first effort to share with historical and genealogical researchers, some of the tools that I have developed to achieve my research goals. I firmly believe that this effort will allow many people to reap the same sorts of treasures that I have.

Our Federal government's General Land Office of the Bureau of Land Management (the "GLO") has given genealogists and historians an incredible gift by virtue of its enormous database housed on its web-site at glorecords.blm.gov. Here, you can search for and find millions of parcels of land purchased by our ancestors in about thirty states.

This GLO web-site is one of the best FREE on-line tools available to family researchers. But, it is not for the faint of heart, nor is it for those unwilling or unable to to sift through and analyze the thousands of records that exist for most counties.

My immediate goal with this series is to spare you the hundreds of hours of work that it would take you to map the Land Patents for this county. Every Lauderdale County homestead or land patent that I have gleaned from public GLO databases is mapped here. Consequently, I can usually show you in an instant, where your ancestor's land is located, as well as the names of nearby land-owners.

Originally, that was my primary goal. But after speaking to other genealogists, it became clear that there was much more that they wanted. Taking their advice set me back almost a full year, but I think you will agree it was worth the wait. Because now, you can learn so much more.

Now, this book answers these sorts of questions:

- Are there any variant spellings for surnames that I have missed in searching GLO records?
- Where is my family's traditional home-place?
- What cemeteries are near Grandma's house?
- My Granddad used to swim in such-and-such-Creek—where is that?
- How close is this little community to that one?
- Are there any other people with the same surname who bought land in the county?
- How about cousins and in-laws—did they buy land in the area?

And these are just for starters!

The rules for using the *Family Maps* books are simple, but the strategies for success are many. Some techniques are apparent on first use, but many are gained with time and experience. Please take the time to notice the roads, cemeteries, creek-names, family names, and unique first-names throughout the whole county. You cannot imagine what YOU might be the first to discover.

I hope to learn that many of you have answered age-old research questions within these pages or that you have discovered relationships previously not even considered. When these sorts of things happen to you, will you please let me hear about it? I would like nothing better. My contact information can always be found at www.arphax.com.

One more thing: please read the "How To Use This Book" chapter; it starts on the next page. This will give you the very best chance to find the treasures that lie within these pages.

My family and I wish you the very best of luck, both in life, and in your research.   Greg Boyd

Family Maps of Lauderdale County, Mississippi

How to Use This Book - A Graphical Summary

Part I
"The Big Picture"

- **Map A** ▸ Counties in the State
- **Map B** ▸ Surrounding Counties
- **Map C** ▸ Congressional Townships (Map Groups) in the County
- **Map D** ▸ Cities & Towns in the County
- **Map E** ▸ Cemeteries in the County
- **Surnames in the County** ▸ Number of Land-Parcels for Each Surname
- **Surname/Township Index** ▸ Directs you to Township Map Groups in Part II

The <u>Surname/Township Index</u> can direct you to any number of **Township Map Groups**

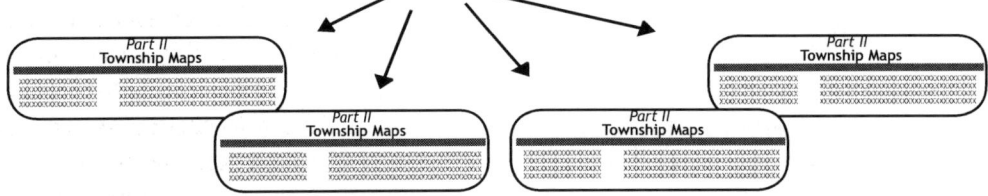

Part II
Township Map Groups
(1 for each Township in the County)

Each Township Map Group contains all four of of the following tools . . .

- **Land Patent Index** ▸ Every-name Index of Patents Mapped in this Township
- **Land Patent Map** ▸ Map of Patents as listed in above Index
- **Road Map** ▸ Map of Roads, City-centers, and Cemeteries in the Township
- **Historical Map** ▸ Map of Railroads, Lakes, Rivers, Creeks, City-Centers, and Cemeteries

Appendices

- **Appendix A** ▸ Congressional Authority enabling Patents within our Maps
- **Appendix B** ▸ Section-Parts / Aliquot Parts (a comprehensive list)
- **Appendix C** ▸ Multi-patentee Groups (Individuals within Buying Groups)

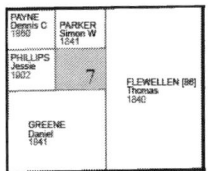

How to Use This Book

The two "Parts" of this *Family Maps* volume seek to answer two different types of questions. Part I deals with broad questions like: what counties surround Lauderdale County, are there any ASHCRAFTs in Lauderdale County, and if so, in which Townships or Maps can I find them? Ultimately, though, Part I should point you to a particular Township Map Group in Part II.

Part II concerns itself with details like: where exactly is this family's land, who else bought land in the area, and what roads and streams run through the land, or are located nearby. The Chart on the opposite page, and the remainder of this chapter attempt to convey to you the particulars of these two "parts", as well as how best to use them to achieve your research goals.

Part I
"The Big Picture"

Within Part I, you will find five "Big Picture" maps and two county-wide surname tools.

These include:

- Map A - Where Lauderdale County lies within the state
- Map B - Counties that surround Lauderdale County
- Map C - Congressional Townships of Lauderdale County (+ Map Group Numbers)
- Map D - Cities & Towns of Lauderdale County (with Index)
- Map E - Cemeteries of Lauderdale County (with Index)
- Surnames in Lauderdale County Patents (with Parcel-counts for each surname)
- Surname/Township Index (with Parcel-counts for each surname by Township)

The five "Big-Picture" Maps are fairly self-explanatory, yet should not be overlooked. This is particularly true of Maps "C", "D", and "E", all of which show Lauderdale County and its Congressional Townships (and their assigned Map Group Numbers).

Let me briefly explain this concept of Map Group Numbers. These are a device completely of our own invention. They were created to help you quickly locate maps without having to remember the full legal name of the various Congressional Townships. It is simply easier to remember "Map Group 1" than a legal name like: "Township 9-North Range 6-West, 5th Principal Meridian." But the fact is that the TRUE legal name for these Townships IS terribly important. These are the designations that others will be familiar with and you will need to accurately record them in your notes. This is why both Map Group numbers AND legal descriptions of Townships are almost always displayed together.

Map "C" will be your first intoduction to "Map Group Numbers", and that is all it contains: legal Township descriptions and their assigned Map Group Numbers. Once you get further into your research, and more immersed in the details, you will likely want to refer back to Map "C" from time to time, in order to regain your bearings on just where in the county you are researching.

Remember, township boundaries are a completely artificial device, created to standardize land descriptions. But do not let them become a boundary in your mind when choosing which townships to research. Your relative's in-laws, children, cousins, siblings, and mamas and papas, might just as easily have lived in the township next to the one your grandfather lived in—rather than in the one where he actually lived. So Map "C" can be your guide to which other Townships/Map Groups you likewise ought to analyze.

3

Of course, the same holds true for County lines; this is the purpose behind Map "B". It shows you surrounding counties that you may want to consider for further reserarch.

Map "D", the Cities and Towns map, is the first map with an index. Map "E" is the second (Cemeteries). Both, Maps "D" and "E" give you broad views of City (or Cemetery) locations in the County. But they go much further by pointing you toward pertinent Township Map Groups so you can locate the patents, roads, and waterways located near a particular city or cemetery.

Once you are familiar with these *Family Maps* volumes and the county you are researching, the "Surnames In Lauderdale County" chapter (or its sister chapter in other volumes) is where you'll likely start your future research sessions. Here, you can quickly scan its few pages and see if anyone in the county possesses the surnames you are researching. The "Surnames in Lauderdale County" list shows only two things: surnames and the number of parcels of land we have located for that surname in Lauderdale County. But whether or not you immediately locate the surnames you are researching, please do not go any further without taking a few moments to scan ALL the surnames in these very few pages.

You cannot imagine how many lost ancestors are waiting to be found by someone willing to take just a little longer to scan the "Surnames In Lauderdale County" list. Misspellings and typographical errors abound in most any index of this sort. Don't miss out on finding your Kinard that was written Rynard or Cox that was written Lox. If it looks funny or wrong, it very often is. And one of those little errors may well be your relative.

Now, armed with a surname and the knowledge that it has one or more entries in this book, you are ready for the "Surname/Township Index." Unlike the "Surnames In Lauderdale County", which has only one line per Surname, the "Surname/Township Index" contains one line-item for each Township Map Group in which each surname is found. In other words, each line represents a different Township Map Group that you will need to review.

Specifically, each line of the Surname/Township Index contains the following four columns of information:

1. Surname
2. Township Map Group Number (these Map Groups are found in Part II)
3. Parcels of Land (number of them with the given Surname within the Township)
4. Meridian/Township/Range (the legal description for this Township Map Group)

The key column here is that of the Township Map Group Number. While you should definitely record the Meridian, Township, and Range, you can do that later. Right now, you need to dig a little deeper. That Map Group Number tells you where in Part II that you need to start digging.

But before you leave the "Surname/Township Index", do the same thing that you did with the "Surnames in Lauderdale County" list: take a moment to scan the pages of the Index and see if there are similarly spelled or misspelled surnames that deserve your attention. Here again, is an easy opportunity to discover grossly misspelled family names with very little effort. Now you are ready to turn to . . .

Part II
"Township Map Groups"

You will normally arrive here in Part II after being directed to do so by one or more "Map Group Numbers" in the Surname/Township Index of Part I.

Each Map Group represents a set of four tools dedicated to a single Congressional Township that is either wholly or partially within the county. If you are trying to learn all that you can about a particular family or their land, then these tools should usually be viewed in the order they are presented.

These four tools include:

1. a Land Patent Index
2. a Land Patent Map
3. a Road Map, and
4. an Historical Map

How to Use This Book

As I mentioned earlier, each grouping of this sort is assigned a Map Group Number. So, let's now move on to a discussion of the four tools that make up one of these Township Map Groups.

Land Patent Index

Each Township Map Group's Index begins with a title, something along these lines:

MAP GROUP 1: Index to Land Patents
Township 16-North Range 5-West (2nd PM)

The Index contains seven (7) columns. They are:

1. ID (a unique ID number for this Individual and a corresponding Parcel of land in this Township)
2. Individual in Patent (name)
3. Sec. (Section), and
4. Sec. Part (Section Part, or Aliquot Part)
5. Date Issued (Patent)
6. Other Counties (often means multiple counties were mentioned in GLO records, or the section lies within multiple counties).
7. For More Info . . . (points to other places within this index or elsewhere in the book where you can find more information)

While most of the seven columns are self-explanatory, I will take a few moments to explain the "Sec. Part." and "For More Info" columns.

The "Sec. Part" column refers to what surveryors and other land professionals refer to as an Aliquot Part. The origins and use of such a term mean little to a non-surveyor, and I have chosen to simply call these sub-sections of land what they are: a "Section Part". No matter what we call them, what we are referring to are things like a quarter-section or half-section or quarter-quarter-section. See Appendix "B" for most of the "Section Parts" you will come across (and many you will not) and what size land-parcel they represent.

The "For More Info" column of the Index may seem like a small appendage to each line, but please recognize quickly that this is not so. And to understand the various items you might find here, you need to become familiar with the Legend that appears at the top of each Land Patent Index.

Here is a sample of the Legend . . .

LEGEND

"For More Info . . . " column

A = Authority (Legislative Act, See Appendix "A")
B = Block or Lot (location in Section unknown)
C = Cancelled Patent
F = Fractional Section
G = Group (Multi-Patentee Patent, see Appendix "C")
V = Overlaps another Parcel
R = Re-Issued (Parcel patented more than once)

Most parcels of land will have only one or two of these items in their "For More Info" columns, but when that is not the case, there is often some valuable information to be gained from further investigation. Below, I will explain what each of these items means to you you as a researcher.

A = Authority
(Legislative Act, See Appendix "A")

All Federal Land Patents were issued because some branch of our government (usually the U.S. Congress) passed a law making such a transfer of title possible. And therefore every patent within these pages will have an "A" item next to it in the index. The number after the "A" indicates which item in Appendix "A" holds the citation to the particular law which authorized the transfer of land to the public. As it stands, most of the Public Land data compiled and released by our government, and which serves as the basis for the patents mapped here, concerns itself with "Cash Sale" homesteads. So in some Counties, the law which authorized cash sales will be the primary, if not the only, entry in the Appendix.

B = Block or Lot (location in Section unknown)
A "B" designation in the Index is a tip-off that the EXACT location of the patent within the map is not apparent from the legal description. This Patent will nonetheless be noted within the proper

Section along with any other Lots purchased in the Section. Given the scope of this project (many states and many Counties are being mapped), trying to locate all relevant plats for Lots (if they even exist) and accurately mapping them would have taken one person several lifetimes. But since our primary goal from the onset has been to establish relationships between neighbors and families, very little is lost to this goal since we can still observe who all lived in which Section.

C = Cancelled Patent

A Cancelled Patent is just that: cancelled. Whether the original Patentee forfeited his or her patent due to fraud, a technicality, non-payment, or whatever, the fact remains that it is significant to know who received patents for what parcels and when. A cancellation may be evidence that the Patentee never physically re-located to the land, but does not in itself prove that point. Further evidence would be required to prove that. *See also,* Re-issued Patents, *below.*

F = Fractional Section

A Fractional Section is one that contains less than 640 acres, almost always because of a body of water. The exact size and shape of land-parcels contained in such sections may not be ascertainable, but we map them nonetheless. Just keep in mind that we are not mapping an actual parcel to scale in such instances. Another point to consider is that we have located some fractional sections that are not so designated by the Bureau of Land Management in their data. This means that not all fractional sections have been so identified in our indexes.

G = Group
(Multi-Patentee Patent, see Appendix "C")

A "G" designation means that the Patent was issued to a GROUP of people (Multi-patentees). The "G" will always be followed by a number. Some such groups were quite large and it was impractical if not impossible to display each individual in our maps without unduly affecting readability. EACH person in the group is named in the Index, but they won't all be found on the Map. You will find the name of the first person in such a Group on the map with the Group number next to it, enclosed in [square brackets].

To find all the members of the Group you can either scan the Index for all people with the same Group Number or you can simply refer to Appendix "C" where all members of the Group are listed next to their number.

O = Overlaps another Parcel

An Overlap is one where PART of a parcel of land gets issued on more than one patent. For genealogical purposes, both transfers of title are important and both Patentees are mapped. If the ENTIRE parcel of land is re-issued, that is what we call it, a Re-Issued Patent (*see below*). The number after the "O" indicates the ID for the overlapping Patent(s) contained within the same Index. Like Re-Issued and Cancelled Patents, Overlaps may cause a map-reader to be confused at first, but for genealogical purposes, all of these parties' relationships to the underlying land is important, and therefore, we map them.

R = Re-Issued (Parcel patented more than once)

The label, "Re-issued Patent" describes Patents which were issued more than once for land with the EXACT SAME LEGAL DESCRIPTION. Whether the original patent was cancelled or not, there were a good many parcels which were patented more than once. The number after the "R" indicates the ID for the other Patent contained within the same Index that was for the same land. A quick glance at the map itself within the relevant Section will be the quickest way to find the other Patentee to whom the Parcel was transferred. They should both be mapped in the same general area.

I have gone to some length describing all sorts of anomalies either in the underlying data or in their representation on the maps and indexes in this book. Most of this will bore the most ardent reseracher, but I do this with all due respect to those researchers who will inevitably (and rightfully) ask: *"Why isn't so-and-so's name on the exact spot that the index says it should be?"*

In most cases it will be due to the existence of a Multi-Patentee Patent, a Re-issued Patent, a Cancelled Patent, or Overlapping Parcels named in separate Patents. I don't pretend that this discussion will answer every question along these lines, but I hope it will at least convince you of the complexity of the subject.

Not to despair, this book's companion web-site will offer a way to further explain "odd-ball" or errant data. Each book (County) will have its own web-page or pages to discuss such situations. You can go to www.arphax.com to find the relevant web-page for Lauderdale County.

Land Patent Map

On the first two-page spread following each Township's Index to Land Patents, you'll find the corresponding Land Patent Map. And here lies the real heart of our work. For the first time anywhere, researchers will be able to observe and analyze, on a grand scale, most of the original land-owners for an area AND see them mapped in proximity to each one another.

We encourage you to make vigorous use of the accompanying Index described above, but then later, to abandon it, and just stare at these maps for a while. This is a great way to catch misspellings or to find collateral kin you'd not known were in the area.

Each Land Patent Map represents one Congressional Township containing approximately 36-square miles. Each of these square miles is labeled by an accompanying Section Number (1 through 36, in most cases). Keep in mind, that this book concerns itself solely with Lauderdale County's patents. Townships which creep into one or more other counties will not be shown in their entirety in any one book. You will need to consult other books, as they become available, in order to view other countys' patents, cities, cemeteries, etc.

But getting back to Lauderdale County: each Land Patent Map contains a Statistical Chart that looks like the following:

Township Statistics

Parcels Mapped	:	173
Number of Patents	:	163
Number of Individuals	:	152
Patentees Identified	:	151
Number of Surnames	:	137
Multi-Patentee Parcels	:	4
Oldest Patent Date	:	11/27/1820
Most Recent Patent	:	9/28/1917
Block/Lot Parcels	:	0
Parcels Re-Issued	:	3
Parcels that Overlap	:	8
Cities and Towns	:	6
Cemeteries	:	6

This information may be of more use to a social statistician or historian than a genealogist, but I think all three will find it interesting.

Most of the statistics are self-explanatory, and what is not, was described in the above discussion of the Index's Legend, but I do want to mention a few of them that may affect your understanding of the Land Patent Maps.

First of all, Patents often contain more than one Parcel of land, so it is common for there to be more Parcels than Patents. Also, the Number of Individuals will more often than not, not match the number of Patentees. A Patentee is literally the person or PERSONS named in a patent. So, a Patent may have a multi-person Patentee or a single-person patentee. Nonetheless, we account for all these individuals in our indexes.

On the lower-righthand side of the Patent Map is a Legend which describes various features in the map, including Section Boundaries, Patent (land) Boundaries, Lots (numbered), and Multi-Patentee Group Numbers. You'll also find a "Helpful Hints" Box that will assist you.

One important note: though the vast majority of Patents mapped in this series will prove to be reasonably accurate representations of their actual locations, we cannot claim this for patents lying along state and county lines, or waterways, or that have been platted (lots).

Shifting boundaries and sparse legal descriptions in the GLO data make this a reality that we have nonetheless tried to overcome by estimating these patents' locations the best that we can.

Road Map

On the two-page spread following each Patent Map you will find a Road Map covering the exact same area (the same Congressional Township).

For me, fully exploring the past means that every once in a while I must leave the library and travel to the actual locations where my ancestors once walked and worked the land. Our Township Road Maps are a great place to begin such a quest.

Keep in mind that the scaling and proportion of these maps was chosen in order to squeeze hundreds of people-names, road-names, and place-names into tinier spaces than you would traditionally see. These are not professional road-maps, and like any secondary genealogical source, should be looked upon as an entry-way to original sources—in this case, original patents and applications, professionally produced maps and surveys, etc.

Both our Road Maps and Historical Maps contain cemeteries and city-centers, along with a listing of these on the left-hand side of the map. I should note that I am showing you city center-points, rather than city-limit boundaries, because in many instances, this will represent a place where settlement began. This may be a good time to mention that many cemeteries are located on private property, Always check with a local historical or genealogical society to see if a particular cemetery is publicly accessible (if it is not obviously so). As a final point, look for your surnames among the road-names. You will often be surprised by what you find.

Historical Map

The third and final map in each Map Group is our attempt to display what each Township might have looked like before the advent of modern roads. In frontier times, people were usually more determined to settle near rivers and creeks than they were near roads, which were often few and far between. As was the case with the Road Map, we've included the same cemeteries and city-centers. We've also included railroads, many of which came along before most roads.

While some may claim "Historical Map" to be a bit of a misnomer for this tool, we settled for this label simply because it was almost as accurate as saying "Railroads, Lakes, Rivers, Cities, and Cemeteries," and it is much easier to remember.

In Closing . . .

By way of example, here is *A Really Good Way to Use a Township Map Group*. First, find the person you are researching in the Township's Index to Land Patents, which will direct you to the proper Section and parcel on the Patent Map. But before leaving the Index, scan all the patents within it, looking for other names of interest. Now, turn to the Patent Map and locate your parcels of land. Pay special attention to the names of patent-holders who own land surrounding your person of interest. Next, turn the page and look at the same Section(s) on the Road Map. Note which roads are closest to your parcels and also the names of nearby towns and cemeteries. Using other resources, you may be able to learn of kin who have been buried here, plus, you may choose to visit these cemeteries the next time you are in the area.

Finally, turn to the Historical Map. Look once more at the same Sections where you found your research subject's land. Note the nearby streams, creeks, and other geographical features. You may be surprised to find family names were used to name them, or you may see a name you haven't heard mentioned in years and years—and a new research possibility is born.

Many more techniques for using these *Family Maps* volumes will no doubt be discovered. If from time to time, you will navigate to Lauderdale County's web-page at www.arphax.com (use the "Research" link), you can learn new tricks as they become known (or you can share ones you have employed). But for now, you are ready to get started. So, go, and good luck.

Part I

The Big Picture

The Big Picture

Map A - Where Lauderdale County, Mississippi Lies Within the State

Copyright © 2006 Boyd IT, Inc. All Rights Reserved

Family Maps of Lauderdale County, Mississippi

Map B - Lauderdale County, Mississippi and Surrounding Counties

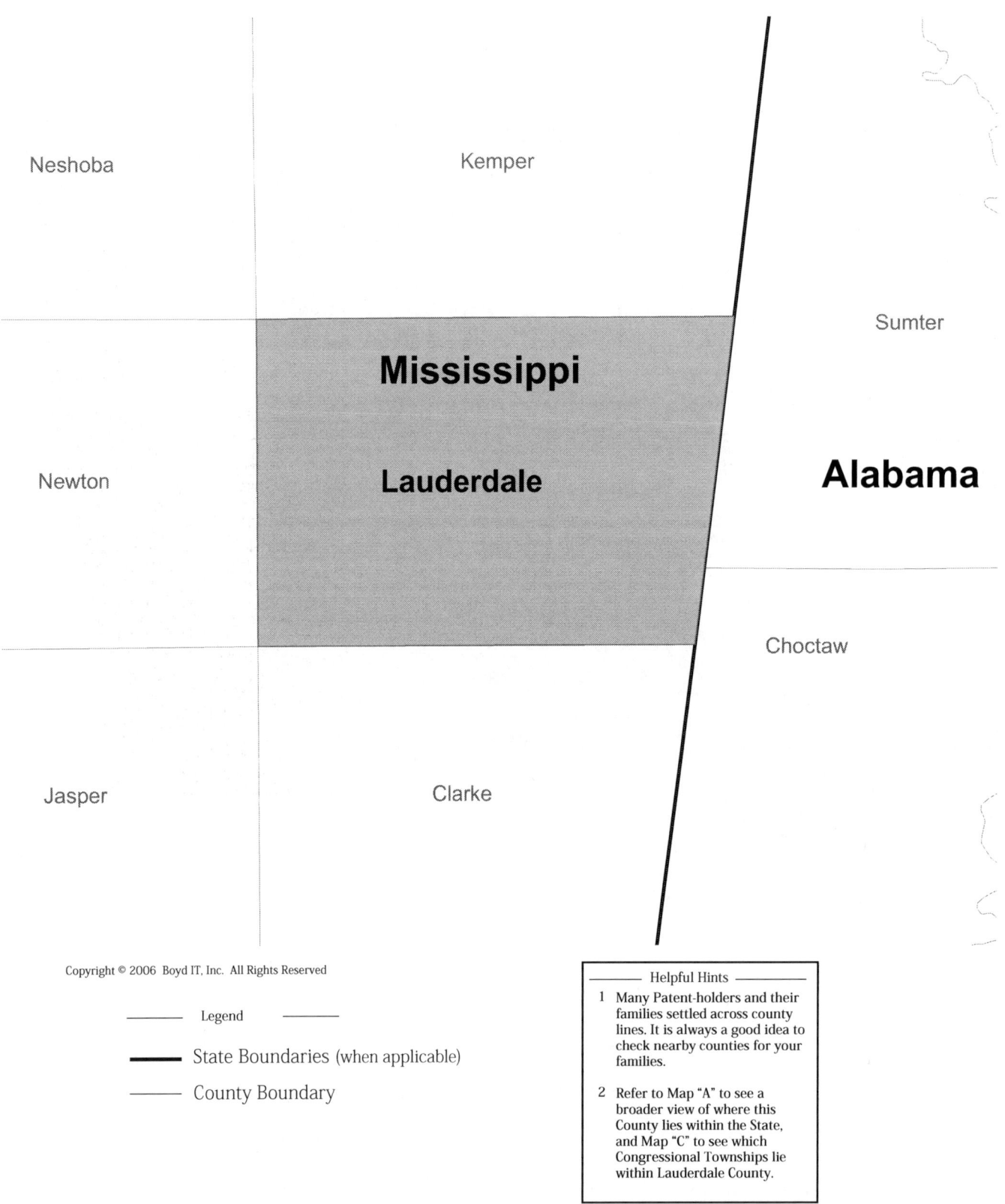

The Big Picture

Map C - Congressional Townships of Lauderdale County, Mississippi

Map Group 1 Township 8-N Range 14-E	Map Group 2 Township 8-N Range 15-E	Map Group 3 Township 8-N Range 16-E	Map Group 4 Township 8-N Range 17-E	Map Group 5 Township 8-N Range 18-E	Map Group 6 Township 8-N Range 19-E
Map Group 7 Township 7-N Range 14-E	Map Group 8 Township 7-N Range 15-E	Map Group 9 Township 7-N Range 16-E	Map Group 10 Township 7-N Range 17-E	Map Group 11 Township 7-N Range 18-E	
Map Group 12 Township 6-N Range 14-E	Map Group 13 Township 6-N Range 15-E	Map Group 14 Township 6-N Range 16-E	Map Group 15 Township 6-N Range 17-E	Map Group 16 Township 6-N Range 18-E	
Map Group 17 Township 5-N Range 14-E	Map Group 18 Township 5-N Range 15-E	Map Group 19 Township 5-N Range 16-E	Map Group 20 Township 5-N Range 17-E	Map Group 21 Township 5-N Range 18-E	

--- Legend ---

▒▒ Lauderdale County, Mississippi

☐ Congressional Townships

Copyright © 2006 Boyd IT, Inc. All Rights Reserved

--- Helpful Hints ---

1. Many Patent-holders and their families settled across county lines. It is always a good idea to check nearby counties for your families (See Map "B").

2. Refer to Map "A" to see a broader view of where this county lies within the State, and Map "B" for a view of the counties surrounding Lauderdale County.

Map D Index: Cities & Towns of Lauderdale County, Mississippi

The following represents the Cities and Towns of Lauderdale County, along with the corresponding Map Group in which each is found. Cities and Towns are displayed in both the Road and Historical maps in the Group.

City/Town	Map Group No.
Alamucha	16
Arunde	18
Bailey	8
Bonita	14
Center Hill	2
Cliff Williams	20
Collinsville	1
Complete	13
Daleville	3
Enzor	19
Graham	12
Hookston	8
Houston	9
Increase	20
Kewanee	11
Lauderdale	4
Lizelia	3
Lockhart	4
Lost Gap	13
Marion	9
Martin	1
Meehan	12
Meridian	14
Moseley	7
Nellieburg	8
Obadiah	2
Pine Springs	8
Pleasant Hill	19
Poplar Springs	9
Russell	10
Savannah Grove	13
Savoy	18
Schamberville	7
Shucktown	2
Smith	11
Sterling	18
Stinson	19
Suqualena	7
Toomsuba	10
Topton	10
Vimville	15
Wolf Springs	19
Zero	19

The Big Picture

Map D - Cities & Towns of Lauderdale County, Mississippi

Map Group 1 — Township 8-N Range 14-E
- Martin
- Collinsville

Map Group 2 — Township 8-N Range 15-E
- Shucktown
- Center Hill
- Obadiah

Map Group 3 — Township 8-N Range 16-E
- Daleville
- Lizelia

Map Group 4 — Township 8-N Range 17-E
- Lauderdale
- Lockhart

Map Group 5 — Township 8-N Range 18-E

Map Group 6 — Township 8-N Range 19-E

Map Group 7 — Township 7-N Range 14-E
- Schamberville
- Moseley
- Suqualena

Map Group 8 — Township 7-N Range 15-E
- Pine Springs
- Bailey
- Hookston
- Nellieburg

Map Group 9 — Township 7-N Range 16-E
- Houston
- Marion
- Poplar Springs

Map Group 10 — Township 7-N Range 17-E
- Topton
- Toomsuba
- Russell

Map Group 11 — Township 7-N Range 18-E
- Smith
- Kewanee

Map Group 12 — Township 6-N Range 14-E
- Graham
- Meehan

Map Group 13 — Township 6-N Range 15-E
- Complete
- Savannah Grove
- Lost Gap

Map Group 14 — Township 6-N Range 16-E
- Meridian
- Bonita

Map Group 15 — Township 6-N Range 17-E
- Vimville

Map Group 16 — Township 6-N Range 18-E
- Alamucha

Map Group 17 — Township 5-N Range 14-E

Map Group 18 — Township 5-N Range 15-E
- Sterling
- Arunde
- Savoy

Map Group 19 — Township 5-N Range 16-E
- Enzor
- Wolf Springs
- Zero
- Stinson
- Pleasant Hill

Map Group 20 — Township 5-N Range 17-E
- Cliff Williams
- Increase

Map Group 21 — Township 5-N Range 18-E

Legend

Lauderdale County, Mississippi

Congressional Townships

Copyright © 2006 Boyd IT, Inc. All Rights Reserved

Helpful Hints

1. Cities and towns are marked only at their center-points as published by the USGS and/or NationalAtlas.gov. This often enables us to more closely approximate where these might have existed when first settled.

2. To see more specifically where these Cities & Towns are located within the county, refer to both the Road and Historical maps in the Map-Group referred to above. See also, the Map "D" Index on the opposite page.

Map E Index: Cemeteries of Lauderdale County, Mississippi

The following represents many of the Cemeteries of Lauderdale County, along with the corresponding Township Map Group in which each is found. Cemeteries are displayed in both the Road and Historical maps in the Map Groups referred to below.

Cemetery	Map Group No.
Andrews Chapel Cem.	3
Anna York Cem.	5
Antioch Cem.	21
Arkadelphia Cem.	2
Barker Cem.	9
Barker Cem.	9
Barrett Cem.	4
Beth Israel Cem.	14
Bethel Cem.	8
Bethel Cem.	11
Burwell Cem.	13
Cedar Grove Cem.	7
Center Ridge Cem.	5
Clayton Cem.	4
Coker Chapel Cem.	15
Collinsville Cem.	1
Collinsville Church Cem.	1
Concord Cem.	17
Confederate Cem.	9
Daleville Cem.	3
Daleville Cem.	3
Fairchild Cem.	13
Fellowship Cem.	2
Forest Lawn Garden	9
Good Hope Cem.	10
Goodwater Cem.	17
Gordon Cem.	4
Gordon Cem.	13
Graham Cem.	18
Gum Log Cem.	2
Hamrick Cem.	1
Hays Cem.	20
Hearn Cem.	10
Hickory Grove Cem.	3
Kelly Cem.	11
Kinard Cem.	9
Lad Cem.	8
Lauderdale Cem.	5
Lauderdale Springs Cem.	5
Lockhart Cem.	4
Long Creek Cem.	19
Magnolia Cem.	14
McLemore Cem.	14
McMurry Cem.	16
Miller Cem.	4
Moore Cem.	3
Mount Carmel Cem.	2
Mount Carmel Cem.	15
Mount Horeb Cem.	19
New Hope Cem.	19
Oak Grove Cem.	8
Oak Grove Cem.	14
Oak Grove Cem.	18
Old Marion Cem.	9
Osborne Cem.	17
Pack Cem.	11
Parker Cem.	10
Payne Cem.	11
Pigford Cem.	15

Cemetery	Map Group No.
Pine Forest Cem.	1
Pine Grove Cem.	1
Pine Grove Cem.	19
Pine Springs Cem.	8
Pleasant Grove Cem.	15
Pleasant Ridge Cem.	1
Rawson Cem.	15
Rose Hill Cem.	13
Sageville Cem.	18
Saint Luke Cem.	14
Saint Marks Cem.	17
Saint Patricks Cem.	13
Salem Cem.	16
Samuel Dale Cem.	3
Segars Cem.	4
Semmes Cem.	13
Shamburger Cem.	16
Stenis Cem.	8
Stinson Cem.	19
Suqualena Cem.	7
Tabernacle Cem.	14
Tinnin Cem.	9
Trussel Cem.	1
Tunnel Hill Cem.	12
West Mount Moriah Cem.	13
White Cem.	10
Wilson Cem.	8

The Big Picture

Map E - Cemeteries of Lauderdale County, Mississippi

Map Group 6
Township 8-N Range 19-E

Map Group 1 — Township 8-N Range 14-E
- Pine Grove Cem.
- Trussel Cem.
- Pine Forest Cem.
- Pleasant Ridge Cem.
- Hamrick Cem.
- Collinsville Cem.
- Collinsville Church Cem.

Map Group 2 — Township 8-N Range 15-E
- Arkadelphia Cem.
- Mount Carmel Cem.
- Fellowship Cem.
- Gum Log Cem.

Map Group 3 — Township 8-N Range 16-E
- Daleville Cem.
- Samuel Dale Cem.
- Daleville Cem.
- Hickory Grove Cem.
- Moore Cem.
- Andrews Chapel Cem.

Map Group 4 — Township 8-N Range 17-E
- Segars Cem.
- Gordon Cem.
- Barrett Cem.
- Clayton Cem.
- Lockhart Cem.
- Miller Cem.

Map Group 5 — Township 8-N Range 18-E
- Center Ridge Cem.
- Anna York Cem.
- Lauderdale Cem.
- Lauderdale Springs Cem.

Map Group 7 — Township 7-N Range 14-E
- Cedar Grove Cem.
- Suqualena Cem.

Map Group 8 — Township 7-N Range 15-E
- Pine Springs Cem.
- Stenis Cem.
- Oak Grove Cem.
- Wilson Cem.
- Lad Cem.
- Bethel Cem.

Map Group 9 — Township 7-N Range 16-E
- Tinnin Cem.
- Forest Lawn Garden
- Kinard Cem.
- Barker Cem.
- Barker Cem.
- Confederate Cem.
- Old Marion Cem.

Map Group 10 — Township 7-N Range 17-E
- Parker Cem.
- White Cem.
- Hearn Cem.
- Good Hope Cem.

Map Group 11 — Township 7-N Range 18-E
- Payne Cem.
- Pack Cem.
- Bethel Cem.
- Kelly Cem.

Map Group 12 — Township 6-N Range 14-E
- Tunnel Hill Cem.

Map Group 13 — Township 6-N Range 15-E
- West Mount Moriah Cem.
- Semmes Cem.
- Gordon Cem.
- Fairchild Cem.
- Saint Patricks Cem.
- Rose Hill Cem.
- Burwell Cem.

Map Group 14 — Township 6-N Range 16-E
- Saint Luke Cem.
- Magnolia Cem.
- Beth Israel Cem.
- Oak Grove Cem.
- McLemore Cem.
- Tabernacle Cem.

Map Group 15 — Township 6-N Range 17-E
- Pigford Cem.
- Rawson Cem.
- Pleasant Grove Cem.
- Mount Carmel Cem.
- Coker Chapel Cem.

Map Group 16 — Township 6-N Range 18-E
- Salem Cem.
- Shamburger Cem.
- McMurry Cem.

Map Group 17 — Township 5-N Range 14-E
- Saint Marks Cem.
- Goodwater Cem.
- Osborne Cem.
- Concord Cem.

Map Group 18 — Township 5-N Range 15-E
- Sageville Cem.
- Graham Cem.
- Oak Grove Cem.

Map Group 19 — Township 5-N Range 16-E
- Long Creek Cem.
- Stinson Cem.
- Mount Horeb Cem.
- Pine Grove Cem.
- New Hope Cem.

Map Group 20 — Township 5-N Range 17-E
- Hays Cem.

Map Group 21 — Township 5-N Range 18-E
- Antioch Cem.

Legend

- Lauderdale County, Mississippi
- Congressional Townships

Copyright © 2006 Boyd IT, Inc. All Rights Reserved

Helpful Hints

1. Cemeteries are marked at locations as published by the USGS and/or NationalAtlas.gov.

2. To see more specifically where these Cemeteries are located, refer to the Road & Historical maps in the Map-Group referred to above. See also, the Map "E" Index on the opposite page to make sure you don't miss any of the Cemeteries located within this Congressional township.

17

Surnames in Lauderdale County, Mississippi Patents

The following list represents the surnames that we have located in Lauderdale County, Mississippi Patents and the number of parcels that we have mapped for each one. Here is a quick way to determine the existence (or not) of Patents to be found in the subsequent indexes and maps of this volume.

Surname	# of Land Parcels	Surname	# of Land Parcels	Surname	# of Land Parcels	Surname	# of Land Parcels
ABERNATHY	1	BOND	2	CARR	5	CUNNINGHAM	1
ADAIR	8	BONEY	1	CARROL	1	CURRIE	2
ADAMS	2	BOON	2	CARSON	2	DALE	9
ADKINSON	1	BORDERS	1	CARTER	7	DANIEL	13
AGLETREE	3	BOSTON	1	CARYE	2	DANIELS	3
AGNEW	3	BOSWELL	4	CASTLES	4	DARDEN	2
AIKINS	2	BOTHWELL	25	CATES	1	DAUCEY	1
ALEXANDER	33	BOUNDS	4	CATLET	2	DAUGHTRY	1
ALFORD	12	BOURDEAUX	5	CHA	1	DAVENPORT	1
ALLBROOKS	1	BOUTWELL	2	CHADICK	1	DAVID	3
ALLEN	17	BOWEN	1	CHAMBLIS	2	DAVIS	19
ALLISON	1	BOWLING	1	CHAMBLISS	1	DAWKINS	10
ALLSOBROOKS	1	BOYD	11	CHANDLER	6	DAY	4
ALVICE	2	BOYKIN	1	CHANEY	1	DEAN	2
ANDERSON	20	BOZEMAN	3	CHESTER	3	DEAR	15
ANDING	2	BOZMAN	3	CHILDE	1	DEARMAN	11
ANDREWS	5	BOZONE	1	CHILES	1	DEASE	2
ARMSTRONG	1	BRAGG	6	CLANTON	8	DEASON	1
ARRANT	2	BRANNAN	1	CLARK	3	DEBOE	1
ARRINGTON	2	BRASSIELL	1	CLAY	13	DEEN	9
ASKEW	3	BRECK	3	CLAYTON	21	DEENS	2
AVARA	2	BREWER	1	CLIET	4	DEER	4
AVERY	1	BREWSTER	2	CLINTON	7	DEERMAN	2
BABERS	1	BRIANT	3	COAKER	1	DEIR	1
BAILES	4	BRIDGES	1	COATES	12	DELK	14
BAILEY	3	BROCK	9	COATS	10	DEMENT	1
BAILIFF	10	BRONSON	1	COBB	2	DENTON	3
BAILY	2	BROOKE	1	COCHRUM	1	DEWITT	1
BAINES	2	BROWER	8	COCKE	1	DICKERSON	1
BAINS	1	BROWN	80	COKER	3	DICKSON	2
BAIRD	1	BRUISTER	5	COLE	13	DOBBS	4
BALL	9	BRUNER	1	COLEMAN	14	DODD	2
BALLARD	5	BRUNSON	4	COLVIN	1	DORMAN	1
BANES	2	BRYAN	2	CONNER	1	DOVE	6
BANYARD	2	BUCHANAN	1	COOK	1	DOWD	1
BARDEN	1	BUCKLEW	1	COOKSEY	1	DRUMOND	2
BAREFOOT	4	BULLARD	4	COOLEY	1	DUBOSE	11
BARFOOT	1	BUNTYN	1	COON	1	DUKE	1
BARNES	3	BUNYARD	8	COOPER	4	DUN	1
BARNET	1	BURKHALTER	2	COOPWOOD	25	DUNAGIN	1
BARNETT	8	BURNS	1	CORLEY	4	DUNBAR	1
BARRETT	1	BURRESS	1	COTES	1	DUNCAN	6
BARRON	1	BURRIS	1	COURTNEY	1	DUNGEN	1
BASTIN	3	BURTON	2	COVINGTON	2	DUNKLEY	2
BATCHEE	1	BURWELL	13	COX	1	DUNN	3
BATTLE	3	BUSBY	11	CRAIG	12	DUPREE	1
BAXTON	1	BUSH	4	CRAIL	1	DURR	23
BEASON	3	BUSON	3	CRAIN	1	DYASS	6
BEEMON	2	BUSTIN	4	CRANE	15	EAKENS	1
BEESON	6	BUTCHEE	7	CRAWFORD	5	EAKIN	1
BELL	2	BUTLER	6	CRAWLEY	1	EAKINS	3
BENNETT	3	BYNUM	5	CREAL	3	EARLONG	1
BERDEAUX	1	CAINS	4	CREEL	9	EASOM	1
BERRIEN	25	CALHOUN	14	CRENSHAW	3	EASTHAM	3
BERRY	2	CALLUM	2	CREWS	5	EASTIS	1
BEVILL	4	CAMMACK	1	CROCKETT	1	EATHAM	2
BICKART	1	CAMPBELL	5	CRORY	2	EAVES	2
BISHOP	6	CANADY	1	CROSBY	4	ECHOLS	2
BLACKMAN	5	CANTERBURY	3	CRUFT	3	EDWARDS	7
BLACKWELL	3	CARAWAY	2	CUBLEY	3	ELAM	3
BLANKS	14	CARMICHAEL	1	CULBRAITH	2	ELLIS	6
BOLDEN	1	CARPENTER	5	CULPEPPER	3	EMBRY	1

Surnames in Lauderdale County, Mississippi Patents

Surname	# of Land Parcels
ENGLISH	1
ERBY	1
ERNEST	1
ESTELL	4
ETHRIDGE	3
ETHRRIDGE	1
ETTURIDGE	1
EUBANKS	13
EVANS	4
EVERETT	4
EVES	1
EWING	1
EZELL	8
FAGIN	1
FAIR	1
FAIRCHILD	6
FAIRCHILDS	5
FALLIN	1
FARCHILES	1
FARISS	3
FARRELL	1
FAULKNER	1
FAY	1
FEDRICK	1
FENDALL	1
FENNER	1
FIKES	6
FINDLEY	1
FINLEY	5
FINNEN	5
FISHER	5
FLETCHER	2
FLEWELLEN	35
FLINN	3
FLOYD	1
FORD	18
FORTSON	1
FOSTER	3
FOUNTAIN	5
FOUNTAINE	1
FOWLER	7
FREELAND	1
FREEMAN	5
FUNK	1
GABEL	1
GADDIS	12
GAINES	3
GAINS	1
GARDNER	6
GARRETT	10
GARY	1
GAY	1
GAYLIE	4
GEE	2
GEORGE	1
GIBSON	6
GILBERT	4
GILHAM	3
GILLASPIE	6
GILLESPIE	7
GILLINDER	2
GILLINER	2
GILMORE	1
GLASCOCK	1
GLASS	1
GLENN	2
GODWIN	1
GOODMAN	1
GOODSON	2
GOODWIN	1

Surname	# of Land Parcels
GOODWYN	3
GORDON	21
GORMAN	1
GOUGH	1
GRAHAM	11
GRANT	59
GRANTLAND	10
GRAVES	2
GRAY	6
GRAYSON	1
GREEN	11
GREENE	5
GREER	1
GRESSETT	1
GRICE	2
GRIFFIN	44
GRIFFITH	7
GRIGSBY	1
GRISSETT	3
GUNN	2
HADEN	1
HAGOOD	1
HAILS	3
HALEY	1
HALL	16
HALTON	1
HAMBRICK	1
HAMLET	3
HAMMOND	2
HAMMONDS	2
HAMRICK	4
HAMSWORTH	1
HAND	6
HANDLEY	1
HANNA	15
HANNAH	3
HANNER	1
HARDEN	3
HARDIN	4
HARDY	36
HARLAN	6
HARLIN	3
HARPER	26
HARRELL	2
HARRINGTON	1
HARRIS	7
HARRISON	11
HART	1
HARVARD	1
HARVEY	4
HARWELL	1
HATCHER	6
HATHORN	2
HAWKINS	5
HAYES	3
HAYS	9
HAYSE	2
HEARN	2
HELVESTON	3
HENDERSON	33
HENDRICK	8
HERNDON	7
HERREN	1
HERRING	1
HERRINGTON	3
HICKS	2
HIGGINBOTHAM	3
HIGHTOWER	4
HILL	16
HINES	4

Surname	# of Land Parcels
HINTON	2
HOBBS	1
HODGES	4
HOGANS	1
HOLDEN	5
HOLDERNESS	2
HOLIMAN	1
HOLLADAY	1
HOLLAM	1
HOLLAND	7
HOLLIMON	2
HOLLINSON	1
HOLLY	2
HOLMES	1
HONAH	1
HOOKS	6
HOPKINS	4
HORSEY	1
HOUSE	28
HOUSTON	5
HOUZE	2
HOWELL	3
HOWSE	2
HUBBARD	18
HUDSON	7
HUEY	1
HUGGINS	2
HUGHES	8
HULET	6
HUMPHREY	1
HUMPHREYS	7
HUMPHRIES	2
HUNT	7
HUNTER	19
HURT	1
HUSSEY	15
HUTTAN	14
HUTTON	13
HUZZA	2
HYDE	1
INGRAM	1
IRBY	8
IVY	2
JACKSON	7
JAMES	6
JARMAN	3
JEMISON	25
JENNINGS	4
JOHNSON	16
JOHNSTON	1
JOINER	1
JOLLY	3
JONES	53
JORDAN	13
JORDON	1
JOSEPH	3
KALSAW	1
KEATH	1
KEELAND	1
KEETON	3
KEITH	4
KELLEY	1
KELLY	6
KENNEDY	2
KENNERLY	1
KERBY	1
KEYS	1
KIDD	1
KIELY	2
KILLENS	3

Surname	# of Land Parcels
KILLILEA	1
KILLIN	4
KILLINGSWORTH	10
KIMBRAUGH	1
KIMBRELL	1
KIMBROUGH	1
KINERD	1
KING	10
KIRKLAND	1
KITERAL	2
KITRELL	4
KITTRELL	6
KNIGHT	4
KNOX	17
KUNERY	1
KYLES	1
LACKEY	4
LACY	12
LAMB	3
LAMBERT	2
LANCASTER	4
LANE	3
LANEY	2
LANG	17
LARD	2
LARKIN	1
LASHLEY	1
LASLEY	3
LATTNEY	1
LEE	19
LEHMAN	1
LEMORE	1
LEWIS	53
LILLEY	2
LILLY	1
LINDLEY	1
LINTON	3
LINZEY	2
LITCHFIELD	1
LITTLE	3
LITTLETON	1
LLOYD	4
LONG	1
LOPER	4
LOTT	1
LOVE	4
LUCAS	1
LUCY	4
LUNSFORD	1
MAASS	2
MABRY	1
MAGRUE	1
MAHAN	4
MALONE	2
MANDEVILLE	1
MANGUM	1
MANN	2
MANNING	1
MARCY	1
MARSH	16
MARSHALL	3
MARTIN	20
MASON	3
MASSENGALL	1
MASTERS	1
MATHEWS	3
MAURY	2
MAXEY	1
MAXWELL	2
MAY	7

Family Maps of Lauderdale County, Mississippi

Surname	# of Land Parcels
MAYATT	1
MAYBERRY	2
MAYERHOFF	1
MCBRIDE	2
MCCALL	1
MCCANN	2
MCCAREY	1
MCCARTY	2
MCCARY	2
MCCOMB	5
MCCOWN	43
MCCRANEY	2
MCCRORY	2
MCCURDY	1
MCDANIEL	3
MCDONALD	8
MCDOUGAL	1
MCDOW	6
MCEACHIN	1
MCELROY	2
MCGAHA	1
MCGAW	2
MCGEE	2
MCGRAW	1
MCGREW	4
MCILWAIN	2
MCINNIS	11
MCKELLAR	6
MCKELVIN	1
MCKINLEY	1
MCKINNON	1
MCKLEVAIN	1
MCLAMORE	6
MCLAREN	1
MCLARRIN	1
MCLAUGHLIN	4
MCLAURIN	4
MCLEAN	1
MCLEMORE	18
MCLIN	3
MCMELLON	1
MCMILLAN	1
MCMILLON	1
MCMORY	2
MCMULLEN	2
MCMULLIN	1
MCMULLUN	1
MCNEILL	8
MCPHAUL	2
MCPHERSON	5
MCRAE	3
MCSHAN	2
MCSHANN	2
MCWHORTER	7
MCWOOTON	1
MEEK	1
MERRITT	4
MESSLEY	1
MIEATT	1
MILES	1
MILLER	17
MILLS	20
MINOR	2
MITCHAM	1
MITCHELL	4
MIXON	1
MOBLEY	1
MOFFET	1
MOFFETT	3
MOLPUS	4

Surname	# of Land Parcels
MONTAGUE	6
MONTGOMERY	1
MOODY	2
MOORE	30
MORGAN	4
MORRIS	4
MORROW	3
MOSELEY	1
MOSLEY	5
MOTEN	1
MOTH	1
MOTT	8
MUNDELL	1
MURPHY	6
MYRICK	2
NASH	8
NAYLOR	1
NEAL	1
NEIGHBOURS	2
NEW	6
NEWBERRY	2
NEWELL	1
NEWMAN	1
NEWSOM	4
NEWTON	1
NICHOL	1
NICHOLAS	8
NICHOLS	2
NICHOLSON	2
NOEL	1
NUNNARY	1
NUTT	3
OBRINK	2
ODAM	2
ODEN	1
ODOM	22
OFALLIN	4
ONEAL	6
OSBORN	3
OTT	1
OWENS	2
PACE	6
PAGE	2
PALMER	1
PARISH	1
PARKER	10
PARKS	7
PARTIN	1
PATHEA	1
PATON	4
PATTON	2
PAYNE	3
PEARL	2
PEARSON	1
PEAVY	5
PERCE	1
PETAGREW	1
PEW	1
PEYTON	1
PHILIPS	6
PHILLIPS	5
PICKARD	7
PIGFORD	5
PISTOLE	2
PITMAN	1
PITTS	1
PLUMMER	2
POGUE	2
POOL	2
POPE	2

Surname	# of Land Parcels
PORTER	1
POWELL	5
PRESTON	2
PRICE	12
PRIER	3
PRINE	1
PRINGLE	17
PRIVETT	1
PRYER	1
PUGH	3
RAGSDALE	6
RAGSDALL	3
RAINER	1
RAMSAY	3
RAMSEY	3
RANDALL	1
RASBERRY	1
RATCLIFF	14
RATSBURG	1
RAWLINGS	1
RAWSON	5
RAY	1
REA	1
REASON	2
REAVES	1
REECE	1
REED	3
REESE	3
REEVES	3
REID	2
REIVES	2
REW	1
REYNOLDS	9
RHODES	3
RICE	1
RICH	3
RICHARDSON	18
RICHE	1
RICHEY	1
RILEY	2
RIVERS	1
ROBBINS	2
ROBERSON	1
ROBERTS	3
ROBERTSON	2
ROBINSON	1
ROBISON	3
ROCHEL	2
RODGERS	5
ROGERS	10
RUNNELS	2
RUSH	1
RUSHING	30
RUSSELL	29
RUSSING	2
RUTLEDGE	2
SADLER	1
SALTER	1
SANDERFORD	2
SANDIFORD	2
SANDYFORD	3
SANFORD	1
SAPP	1
SATCHER	1
SAVELL	2
SCARBOROUGH	13
SCARBROUGH	4
SCHANROCK	1
SCOTT	5
SCRUGGS	1

Surname	# of Land Parcels
SCRUGS	1
SEALE	3
SECREST	1
SEGARS	1
SELLERS	2
SEMMES	3
SHACKELFORD	3
SHAMBERGER	1
SHAMBURGER	6
SHANBURGER	1
SHANNON	1
SHAW	8
SHELBY	1
SHOWS	2
SHUMATE	7
SHURDEN	2
SIGRIS	1
SIKES	6
SILLIMAN	1
SIMMONS	10
SIMPSON	1
SIMS	5
SKELTON	2
SKINNER	3
SKIPPER	3
SMITH	52
SMITHWICK	3
SMITT	1
SNOWDEN	12
SPEED	3
SPINKS	30
STAFFORD	11
STALLINGS	1
STANSEL	1
STANTON	5
STEDIVANT	1
STEED	2
STEEDLEY	1
STEELE	1
STEINWINDER	3
STEPHENS	4
STEPHENSON	3
STEWART	1
STINSON	5
STOKES	5
STONE	6
STRAIT	20
STRANG	1
STRANGE	2
STRINGER	2
STROUD	7
STUCKEY	3
SUGGS	2
SULLIVAN	1
SUMMERS	1
SUTTLE	4
SUTTLES	4
SWAIN	3
SWIFT	5
SWILLEY	5
SWILLY	1
TABB	29
TALBERT	2
TAPPAN	18
TARRANT	1
TARTT	2
TATUM	1
TAYLOR	18
TEMPLETON	1
TERRAL	1

Surnames in Lauderdale County, Mississippi Patents

Surname	# of Land Parcels
TERRELL	3
THOMAS	6
THOMPSON	20
THOMSON	1
THORNTON	4
TILL	1
TILLMAN	8
TIMBREL	1
TIMMS	2
TIMS	1
TINNIN	5
TODD	3
TOLSON	3
TOOL	3
TOOMEY	1
TORRANS	1
TOUCHSTONE	1
TOURELL	1
TOWNSEND	2
TRAWICK	7
TRUSSEL	1
TRUSSELL	54
TRUST	1
TUCKER	3
TUGGLE	1
TURNER	3
TUTON	7
TUTT	30
TWILLEY	1
ULMER	1
ULRIC	1
ULRICK	3
UPCHURCH	2
URY	1
VANCE	4
VAUGHAN	2
VAUGHN	12
VINSON	3
WABINGTON	2
WADE	1
WADKINS	2
WAGGENER	1
WAGGONER	1
WAGONER	2
WAITS	2
WALKER	42
WALLACE	21
WALTERER	1
WALTERS	2
WALTHALL	13
WARBINGTON	6
WARD	8
WARLINGTON	1
WARREN	14
WASHINGTON	3
WATERS	2
WATES	1
WATKINS	5
WATSON	4
WATTERS	52
WATTS	6
WEATHERFORD	1
WEATHERS	1
WEBB	1
WEBSTER	1
WEIR	2
WEISSINGER	1
WELCH	6
WELLBORN	4
WELLS	3
WELSH	1
WEST	5
WESTBROOK	1
WHEAT	2
WHITE	62
WHITEHEAD	11
WHITFIELD	4
WHITIKER	1
WHITLOCK	6
WHITSETT	10
WIER	2
WIGGINS	2
WILKERSON	1
WILKINSON	12
WILLARD	1
WILLIAMS	16
WILLIAMSON	6
WILLIS	1
WILSON	13
WINDHAM	2
WINGATE	3
WINN	1
WINNINGHAM	5
WITHERSPOON	2
WITT	2
WOLF	2
WOMACK	2
WOOD	3
WOODS	1
WOODWARD	1
WOOTAN	1
WOOTON	1
WORBINGTON	4
WORMACK	1
WRIGHT	13
YARBOROUGH	2
YARBROUGH	3
YARRELL	2
YATES	2
YEAGER	3
YOES	1
YOUNG	5

Surname/Township Index

This Index allows you to determine which *Township Map Group(s)* contain individuals with the following surnames. Each *Map Group* has a corresponding full-name index of all individuals who obtained patents for land within its Congressional township's borders. After each index you will find the Patent Map to which it refers, and just thereafter, you can view the township's Road Map and Historical Map, with the latter map displaying streams, railroads, and more.

So, once you find your Surname here, proceed to the Index at the beginning of the **Map Group** indicated below.

Surname	Map Group	Parcels of Land	Meridian/Township/Range
ABERNATHY	**11**	1	Choctaw 7-N 18-E
ADAIR	**2**	6	Choctaw 8-N 15-E
" "	**3**	2	Choctaw 8-N 16-E
ADAMS	**12**	1	Choctaw 6-N 14-E
" "	**10**	1	Choctaw 7-N 17-E
ADKINSON	**14**	1	Choctaw 6-N 16-E
AGLETREE	**17**	3	CHOCTAW 5-N 14-E
AGNEW	**10**	3	Choctaw 7-N 17-E
AIKINS	**5**	2	Choctaw 8-N 18-E
ALEXANDER	**9**	24	Choctaw 7-N 16-E
" "	**11**	3	Choctaw 7-N 18-E
" "	**14**	2	Choctaw 6-N 16-E
" "	**10**	2	Choctaw 7-N 17-E
" "	**19**	1	Choctaw 5-N 16-E
" "	**3**	1	Choctaw 8-N 16-E
ALFORD	**2**	4	Choctaw 8-N 15-E
" "	**9**	3	Choctaw 7-N 16-E
" "	**14**	2	Choctaw 6-N 16-E
" "	**8**	2	Choctaw 7-N 15-E
" "	**1**	1	Choctaw 8-N 14-E
ALLBROOKS	**14**	1	Choctaw 6-N 16-E
ALLEN	**5**	8	Choctaw 8-N 18-E
" "	**4**	4	Choctaw 8-N 17-E
" "	**17**	3	Choctaw 5-N 14-E
" "	**11**	2	Choctaw 7-N 18-E
ALLISON	**8**	1	Choctaw 7-N 15-E
ALLSOBROOKS	**19**	1	Choctaw 5-N 16-E
ALVICE	**3**	2	Choctaw 8-N 16-E
ANDERSON	**2**	8	Choctaw 8-N 15-E
" "	**5**	4	Choctaw 8-N 18-E
" "	**19**	2	Choctaw 5-N 16-E
" "	**16**	2	Choctaw 6-N 18-E
" "	**3**	2	Choctaw 8-N 16-E
" "	**17**	1	CHOCTAW 5-N 14-E
" "	**20**	1	Choctaw 5-N 17-E
ANDING	**10**	1	Choctaw 7-N 17-E
" "	**4**	1	Choctaw 8-N 17-E
ANDREWS	**3**	4	Choctaw 8-N 16-E
" "	**10**	1	Choctaw 7-N 17-E
ARMSTRONG	**3**	1	Choctaw 8-N 16-E
ARRANT	**17**	2	Choctaw 5-N 14-E
ARRINGTON	**17**	1	Choctaw 5-N 14-E
" "	**9**	1	Choctaw 7-N 16-E
ASKEW	**10**	3	Choctaw 7-N 17-E

Family Maps of Lauderdale County, Mississippi

Surname	Map Group	Parcels of Land	Meridian/Township/Range
AVARA	2	2	Choctaw 8-N 15-E
AVERY	2	1	Choctaw 8-N 15-E
BABERS	8	1	Choctaw 7-N 15-E
BAILES	2	3	Choctaw 8-N 15-E
" "	1	1	Choctaw 8-N 14-E
BAILEY	21	2	Choctaw 5-N 18-E
" "	5	1	Choctaw 8-N 18-E
BAILIFF	3	8	Choctaw 8-N 16-E
" "	16	2	Choctaw 6-N 18-E
BAILY	8	2	Choctaw 7-N 15-E
BAINES	2	2	Choctaw 8-N 15-E
BAINS	2	1	Choctaw 8-N 15-E
BAIRD	1	1	Choctaw 8-N 14-E
BALL	14	5	Choctaw 6-N 16-E
" "	13	2	Choctaw 6-N 15-E
" "	7	1	Choctaw 7-N 14-E
" "	5	1	Choctaw 8-N 18-E
BALLARD	18	2	Choctaw 5-N 15-E
" "	9	2	Choctaw 7-N 16-E
" "	10	1	Choctaw 7-N 17-E
BANES	18	2	Choctaw 5-N 15-E
BANYARD	20	2	Choctaw 5-N 17-E
BARDEN	8	1	Choctaw 7-N 15-E
BAREFOOT	4	4	Choctaw 8-N 17-E
BARFOOT	4	1	Choctaw 8-N 17-E
BARNES	7	1	Choctaw 7-N 14-E
" "	10	1	Choctaw 7-N 17-E
" "	3	1	Choctaw 8-N 16-E
BARNET	8	1	Choctaw 7-N 15-E
BARNETT	13	5	Choctaw 6-N 15-E
" "	8	2	Choctaw 7-N 15-E
" "	3	1	Choctaw 8-N 16-E
BARRETT	17	1	Choctaw 5-N 14-E
BARRON	19	1	Choctaw 5-N 16-E
BASTIN	11	3	Choctaw 7-N 18-E
BATCHEE	10	1	Choctaw 7-N 17-E
BATTLE	8	3	Choctaw 7-N 15-E
BAXTON	18	1	CHOCTAW 5-N 15-E
BEASON	8	2	Choctaw 7-N 15-E
" "	9	1	Choctaw 7-N 16-E
BEEMON	1	2	Choctaw 8-N 14-E
BEESON	5	6	Choctaw 8-N 18-E
BELL	16	1	Choctaw 6-N 18-E
" "	11	1	Choctaw 7-N 18-E
BENNETT	17	2	Choctaw 5-N 14-E
" "	18	1	Choctaw 5-N 15-E
BERDEAUX	11	1	Choctaw 7-N 18-E
BERRIEN	14	13	Choctaw 6-N 16-E
" "	9	12	Choctaw 7-N 16-E
BERRY	18	1	Choctaw 5-N 15-E
" "	13	1	Choctaw 6-N 15-E
BEVILL	5	4	Choctaw 8-N 18-E
BICKART	14	1	Choctaw 6-N 16-E
BISHOP	9	3	Choctaw 7-N 16-E
" "	14	2	Choctaw 6-N 16-E
" "	13	1	Choctaw 6-N 15-E
BLACKMAN	7	4	Choctaw 7-N 14-E
" "	14	1	CHOCTAW 6-N 16-E
BLACKWELL	19	2	Choctaw 5-N 16-E
" "	15	1	Choctaw 6-N 17-E

Surname/Township Index

Surname	Map Group	Parcels of Land	Meridian/Township/Range
BLANKS	21	11	Choctaw 5-N 18-E
" "	3	2	Choctaw 8-N 16-E
" "	8	1	Choctaw 7-N 15-E
BOLDEN	15	1	Choctaw 6-N 17-E
BOND	5	2	Choctaw 8-N 18-E
BONEY	15	1	Choctaw 6-N 17-E
BOON	13	1	Choctaw 6-N 15-E
" "	8	1	Choctaw 7-N 15-E
BORDERS	14	1	Choctaw 6-N 16-E
BOSTON	3	1	Choctaw 8-N 16-E
BOSWELL	16	3	Choctaw 6-N 18-E
" "	10	1	Choctaw 7-N 17-E
BOTHWELL	14	13	Choctaw 6-N 16-E
" "	9	12	Choctaw 7-N 16-E
BOUNDS	12	4	Choctaw 6-N 14-E
BOURDEAUX	11	5	Choctaw 7-N 18-E
BOUTWELL	20	2	Choctaw 5-N 17-E
BOWEN	1	1	Choctaw 8-N 14-E
BOWLING	15	1	Choctaw 6-N 17-E
BOYD	5	11	Choctaw 8-N 18-E
BOYKIN	10	1	Choctaw 7-N 17-E
BOZEMAN	18	1	Choctaw 5-N 15-E
" "	10	1	Choctaw 7-N 17-E
" "	11	1	Choctaw 7-N 18-E
BOZMAN	11	3	Choctaw 7-N 18-E
BOZONE	14	1	Choctaw 6-N 16-E
BRAGG	11	4	Choctaw 7-N 18-E
" "	16	2	Choctaw 6-N 18-E
BRANNAN	10	1	Choctaw 7-N 17-E
BRASSIELL	10	1	Choctaw 7-N 17-E
BRECK	1	3	Choctaw 8-N 14-E
BREWER	13	1	Choctaw 6-N 15-E
BREWSTER	21	2	Choctaw 5-N 18-E
BRIANT	17	2	Choctaw 5-N 14-E
" "	15	1	Choctaw 6-N 17-E
BRIDGES	13	1	Choctaw 6-N 15-E
BROCK	21	9	CHOCTAW 5-N 18-E
BRONSON	21	1	Choctaw 5-N 18-E
BROOKE	2	1	Choctaw 8-N 15-E
BROWER	11	4	Choctaw 7-N 18-E
" "	16	3	Choctaw 6-N 18-E
" "	14	1	Choctaw 6-N 16-E
BROWN	8	20	Choctaw 7-N 15-E
" "	2	12	Choctaw 8-N 15-E
" "	3	9	Choctaw 8-N 16-E
" "	10	8	Choctaw 7-N 17-E
" "	11	8	Choctaw 7-N 18-E
" "	4	8	Choctaw 8-N 17-E
" "	17	5	Choctaw 5-N 14-E
" "	9	3	Choctaw 7-N 16-E
" "	18	2	Choctaw 5-N 15-E
" "	21	2	Choctaw 5-N 18-E
" "	12	1	Choctaw 6-N 14-E
" "	13	1	Choctaw 6-N 15-E
" "	5	1	Choctaw 8-N 18-E
BRUISTER	15	4	Choctaw 6-N 17-E
" "	20	1	Choctaw 5-N 17-E
BRUNER	10	1	Choctaw 7-N 17-E
BRUNSON	11	2	CHOCTAW 7-N 18-E
" "	18	1	Choctaw 5-N 15-E

25

Surname	Map Group	Parcels of Land	Meridian/Township/Range
BRUNSON (Cont'd)	19	1	Choctaw 5-N 16-E
BRYAN	21	2	Choctaw 5-N 18-E
BUCHANAN	13	1	Choctaw 6-N 15-E
BUCKLEW	21	1	Choctaw 5-N 18-E
BULLARD	18	2	Choctaw 5-N 15-E
" "	9	1	Choctaw 7-N 16-E
" "	3	1	Choctaw 8-N 16-E
BUNTYN	14	1	Choctaw 6-N 16-E
BUNYARD	19	4	Choctaw 5-N 16-E
" "	20	3	Choctaw 5-N 17-E
" "	10	1	Choctaw 7-N 17-E
BURKHALTER	15	1	Choctaw 6-N 17-E
" "	10	1	Choctaw 7-N 17-E
BURNS	4	1	Choctaw 8-N 17-E
BURRESS	3	1	Choctaw 8-N 16-E
BURRIS	3	1	Choctaw 8-N 16-E
BURTON	5	1	Choctaw 8-N 18-E
" "	6	1	Choctaw 8-N 19-E
BURWELL	13	12	Choctaw 6-N 15-E
" "	14	1	Choctaw 6-N 16-E
BUSBY	10	6	Choctaw 7-N 17-E
" "	9	2	Choctaw 7-N 16-E
" "	4	2	Choctaw 8-N 17-E
" "	5	1	Choctaw 8-N 18-E
BUSH	4	4	Choctaw 8-N 17-E
BUSON	5	2	Choctaw 8-N 18-E
" "	15	1	Choctaw 6-N 17-E
BUSTIN	11	3	Choctaw 7-N 18-E
" "	5	1	Choctaw 8-N 18-E
BUTCHEE	4	6	Choctaw 8-N 17-E
" "	10	1	Choctaw 7-N 17-E
BUTLER	2	4	Choctaw 8-N 15-E
" "	20	2	CHOCTAW 5-N 17-E
BYNUM	18	2	Choctaw 5-N 15-E
" "	19	1	Choctaw 5-N 16-E
" "	14	1	Choctaw 6-N 16-E
" "	11	1	Choctaw 7-N 18-E
CAINS	10	3	Choctaw 7-N 17-E
" "	9	1	Choctaw 7-N 16-E
CALHOUN	8	9	Choctaw 7-N 15-E
" "	7	2	Choctaw 7-N 14-E
" "	4	2	Choctaw 8-N 17-E
" "	9	1	Choctaw 7-N 16-E
CALLUM	1	2	Choctaw 8-N 14-E
CAMMACK	16	1	Choctaw 6-N 18-E
CAMPBELL	8	2	Choctaw 7-N 15-E
" "	11	1	Choctaw 7-N 18-E
" "	3	1	Choctaw 8-N 16-E
" "	5	1	Choctaw 8-N 18-E
CANADY	8	1	Choctaw 7-N 15-E
CANTERBURY	5	3	CHOCTAW 8-N 18-E
CARAWAY	18	2	Choctaw 5-N 15-E
CARMICHAEL	10	1	Choctaw 7-N 17-E
CARPENTER	7	2	Choctaw 7-N 14-E
" "	3	2	Choctaw 8-N 16-E
" "	10	1	Choctaw 7-N 17-E
CARR	21	4	CHOCTAW 5-N 18-E
" "	20	1	Choctaw 5-N 17-E
CARROL	9	1	Choctaw 7-N 16-E
CARSON	18	2	Choctaw 5-N 15-E

Surname/Township Index

Surname	Map Group	Parcels of Land	Meridian/Township/Range
CARTER	18	3	Choctaw 5-N 15-E
" "	11	2	Choctaw 7-N 18-E
" "	17	1	Choctaw 5-N 14-E
" "	3	1	Choctaw 8-N 16-E
CARYE	18	2	Choctaw 5-N 15-E
CASTLES	8	3	Choctaw 7-N 15-E
" "	2	1	Choctaw 8-N 15-E
CATES	3	1	Choctaw 8-N 16-E
CATLET	16	2	Choctaw 6-N 18-E
CHA	7	1	CHOCTAW 7-N 14-E
CHADICK	17	1	Choctaw 5-N 14-E
CHAMBLIS	17	2	Choctaw 5-N 14-E
CHAMBLISS	17	1	CHOCTAW 5-N 14-E
CHANDLER	9	6	Choctaw 7-N 16-E
CHANEY	9	1	Choctaw 7-N 16-E
CHESTER	9	3	Choctaw 7-N 16-E
CHILDE	4	1	Choctaw 8-N 17-E
CHILES	5	1	Choctaw 8-N 18-E
CLANTON	11	8	CHOCTAW 7-N 18-E
CLARK	17	1	Choctaw 5-N 14-E
" "	21	1	Choctaw 5-N 18-E
" "	16	1	Choctaw 6-N 18-E
CLAY	5	10	Choctaw 8-N 18-E
" "	8	2	Choctaw 7-N 15-E
" "	13	1	Choctaw 6-N 15-E
CLAYTON	4	20	Choctaw 8-N 17-E
" "	10	1	Choctaw 7-N 17-E
CLIET	21	4	CHOCTAW 5-N 18-E
CLINTON	11	3	Choctaw 7-N 18-E
" "	14	2	Choctaw 6-N 16-E
" "	9	2	Choctaw 7-N 16-E
COAKER	15	1	Choctaw 6-N 17-E
COATES	3	10	Choctaw 8-N 16-E
" "	18	1	Choctaw 5-N 15-E
" "	13	1	Choctaw 6-N 15-E
COATS	18	4	Choctaw 5-N 15-E
" "	8	3	Choctaw 7-N 15-E
" "	3	2	Choctaw 8-N 16-E
" "	7	1	Choctaw 7-N 14-E
COBB	6	2	Choctaw 8-N 19-E
COCHRUM	7	1	Choctaw 7-N 14-E
COCKE	4	1	Choctaw 8-N 17-E
COKER	15	3	Choctaw 6-N 17-E
COLE	3	7	Choctaw 8-N 16-E
" "	8	2	Choctaw 7-N 15-E
" "	17	1	Choctaw 5-N 14-E
" "	18	1	Choctaw 5-N 15-E
" "	13	1	Choctaw 6-N 15-E
" "	14	1	Choctaw 6-N 16-E
COLEMAN	2	12	Choctaw 8-N 15-E
" "	14	1	Choctaw 6-N 16-E
" "	3	1	Choctaw 8-N 16-E
COLVIN	9	1	Choctaw 7-N 16-E
CONNER	13	1	Choctaw 6-N 15-E
COOK	15	1	Choctaw 6-N 17-E
COOKSEY	1	1	Choctaw 8-N 14-E
COOLEY	12	1	Choctaw 6-N 14-E
COON	12	1	Choctaw 6-N 14-E
COOPER	18	2	Choctaw 5-N 15-E
" "	7	1	Choctaw 7-N 14-E

Surname	Map Group	Parcels of Land	Meridian/Township/Range
COOPER (Cont'd)	2	1	Choctaw 8-N 15-E
COOPWOOD	4	19	Choctaw 8-N 17-E
" "	5	6	Choctaw 8-N 18-E
CORLEY	17	1	Choctaw 5-N 14-E
" "	18	1	Choctaw 5-N 15-E
" "	19	1	Choctaw 5-N 16-E
" "	16	1	Choctaw 6-N 18-E
COTES	3	1	Choctaw 8-N 16-E
COURTNEY	10	1	Choctaw 7-N 17-E
COVINGTON	19	1	Choctaw 5-N 16-E
" "	9	1	Choctaw 7-N 16-E
COX	12	1	Choctaw 6-N 14-E
CRAIG	2	12	Choctaw 8-N 15-E
CRAIL	3	1	Choctaw 8-N 16-E
CRAIN	8	1	Choctaw 7-N 15-E
CRANE	9	7	Choctaw 7-N 16-E
" "	8	4	Choctaw 7-N 15-E
" "	3	4	Choctaw 8-N 16-E
CRAWFORD	2	3	Choctaw 8-N 15-E
" "	19	1	Choctaw 5-N 16-E
" "	16	1	Choctaw 6-N 18-E
CRAWLEY	2	1	Choctaw 8-N 15-E
CREAL	14	2	Choctaw 6-N 16-E
" "	13	1	Choctaw 6-N 15-E
CREEL	14	5	Choctaw 6-N 16-E
" "	19	2	Choctaw 5-N 16-E
" "	13	2	Choctaw 6-N 15-E
CRENSHAW	15	1	Choctaw 6-N 17-E
" "	16	1	Choctaw 6-N 18-E
" "	5	1	Choctaw 8-N 18-E
CREWS	11	5	Choctaw 7-N 18-E
CROCKETT	3	1	Choctaw 8-N 16-E
CRORY	14	1	Choctaw 6-N 16-E
" "	7	1	Choctaw 7-N 14-E
CROSBY	19	4	Choctaw 5-N 16-E
CRUFT	1	3	Choctaw 8-N 14-E
CUBLEY	18	3	Choctaw 5-N 15-E
CULBRAITH	10	2	Choctaw 7-N 17-E
CULPEPPER	16	3	Choctaw 6-N 18-E
CUNNINGHAM	9	1	Choctaw 7-N 16-E
CURRIE	12	2	Choctaw 6-N 14-E
DALE	3	9	Choctaw 8-N 16-E
DANIEL	13	4	Choctaw 6-N 15-E
" "	1	4	Choctaw 8-N 14-E
" "	8	2	Choctaw 7-N 15-E
" "	3	2	Choctaw 8-N 16-E
" "	18	1	Choctaw 5-N 15-E
DANIELS	13	2	Choctaw 6-N 15-E
" "	14	1	Choctaw 6-N 16-E
DARDEN	4	2	Choctaw 8-N 17-E
DAUCEY	17	1	Choctaw 5-N 14-E
DAUGHTRY	16	1	Choctaw 6-N 18-E
DAVENPORT	14	1	Choctaw 6-N 16-E
DAVID	13	3	Choctaw 6-N 15-E
DAVIS	2	7	Choctaw 8-N 15-E
" "	4	5	Choctaw 8-N 17-E
" "	3	2	Choctaw 8-N 16-E
" "	13	1	Choctaw 6-N 15-E
" "	14	1	Choctaw 6-N 16-E
" "	15	1	Choctaw 6-N 17-E

Surname/Township Index

Surname	Map Group	Parcels of Land	Meridian/Township/Range
DAVIS (Cont'd)	1	1	Choctaw 8-N 14-E
" "	5	1	Choctaw 8-N 18-E
DAWKINS	3	8	Choctaw 8-N 16-E
" "	2	2	Choctaw 8-N 15-E
DAY	21	4	Choctaw 5-N 18-E
DEAN	14	2	Choctaw 6-N 16-E
DEAR	17	13	Choctaw 5-N 14-E
" "	12	2	Choctaw 6-N 14-E
DEARMAN	14	7	Choctaw 6-N 16-E
" "	13	3	Choctaw 6-N 15-E
" "	15	1	Choctaw 6-N 17-E
DEASE	18	2	Choctaw 5-N 15-E
DEASON	16	1	Choctaw 6-N 18-E
DEBOE	17	1	Choctaw 5-N 14-E
DEEN	8	5	Choctaw 7-N 15-E
" "	14	3	Choctaw 6-N 16-E
" "	9	1	Choctaw 7-N 16-E
DEENS	14	2	Choctaw 6-N 16-E
DEER	17	4	Choctaw 5-N 14-E
DEERMAN	13	2	Choctaw 6-N 15-E
DEIR	17	1	Choctaw 5-N 14-E
DELK	5	9	Choctaw 8-N 18-E
" "	17	3	CHOCTAW 5-N 14-E
" "	4	2	Choctaw 8-N 17-E
DEMENT	4	1	Choctaw 8-N 17-E
DENTON	2	2	Choctaw 8-N 15-E
" "	3	1	Choctaw 8-N 16-E
DEWITT	11	1	Choctaw 7-N 18-E
DICKERSON	18	1	Choctaw 5-N 15-E
DICKSON	1	2	Choctaw 8-N 14-E
DOBBS	1	2	Choctaw 8-N 14-E
" "	3	2	Choctaw 8-N 16-E
DODD	16	2	Choctaw 6-N 18-E
DORMAN	2	1	Choctaw 8-N 15-E
DOVE	5	4	Choctaw 8-N 18-E
" "	10	2	Choctaw 7-N 17-E
DOWD	3	1	Choctaw 8-N 16-E
DRUMOND	12	2	Choctaw 6-N 14-E
DUBOSE	18	8	Choctaw 5-N 15-E
" "	10	2	Choctaw 7-N 17-E
" "	14	1	Choctaw 6-N 16-E
DUKE	3	1	Choctaw 8-N 16-E
DUN	10	1	Choctaw 7-N 17-E
DUNAGIN	11	1	Choctaw 7-N 18-E
DUNBAR	21	1	Choctaw 5-N 18-E
DUNCAN	13	6	Choctaw 6-N 15-E
DUNGEN	14	1	Choctaw 6-N 16-E
DUNKLEY	1	2	Choctaw 8-N 14-E
DUNN	17	2	Choctaw 5-N 14-E
" "	9	1	Choctaw 7-N 16-E
DUPREE	10	1	Choctaw 7-N 17-E
DURR	10	18	Choctaw 7-N 17-E
" "	15	3	CHOCTAW 6-N 17-E
" "	14	2	CHOCTAW 6-N 16-E
DYASS	17	5	Choctaw 5-N 14-E
" "	13	1	Choctaw 6-N 15-E
EAKENS	11	1	Choctaw 7-N 18-E
EAKIN	11	1	Choctaw 7-N 18-E
EAKINS	5	2	Choctaw 8-N 18-E
" "	10	1	Choctaw 7-N 17-E

Surname	Map Group	Parcels of Land	Meridian/Township/Range
EARLONG	13	1	Choctaw 6-N 15-E
EASOM	3	1	Choctaw 8-N 16-E
EASTHAM	4	3	Choctaw 8-N 17-E
EASTIS	16	1	Choctaw 6-N 18-E
EATHAM	4	2	Choctaw 8-N 17-E
EAVES	16	2	Choctaw 6-N 18-E
ECHOLS	10	2	Choctaw 7-N 17-E
EDWARDS	14	3	Choctaw 6-N 16-E
" "	5	3	Choctaw 8-N 18-E
" "	7	1	Choctaw 7-N 14-E
ELAM	17	3	Choctaw 5-N 14-E
ELLIS	11	4	Choctaw 7-N 18-E
" "	18	2	Choctaw 5-N 15-E
EMBRY	15	1	Choctaw 6-N 17-E
ENGLISH	3	1	Choctaw 8-N 16-E
ERBY	20	1	Choctaw 5-N 17-E
ERNEST	17	1	Choctaw 5-N 14-E
ESTELL	5	4	Choctaw 8-N 18-E
ETHRIDGE	1	2	Choctaw 8-N 14-E
" "	2	1	Choctaw 8-N 15-E
ETHRRIDGE	19	1	Choctaw 5-N 16-E
ETTURIDGE	19	1	Choctaw 5-N 16-E
EUBANKS	8	7	Choctaw 7-N 15-E
" "	13	3	Choctaw 6-N 15-E
" "	10	2	Choctaw 7-N 17-E
" "	9	1	Choctaw 7-N 16-E
EVANS	14	2	Choctaw 6-N 16-E
" "	8	1	Choctaw 7-N 15-E
" "	11	1	Choctaw 7-N 18-E
EVERETT	17	3	CHOCTAW 5-N 14-E
" "	1	1	Choctaw 8-N 14-E
EVES	16	1	Choctaw 6-N 18-E
EWING	16	1	Choctaw 6-N 18-E
EZELL	5	5	Choctaw 8-N 18-E
" "	11	3	Choctaw 7-N 18-E
FAGIN	14	1	Choctaw 6-N 16-E
FAIR	10	1	Choctaw 7-N 17-E
FAIRCHILD	13	3	Choctaw 6-N 15-E
" "	18	2	Choctaw 5-N 15-E
" "	14	1	Choctaw 6-N 16-E
FAIRCHILDS	19	3	Choctaw 5-N 16-E
" "	18	2	Choctaw 5-N 15-E
FALLIN	11	1	Choctaw 7-N 18-E
FARCHILES	14	1	Choctaw 6-N 16-E
FARISS	14	3	Choctaw 6-N 16-E
FARRELL	13	1	Choctaw 6-N 15-E
FAULKNER	13	1	Choctaw 6-N 15-E
FAY	18	1	Choctaw 5-N 15-E
FEDRICK	4	1	Choctaw 8-N 17-E
FENDALL	13	1	Choctaw 6-N 15-E
FENNER	14	1	Choctaw 6-N 16-E
FIKES	17	6	Choctaw 5-N 14-E
FINDLEY	4	1	Choctaw 8-N 17-E
FINLEY	3	3	Choctaw 8-N 16-E
" "	4	2	Choctaw 8-N 17-E
FINNEN	9	5	Choctaw 7-N 16-E
FISHER	19	5	Choctaw 5-N 16-E
FLETCHER	16	1	Choctaw 6-N 18-E
" "	10	1	Choctaw 7-N 17-E
FLEWELLEN	4	28	Choctaw 8-N 17-E

Surname/Township Index

Surname	Map Group	Parcels of Land	Meridian/Township/Range
FLEWELLEN (Cont'd)	5	7	Choctaw 8-N 18-E
FLINN	18	3	Choctaw 5-N 15-E
FLOYD	9	1	Choctaw 7-N 16-E
FORD	7	10	Choctaw 7-N 14-E
" "	9	4	Choctaw 7-N 16-E
" "	8	3	Choctaw 7-N 15-E
" "	17	1	Choctaw 5-N 14-E
FORTSON	14	1	Choctaw 6-N 16-E
FOSTER	10	2	Choctaw 7-N 17-E
" "	1	1	Choctaw 8-N 14-E
FOUNTAIN	13	3	Choctaw 6-N 15-E
" "	8	2	Choctaw 7-N 15-E
FOUNTAINE	13	1	Choctaw 6-N 15-E
FOWLER	4	4	Choctaw 8-N 17-E
" "	12	3	Choctaw 6-N 14-E
FREELAND	9	1	Choctaw 7-N 16-E
FREEMAN	15	5	Choctaw 6-N 17-E
FUNK	14	1	Choctaw 6-N 16-E
GABEL	3	1	Choctaw 8-N 16-E
GADDIS	18	5	Choctaw 5-N 15-E
" "	13	4	Choctaw 6-N 15-E
" "	12	2	Choctaw 6-N 14-E
" "	17	1	Choctaw 5-N 14-E
GAINES	5	2	Choctaw 8-N 18-E
" "	10	1	Choctaw 7-N 17-E
GAINS	8	1	Choctaw 7-N 15-E
GARDNER	21	3	Choctaw 5-N 18-E
" "	19	1	Choctaw 5-N 16-E
" "	12	1	Choctaw 6-N 14-E
" "	13	1	Choctaw 6-N 15-E
GARRETT	14	5	Choctaw 6-N 16-E
" "	9	3	Choctaw 7-N 16-E
" "	13	2	Choctaw 6-N 15-E
GARY	14	1	Choctaw 6-N 16-E
GAY	19	1	Choctaw 5-N 16-E
GAYLIE	5	4	Choctaw 8-N 18-E
GEE	5	2	Choctaw 8-N 18-E
GEORGE	17	1	Choctaw 5-N 14-E
GIBSON	1	6	Choctaw 8-N 14-E
GILBERT	12	1	Choctaw 6-N 14-E
" "	13	1	Choctaw 6-N 15-E
" "	8	1	Choctaw 7-N 15-E
" "	9	1	Choctaw 7-N 16-E
GILHAM	4	3	Choctaw 8-N 17-E
GILLASPIE	9	5	Choctaw 7-N 16-E
" "	8	1	Choctaw 7-N 15-E
GILLESPIE	3	7	Choctaw 8-N 16-E
GILLINDER	14	2	Choctaw 6-N 16-E
GILLINER	19	2	Choctaw 5-N 16-E
GILMORE	20	1	Choctaw 5-N 17-E
GLASCOCK	4	1	Choctaw 8-N 17-E
GLASS	3	1	Choctaw 8-N 16-E
GLENN	9	2	Choctaw 7-N 16-E
GODWIN	20	1	Choctaw 5-N 17-E
GOODMAN	21	1	Choctaw 5-N 18-E
GOODSON	13	2	Choctaw 6-N 15-E
GOODWIN	12	1	Choctaw 6-N 14-E
GOODWYN	16	2	Choctaw 6-N 18-E
" "	21	1	Choctaw 5-N 18-E
GORDON	15	13	Choctaw 6-N 17-E

Surname	Map Group	Parcels of Land	Meridian/Township/Range
GORDON (Cont'd)	10	3	Choctaw 7-N 17-E
" "	18	2	Choctaw 5-N 15-E
" "	3	2	Choctaw 8-N 16-E
" "	14	1	Choctaw 6-N 16-E
GORMAN	3	1	Choctaw 8-N 16-E
GOUGH	17	1	Choctaw 5-N 14-E
GRAHAM	11	4	Choctaw 7-N 18-E
" "	18	2	Choctaw 5-N 15-E
" "	19	2	Choctaw 5-N 16-E
" "	13	2	Choctaw 6-N 15-E
" "	4	1	CHOCTAW 8-N 17-E
GRANT	4	32	Choctaw 8-N 17-E
" "	5	15	Choctaw 8-N 18-E
" "	9	6	Choctaw 7-N 16-E
" "	18	2	Choctaw 5-N 15-E
" "	3	2	Choctaw 8-N 16-E
" "	14	1	Choctaw 6-N 16-E
" "	15	1	Choctaw 6-N 17-E
GRANTLAND	2	8	Choctaw 8-N 15-E
" "	3	2	Choctaw 8-N 16-E
GRAVES	2	2	Choctaw 8-N 15-E
GRAY	17	2	Choctaw 5-N 14-E
" "	19	2	Choctaw 5-N 16-E
" "	9	2	Choctaw 7-N 16-E
GRAYSON	16	1	Choctaw 6-N 18-E
GREEN	18	5	Choctaw 5-N 15-E
" "	4	3	CHOCTAW 8-N 17-E
" "	14	1	Choctaw 6-N 16-E
" "	1	1	Choctaw 8-N 14-E
" "	3	1	Choctaw 8-N 16-E
GREENE	5	5	Choctaw 8-N 18-E
GREER	14	1	Choctaw 6-N 16-E
GRESSETT	12	1	Choctaw 6-N 14-E
GRICE	17	2	Choctaw 5-N 14-E
GRIFFIN	5	37	Choctaw 8-N 18-E
" "	4	7	Choctaw 8-N 17-E
GRIFFITH	10	3	Choctaw 7-N 17-E
" "	15	2	Choctaw 6-N 17-E
" "	4	2	Choctaw 8-N 17-E
GRIGSBY	3	1	Choctaw 8-N 16-E
GRISSETT	12	3	Choctaw 6-N 14-E
GUNN	21	2	Choctaw 5-N 18-E
HADEN	17	1	Choctaw 5-N 14-E
HAGOOD	21	1	Choctaw 5-N 18-E
HAILS	9	3	Choctaw 7-N 16-E
HALEY	17	1	Choctaw 5-N 14-E
HALL	18	3	Choctaw 5-N 15-E
" "	14	3	Choctaw 6-N 16-E
" "	9	3	Choctaw 7-N 16-E
" "	11	3	Choctaw 7-N 18-E
" "	15	2	Choctaw 6-N 17-E
" "	16	2	Choctaw 6-N 18-E
HALTON	16	1	Choctaw 6-N 18-E
HAMBRICK	8	1	Choctaw 7-N 15-E
HAMLET	3	3	Choctaw 8-N 16-E
HAMMOND	10	2	Choctaw 7-N 17-E
HAMMONDS	10	2	Choctaw 7-N 17-E
HAMRICK	8	3	Choctaw 7-N 15-E
" "	13	1	Choctaw 6-N 15-E
HAMSWORTH	11	1	Choctaw 7-N 18-E

Surname/Township Index

Surname	Map Group	Parcels of Land	Meridian/Township/Range
HAND	5	4	Choctaw 8-N 18-E
" "	18	2	Choctaw 5-N 15-E
HANDLEY	17	1	CHOCTAW 5-N 14-E
HANNA	5	8	Choctaw 8-N 18-E
" "	4	7	Choctaw 8-N 17-E
HANNAH	4	3	Choctaw 8-N 17-E
HANNER	5	1	Choctaw 8-N 18-E
HARDEN	16	3	Choctaw 6-N 18-E
HARDIN	16	4	Choctaw 6-N 18-E
HARDY	4	28	Choctaw 8-N 17-E
" "	5	7	Choctaw 8-N 18-E
" "	18	1	Choctaw 5-N 15-E
HARLAN	4	4	Choctaw 8-N 17-E
" "	5	2	Choctaw 8-N 18-E
HARLIN	4	3	Choctaw 8-N 17-E
HARPER	8	16	Choctaw 7-N 15-E
" "	15	7	Choctaw 6-N 17-E
" "	9	2	Choctaw 7-N 16-E
" "	14	1	Choctaw 6-N 16-E
HARRELL	21	2	Choctaw 5-N 18-E
HARRINGTON	17	1	Choctaw 5-N 14-E
HARRIS	17	4	CHOCTAW 5-N 14-E
" "	3	2	Choctaw 8-N 16-E
" "	5	1	Choctaw 8-N 18-E
HARRISON	13	7	Choctaw 6-N 15-E
" "	21	4	Choctaw 5-N 18-E
HART	5	1	Choctaw 8-N 18-E
HARVARD	10	1	Choctaw 7-N 17-E
HARVEY	17	2	Choctaw 5-N 14-E
" "	18	2	Choctaw 5-N 15-E
HARWELL	17	1	Choctaw 5-N 14-E
HATCHER	8	6	Choctaw 7-N 15-E
HATHORN	16	2	Choctaw 6-N 18-E
HAWKINS	18	2	Choctaw 5-N 15-E
" "	3	2	Choctaw 8-N 16-E
" "	19	1	Choctaw 5-N 16-E
HAYES	9	2	Choctaw 7-N 16-E
" "	20	1	Choctaw 5-N 17-E
HAYS	9	5	Choctaw 7-N 16-E
" "	20	4	Choctaw 5-N 17-E
HAYSE	20	2	Choctaw 5-N 17-E
HEARN	16	1	Choctaw 6-N 18-E
" "	10	1	Choctaw 7-N 17-E
HELVESTON	19	3	Choctaw 5-N 16-E
HENDERSON	14	10	Choctaw 6-N 16-E
" "	9	7	Choctaw 7-N 16-E
" "	10	7	Choctaw 7-N 17-E
" "	15	3	Choctaw 6-N 17-E
" "	5	3	Choctaw 8-N 18-E
" "	3	2	Choctaw 8-N 16-E
" "	17	1	Choctaw 5-N 14-E
HENDRICK	4	6	CHOCTAW 8-N 17-E
" "	5	2	Choctaw 8-N 18-E
HERNDON	9	4	Choctaw 7-N 16-E
" "	18	2	Choctaw 5-N 15-E
" "	14	1	Choctaw 6-N 16-E
HERREN	18	1	Choctaw 5-N 15-E
HERRING	17	1	Choctaw 5-N 14-E
HERRINGTON	17	2	Choctaw 5-N 14-E
" "	1	1	Choctaw 8-N 14-E

Surname	Map Group	Parcels of Land	Meridian/Township/Range
HICKS	16	2	Choctaw 6-N 18-E
HIGGINBOTHAM	5	3	Choctaw 8-N 18-E
HIGHTOWER	3	4	Choctaw 8-N 16-E
HILL	3	10	Choctaw 8-N 16-E
" "	14	2	Choctaw 6-N 16-E
" "	1	2	Choctaw 8-N 14-E
" "	19	1	Choctaw 5-N 16-E
" "	15	1	Choctaw 6-N 17-E
HINES	17	2	Choctaw 5-N 14-E
" "	5	2	Choctaw 8-N 18-E
HINTON	5	2	Choctaw 8-N 18-E
HOBBS	14	1	Choctaw 6-N 16-E
HODGES	5	3	Choctaw 8-N 18-E
" "	17	1	Choctaw 5-N 14-E
HOGANS	13	1	Choctaw 6-N 15-E
HOLDEN	15	4	Choctaw 6-N 17-E
" "	14	1	Choctaw 6-N 16-E
HOLDERNESS	4	2	Choctaw 8-N 17-E
HOLIMAN	7	1	Choctaw 7-N 14-E
HOLLADAY	3	1	Choctaw 8-N 16-E
HOLLAM	16	1	Choctaw 6-N 18-E
HOLLAND	16	3	Choctaw 6-N 18-E
" "	2	3	Choctaw 8-N 15-E
" "	8	1	Choctaw 7-N 15-E
HOLLIMON	17	2	Choctaw 5-N 14-E
HOLLINSON	16	1	Choctaw 6-N 18-E
HOLLY	18	2	Choctaw 5-N 15-E
HOLMES	9	1	Choctaw 7-N 16-E
HONAH	7	1	CHOCTAW 7-N 14-E
HOOKS	9	3	Choctaw 7-N 16-E
" "	19	2	Choctaw 5-N 16-E
" "	16	1	Choctaw 6-N 18-E
HOPKINS	18	4	Choctaw 5-N 15-E
HORSEY	8	1	Choctaw 7-N 15-E
HOUSE	1	28	Choctaw 8-N 14-E
HOUSTON	12	2	Choctaw 6-N 14-E
" "	5	2	Choctaw 8-N 18-E
" "	8	1	Choctaw 7-N 15-E
HOUZE	19	1	Choctaw 5-N 16-E
" "	9	1	Choctaw 7-N 16-E
HOWELL	8	2	Choctaw 7-N 15-E
" "	9	1	Choctaw 7-N 16-E
HOWSE	17	2	CHOCTAW 5-N 14-E
HUBBARD	9	5	Choctaw 7-N 16-E
" "	5	5	Choctaw 8-N 18-E
" "	2	3	Choctaw 8-N 15-E
" "	3	3	Choctaw 8-N 16-E
" "	8	2	Choctaw 7-N 15-E
HUDSON	5	7	Choctaw 8-N 18-E
HUEY	12	1	Choctaw 6-N 14-E
HUGGINS	7	1	Choctaw 7-N 14-E
" "	5	1	Choctaw 8-N 18-E
HUGHES	3	8	Choctaw 8-N 16-E
HULET	5	6	Choctaw 8-N 18-E
HUMPHREY	5	1	Choctaw 8-N 18-E
HUMPHREYS	16	5	Choctaw 6-N 18-E
" "	15	1	Choctaw 6-N 17-E
" "	10	1	Choctaw 7-N 17-E
HUMPHRIES	4	2	Choctaw 8-N 17-E
HUNT	5	3	Choctaw 8-N 18-E

Surname/Township Index

Surname	Map Group	Parcels of Land	Meridian/Township/Range
HUNT (Cont'd)	15	2	Choctaw 6-N 17-E
" "	16	2	Choctaw 6-N 18-E
HUNTER	21	14	Choctaw 5-N 18-E
" "	8	4	Choctaw 7-N 15-E
" "	11	1	Choctaw 7-N 18-E
HURT	16	1	Choctaw 6-N 18-E
HUSSEY	9	7	Choctaw 7-N 16-E
" "	10	4	Choctaw 7-N 17-E
" "	4	3	Choctaw 8-N 17-E
" "	3	1	Choctaw 8-N 16-E
HUTTAN	2	14	Choctaw 8-N 15-E
HUTTON	2	11	Choctaw 8-N 15-E
" "	15	1	Choctaw 6-N 17-E
" "	8	1	Choctaw 7-N 15-E
HUZZA	3	2	Choctaw 8-N 16-E
HYDE	19	1	Choctaw 5-N 16-E
INGRAM	10	1	Choctaw 7-N 17-E
IRBY	15	4	Choctaw 6-N 17-E
" "	14	3	Choctaw 6-N 16-E
" "	19	1	Choctaw 5-N 16-E
IVY	5	2	Choctaw 8-N 18-E
JACKSON	13	3	Choctaw 6-N 15-E
" "	8	2	Choctaw 7-N 15-E
" "	15	1	Choctaw 6-N 17-E
" "	4	1	Choctaw 8-N 17-E
JAMES	1	3	Choctaw 8-N 14-E
" "	19	1	Choctaw 5-N 16-E
" "	13	1	Choctaw 6-N 15-E
" "	10	1	Choctaw 7-N 17-E
JARMAN	5	3	Choctaw 8-N 18-E
JEMISON	18	16	Choctaw 5-N 15-E
" "	11	7	Choctaw 7-N 18-E
" "	4	2	Choctaw 8-N 17-E
JENNINGS	17	4	CHOCTAW 5-N 14-E
JOHNSON	3	4	Choctaw 8-N 16-E
" "	15	3	Choctaw 6-N 17-E
" "	13	2	Choctaw 6-N 15-E
" "	14	2	Choctaw 6-N 16-E
" "	18	1	Choctaw 5-N 15-E
" "	21	1	Choctaw 5-N 18-E
" "	9	1	Choctaw 7-N 16-E
" "	10	1	Choctaw 7-N 17-E
" "	4	1	Choctaw 8-N 17-E
JOHNSTON	4	1	Choctaw 8-N 17-E
JOINER	14	1	Choctaw 6-N 16-E
JOLLY	8	1	Choctaw 7-N 15-E
" "	2	1	Choctaw 8-N 15-E
" "	3	1	Choctaw 8-N 16-E
JONES	10	8	Choctaw 7-N 17-E
" "	4	7	Choctaw 8-N 17-E
" "	5	6	Choctaw 8-N 18-E
" "	17	5	Choctaw 5-N 14-E
" "	19	5	Choctaw 5-N 16-E
" "	13	4	Choctaw 6-N 15-E
" "	12	3	Choctaw 6-N 14-E
" "	15	3	Choctaw 6-N 17-E
" "	16	3	CHOCTAW 6-N 18-E
" "	3	3	Choctaw 8-N 16-E
" "	14	2	Choctaw 6-N 16-E
" "	7	2	Choctaw 7-N 14-E

Surname	Map Group	Parcels of Land	Meridian/Township/Range
JONES (Cont'd)	20	1	CHOCTAW 5-N 17-E
" "	21	1	Choctaw 5-N 18-E
JORDAN	9	12	Choctaw 7-N 16-E
" "	15	1	Choctaw 6-N 17-E
JORDON	9	1	Choctaw 7-N 16-E
JOSEPH	13	3	Choctaw 6-N 15-E
KALSAW	8	1	Choctaw 7-N 15-E
KEATH	12	1	Choctaw 6-N 14-E
KEELAND	5	1	Choctaw 8-N 18-E
KEETON	9	3	Choctaw 7-N 16-E
KEITH	12	4	Choctaw 6-N 14-E
KELLEY	17	1	Choctaw 5-N 14-E
KELLY	11	5	Choctaw 7-N 18-E
" "	19	1	Choctaw 5-N 16-E
KENNEDY	8	2	Choctaw 7-N 15-E
KENNERLY	20	1	Choctaw 5-N 17-E
KERBY	8	1	Choctaw 7-N 15-E
KEYS	13	1	Choctaw 6-N 15-E
KIDD	16	1	Choctaw 6-N 18-E
KIELY	14	2	Choctaw 6-N 16-E
KILLENS	8	2	Choctaw 7-N 15-E
" "	13	1	Choctaw 6-N 15-E
KILLILEA	13	1	Choctaw 6-N 15-E
KILLIN	8	4	Choctaw 7-N 15-E
KILLINGSWORTH	10	5	Choctaw 7-N 17-E
" "	5	4	Choctaw 8-N 18-E
" "	4	1	Choctaw 8-N 17-E
KIMBRAUGH	17	1	Choctaw 5-N 14-E
KIMBRELL	15	1	Choctaw 6-N 17-E
KIMBROUGH	17	1	Choctaw 5-N 14-E
KINERD	9	1	Choctaw 7-N 16-E
KING	17	5	Choctaw 5-N 14-E
" "	18	3	Choctaw 5-N 15-E
" "	1	2	Choctaw 8-N 14-E
KIRKLAND	17	1	Choctaw 5-N 14-E
KITERAL	3	2	Choctaw 8-N 16-E
KITRELL	3	4	Choctaw 8-N 16-E
KITTRELL	3	5	Choctaw 8-N 16-E
" "	9	1	Choctaw 7-N 16-E
KNIGHT	17	3	CHOCTAW 5-N 14-E
" "	15	1	Choctaw 6-N 17-E
KNOX	5	12	Choctaw 8-N 18-E
" "	14	2	Choctaw 6-N 16-E
" "	15	2	Choctaw 6-N 17-E
" "	2	1	Choctaw 8-N 15-E
KUNERY	5	1	Choctaw 8-N 18-E
KYLES	8	1	Choctaw 7-N 15-E
LACKEY	3	3	Choctaw 8-N 16-E
" "	15	1	Choctaw 6-N 17-E
LACY	3	9	Choctaw 8-N 16-E
" "	12	3	Choctaw 6-N 14-E
LAMB	4	3	Choctaw 8-N 17-E
LAMBERT	10	2	Choctaw 7-N 17-E
LANCASTER	5	4	Choctaw 8-N 18-E
LANE	5	2	Choctaw 8-N 18-E
" "	15	1	Choctaw 6-N 17-E
LANEY	18	2	Choctaw 5-N 15-E
LANG	13	15	Choctaw 6-N 15-E
" "	17	1	Choctaw 5-N 14-E
" "	14	1	Choctaw 6-N 16-E

Surname/Township Index

Surname	Map Group	Parcels of Land	Meridian/Township/Range
LARD	16	1	Choctaw 6-N 18-E
" "	6	1	Choctaw 8-N 19-E
LARKIN	15	1	Choctaw 6-N 17-E
LASHLEY	15	1	Choctaw 6-N 17-E
LASLEY	15	2	Choctaw 6-N 17-E
" "	19	1	Choctaw 5-N 16-E
LATTNEY	12	1	Choctaw 6-N 14-E
LEE	8	9	Choctaw 7-N 15-E
" "	14	3	Choctaw 6-N 16-E
" "	15	3	Choctaw 6-N 17-E
" "	9	2	Choctaw 7-N 16-E
" "	17	1	Choctaw 5-N 14-E
" "	21	1	Choctaw 5-N 18-E
LEHMAN	14	1	Choctaw 6-N 16-E
LEMORE	19	1	CHOCTAW 5-N 16-E
LEWIS	5	15	Choctaw 8-N 18-E
" "	9	8	Choctaw 7-N 16-E
" "	17	5	Choctaw 5-N 14-E
" "	8	4	Choctaw 7-N 15-E
" "	13	3	Choctaw 6-N 15-E
" "	15	3	Choctaw 6-N 17-E
" "	16	3	Choctaw 6-N 18-E
" "	2	3	Choctaw 8-N 15-E
" "	3	3	Choctaw 8-N 16-E
" "	4	3	Choctaw 8-N 17-E
" "	6	2	Choctaw 8-N 19-E
" "	11	1	Choctaw 7-N 18-E
LILLEY	20	2	Choctaw 5-N 17-E
LILLY	20	1	Choctaw 5-N 17-E
LINDLEY	14	1	Choctaw 6-N 16-E
LINTON	19	3	Choctaw 5-N 16-E
LINZEY	11	2	CHOCTAW 7-N 18-E
LITCHFIELD	1	1	Choctaw 8-N 14-E
LITTLE	17	3	Choctaw 5-N 14-E
LITTLETON	13	1	Choctaw 6-N 15-E
LLOYD	17	3	Choctaw 5-N 14-E
" "	2	1	Choctaw 8-N 15-E
LONG	2	1	Choctaw 8-N 15-E
LOPER	13	4	Choctaw 6-N 15-E
LOTT	13	1	Choctaw 6-N 15-E
LOVE	10	3	Choctaw 7-N 17-E
" "	3	1	Choctaw 8-N 16-E
LUCAS	10	1	Choctaw 7-N 17-E
LUCY	16	4	Choctaw 6-N 18-E
LUNSFORD	4	1	Choctaw 8-N 17-E
MAASS	14	2	Choctaw 6-N 16-E
MABRY	12	1	Choctaw 6-N 14-E
MAGRUE	10	1	Choctaw 7-N 17-E
MAHAN	7	4	Choctaw 7-N 14-E
MALONE	9	2	Choctaw 7-N 16-E
MANDEVILLE	18	1	Choctaw 5-N 15-E
MANGUM	3	1	Choctaw 8-N 16-E
MANN	9	1	Choctaw 7-N 16-E
" "	10	1	Choctaw 7-N 17-E
MANNING	14	1	Choctaw 6-N 16-E
MARCY	13	1	Choctaw 6-N 15-E
MARSH	16	14	Choctaw 6-N 18-E
" "	11	2	Choctaw 7-N 18-E
MARSHALL	8	2	Choctaw 7-N 15-E
" "	15	1	Choctaw 6-N 17-E

Surname	Map Group	Parcels of Land	Meridian/Township/Range
MARTIN	19	13	Choctaw 5-N 16-E
" "	20	2	Choctaw 5-N 17-E
" "	1	2	Choctaw 8-N 14-E
" "	13	1	Choctaw 6-N 15-E
" "	11	1	Choctaw 7-N 18-E
" "	3	1	Choctaw 8-N 16-E
MASON	9	2	Choctaw 7-N 16-E
" "	1	1	Choctaw 8-N 14-E
MASSENGALL	13	1	Choctaw 6-N 15-E
MASTERS	15	1	Choctaw 6-N 17-E
MATHEWS	13	1	Choctaw 6-N 15-E
" "	16	1	Choctaw 6-N 18-E
" "	8	1	Choctaw 7-N 15-E
MAURY	5	2	Choctaw 8-N 18-E
MAXEY	12	1	Choctaw 6-N 14-E
MAXWELL	12	2	Choctaw 6-N 14-E
MAY	1	5	Choctaw 8-N 14-E
" "	16	2	Choctaw 6-N 18-E
MAYATT	1	1	Choctaw 8-N 14-E
MAYBERRY	12	2	Choctaw 6-N 14-E
MAYERHOFF	17	1	Choctaw 5-N 14-E
MCBRIDE	10	2	Choctaw 7-N 17-E
MCCALL	17	1	Choctaw 5-N 14-E
MCCANN	9	2	Choctaw 7-N 16-E
MCCAREY	14	1	Choctaw 6-N 16-E
MCCARTY	3	2	Choctaw 8-N 16-E
MCCARY	18	1	Choctaw 5-N 15-E
" "	14	1	Choctaw 6-N 16-E
MCCOMB	16	3	Choctaw 6-N 18-E
" "	11	2	Choctaw 7-N 18-E
MCCOWN	4	23	Choctaw 8-N 17-E
" "	5	11	Choctaw 8-N 18-E
" "	6	9	Choctaw 8-N 19-E
MCCRANEY	15	1	Choctaw 6-N 17-E
" "	16	1	Choctaw 6-N 18-E
MCCRORY	2	2	Choctaw 8-N 15-E
MCCURDY	16	1	Choctaw 6-N 18-E
MCDANIEL	14	2	Choctaw 6-N 16-E
" "	4	1	Choctaw 8-N 17-E
MCDONALD	8	3	Choctaw 7-N 15-E
" "	19	2	Choctaw 5-N 16-E
" "	17	1	Choctaw 5-N 14-E
" "	18	1	Choctaw 5-N 15-E
" "	9	1	Choctaw 7-N 16-E
MCDOUGAL	20	1	Choctaw 5-N 17-E
MCDOW	9	6	Choctaw 7-N 16-E
MCEACHIN	19	1	Choctaw 5-N 16-E
MCELROY	15	1	Choctaw 6-N 17-E
" "	4	1	Choctaw 8-N 17-E
MCGAHA	16	1	Choctaw 6-N 18-E
MCGAW	19	2	Choctaw 5-N 16-E
MCGEE	8	1	Choctaw 7-N 15-E
" "	3	1	Choctaw 8-N 16-E
MCGRAW	5	1	Choctaw 8-N 18-E
MCGREW	5	3	Choctaw 8-N 18-E
" "	4	1	Choctaw 8-N 17-E
MCILWAIN	10	2	Choctaw 7-N 17-E
MCINNIS	19	6	Choctaw 5-N 16-E
" "	20	3	Choctaw 5-N 17-E
" "	14	2	Choctaw 6-N 16-E

Surname/Township Index

Surname	Map Group	Parcels of Land	Meridian/Township/Range
MCKELLAR	3	6	Choctaw 8-N 16-E
MCKELVIN	20	1	CHOCTAW 5-N 17-E
MCKINLEY	10	1	Choctaw 7-N 17-E
MCKINNON	16	1	Choctaw 6-N 18-E
MCKLEVAIN	20	1	CHOCTAW 5-N 17-E
MCLAMORE	14	3	Choctaw 6-N 16-E
" "	9	2	Choctaw 7-N 16-E
" "	2	1	Choctaw 8-N 15-E
MCLAREN	12	1	Choctaw 6-N 14-E
MCLARRIN	11	1	Choctaw 7-N 18-E
MCLAUGHLIN	18	4	Choctaw 5-N 15-E
MCLAURIN	10	4	Choctaw 7-N 17-E
MCLEAN	4	1	Choctaw 8-N 17-E
MCLEMORE	14	14	Choctaw 6-N 16-E
" "	19	2	CHOCTAW 5-N 16-E
" "	13	1	Choctaw 6-N 15-E
" "	15	1	Choctaw 6-N 17-E
MCLIN	10	3	Choctaw 7-N 17-E
MCMELLON	8	1	Choctaw 7-N 15-E
MCMILLAN	4	1	Choctaw 8-N 17-E
MCMILLON	9	1	Choctaw 7-N 16-E
MCMORY	15	2	Choctaw 6-N 17-E
MCMULLEN	15	2	Choctaw 6-N 17-E
MCMULLIN	14	1	Choctaw 6-N 16-E
MCMULLUN	9	1	Choctaw 7-N 16-E
MCNEILL	4	5	Choctaw 8-N 17-E
" "	5	2	Choctaw 8-N 18-E
" "	10	1	Choctaw 7-N 17-E
MCPHAUL	4	2	Choctaw 8-N 17-E
MCPHERSON	5	5	Choctaw 8-N 18-E
MCRAE	21	3	Choctaw 5-N 18-E
MCSHAN	15	2	Choctaw 6-N 17-E
MCSHANN	14	2	Choctaw 6-N 16-E
MCWHORTER	4	7	Choctaw 8-N 17-E
MCWOOTON	8	1	Choctaw 7-N 15-E
MEEK	1	1	Choctaw 8-N 14-E
MERRITT	15	4	Choctaw 6-N 17-E
MESSLEY	13	1	Choctaw 6-N 15-E
MIEATT	1	1	Choctaw 8-N 14-E
MILES	2	1	Choctaw 8-N 15-E
MILLER	11	9	Choctaw 7-N 18-E
" "	9	3	Choctaw 7-N 16-E
" "	4	2	Choctaw 8-N 17-E
" "	13	1	Choctaw 6-N 15-E
" "	3	1	Choctaw 8-N 16-E
" "	5	1	Choctaw 8-N 18-E
MILLS	1	6	Choctaw 8-N 14-E
" "	4	4	Choctaw 8-N 17-E
" "	14	3	Choctaw 6-N 16-E
" "	10	3	Choctaw 7-N 17-E
" "	2	3	Choctaw 8-N 15-E
" "	3	1	Choctaw 8-N 16-E
MINOR	16	2	Choctaw 6-N 18-E
MITCHAM	19	1	Choctaw 5-N 16-E
MITCHELL	18	1	Choctaw 5-N 15-E
" "	20	1	Choctaw 5-N 17-E
" "	13	1	Choctaw 6-N 15-E
" "	8	1	Choctaw 7-N 15-E
MIXON	11	1	CHOCTAW 7-N 18-E
MOBLEY	8	1	Choctaw 7-N 15-E

Surname	Map Group	Parcels of Land	Meridian/Township/Range		
MOFFET	14	1	Choctaw	6-N	16-E
MOFFETT	12	3	Choctaw	6-N	14-E
MOLPUS	19	4	Choctaw	5-N	16-E
MONTAGUE	13	3	Choctaw	6-N	15-E
" "	1	3	Choctaw	8-N	14-E
MONTGOMERY	4	1	Choctaw	8-N	17-E
MOODY	16	1	Choctaw	6-N	18-E
" "	10	1	Choctaw	7-N	17-E
MOORE	4	9	Choctaw	8-N	17-E
" "	5	7	Choctaw	8-N	18-E
" "	17	6	CHOCTAW	5-N	14-E
" "	13	3	Choctaw	6-N	15-E
" "	6	2	Choctaw	8-N	19-E
" "	18	1	Choctaw	5-N	15-E
" "	15	1	Choctaw	6-N	17-E
" "	1	1	Choctaw	8-N	14-E
MORGAN	5	3	Choctaw	8-N	18-E
" "	14	1	Choctaw	6-N	16-E
MORRIS	17	2	Choctaw	5-N	14-E
" "	14	1	Choctaw	6-N	16-E
" "	2	1	Choctaw	8-N	15-E
MORROW	1	3	Choctaw	8-N	14-E
MOSELEY	3	1	Choctaw	8-N	16-E
MOSLEY	2	5	Choctaw	8-N	15-E
MOTEN	15	1	Choctaw	6-N	17-E
MOTH	8	1	Choctaw	7-N	15-E
MOTT	8	6	Choctaw	7-N	15-E
" "	7	1	Choctaw	7-N	14-E
" "	9	1	Choctaw	7-N	16-E
MUNDELL	17	1	Choctaw	5-N	14-E
MURPHY	5	2	Choctaw	8-N	18-E
" "	17	1	Choctaw	5-N	14-E
" "	9	1	Choctaw	7-N	16-E
" "	2	1	Choctaw	8-N	15-E
" "	3	1	Choctaw	8-N	16-E
MYRICK	10	2	Choctaw	7-N	17-E
NASH	2	7	Choctaw	8-N	15-E
" "	5	1	Choctaw	8-N	18-E
NAYLOR	9	1	Choctaw	7-N	16-E
NEAL	14	1	Choctaw	6-N	16-E
NEIGHBOURS	2	2	Choctaw	8-N	15-E
NEW	4	3	Choctaw	8-N	17-E
" "	17	2	Choctaw	5-N	14-E
" "	12	1	Choctaw	6-N	14-E
NEWBERRY	5	2	Choctaw	8-N	18-E
NEWELL	10	1	Choctaw	7-N	17-E
NEWMAN	18	1	Choctaw	5-N	15-E
NEWSOM	13	2	Choctaw	6-N	15-E
" "	14	2	Choctaw	6-N	16-E
NEWTON	15	1	Choctaw	6-N	17-E
NICHOL	14	1	Choctaw	6-N	16-E
NICHOLAS	18	8	Choctaw	5-N	15-E
NICHOLS	14	1	Choctaw	6-N	16-E
" "	10	1	Choctaw	7-N	17-E
NICHOLSON	9	2	Choctaw	7-N	16-E
NOEL	4	1	Choctaw	8-N	17-E
NUNNARY	5	1	Choctaw	8-N	18-E
NUTT	16	2	Choctaw	6-N	18-E
" "	11	1	Choctaw	7-N	18-E
OBRINK	3	2	Choctaw	8-N	16-E

Surname/Township Index

Surname	Map Group	Parcels of Land	Meridian/Township/Range
ODAM	3	2	Choctaw 8-N 16-E
ODEN	18	1	Choctaw 5-N 15-E
ODOM	9	8	Choctaw 7-N 16-E
" "	4	8	Choctaw 8-N 17-E
" "	14	2	Choctaw 6-N 16-E
" "	16	2	Choctaw 6-N 18-E
" "	10	1	Choctaw 7-N 17-E
" "	11	1	Choctaw 7-N 18-E
OFALLIN	4	4	CHOCTAW 8-N 17-E
ONEAL	16	5	Choctaw 6-N 18-E
" "	4	1	Choctaw 8-N 17-E
OSBORN	7	2	Choctaw 7-N 14-E
" "	17	1	Choctaw 5-N 14-E
OTT	10	1	Choctaw 7-N 17-E
OWENS	3	2	Choctaw 8-N 16-E
PACE	14	3	Choctaw 6-N 16-E
" "	8	2	Choctaw 7-N 15-E
" "	9	1	Choctaw 7-N 16-E
PAGE	10	2	Choctaw 7-N 17-E
PALMER	9	1	Choctaw 7-N 16-E
PARISH	16	1	Choctaw 6-N 18-E
PARKER	3	5	Choctaw 8-N 16-E
" "	17	1	Choctaw 5-N 14-E
" "	18	1	Choctaw 5-N 15-E
" "	13	1	Choctaw 6-N 15-E
" "	10	1	Choctaw 7-N 17-E
" "	4	1	Choctaw 8-N 17-E
PARKS	1	7	Choctaw 8-N 14-E
PARTIN	3	1	Choctaw 8-N 16-E
PATHEA	10	1	Choctaw 7-N 17-E
PATON	11	4	Choctaw 7-N 18-E
PATTON	10	2	Choctaw 7-N 17-E
PAYNE	11	1	Choctaw 7-N 18-E
" "	1	1	Choctaw 8-N 14-E
" "	5	1	Choctaw 8-N 18-E
PEARL	1	2	Choctaw 8-N 14-E
PEARSON	10	1	Choctaw 7-N 17-E
PEAVY	11	4	Choctaw 7-N 18-E
" "	5	1	Choctaw 8-N 18-E
PERCE	10	1	Choctaw 7-N 17-E
PETAGREW	14	1	Choctaw 6-N 16-E
PEW	13	1	Choctaw 6-N 15-E
PEYTON	5	1	Choctaw 8-N 18-E
PHILIPS	9	4	Choctaw 7-N 16-E
" "	11	2	Choctaw 7-N 18-E
PHILLIPS	11	4	Choctaw 7-N 18-E
" "	3	1	Choctaw 8-N 16-E
PICKARD	14	4	Choctaw 6-N 16-E
" "	19	3	Choctaw 5-N 16-E
PIGFORD	15	2	Choctaw 6-N 17-E
" "	10	2	Choctaw 7-N 17-E
" "	5	1	Choctaw 8-N 18-E
PISTOLE	4	2	Choctaw 8-N 17-E
PITMAN	17	1	Choctaw 5-N 14-E
PITTS	5	1	Choctaw 8-N 18-E
PLUMMER	5	2	Choctaw 8-N 18-E
POGUE	20	1	Choctaw 5-N 17-E
" "	15	1	Choctaw 6-N 17-E
POOL	8	2	Choctaw 7-N 15-E
POPE	16	2	Choctaw 6-N 18-E

Family Maps of Lauderdale County, Mississippi

Surname	Map Group	Parcels of Land	Meridian/Township/Range
PORTER	16	1	Choctaw 6-N 18-E
POWELL	1	4	Choctaw 8-N 14-E
" "	21	1	Choctaw 5-N 18-E
PRESTON	19	1	Choctaw 5-N 16-E
" "	14	1	Choctaw 6-N 16-E
PRICE	11	7	Choctaw 7-N 18-E
" "	17	4	Choctaw 5-N 14-E
" "	2	1	Choctaw 8-N 15-E
PRIER	8	2	Choctaw 7-N 15-E
" "	2	1	Choctaw 8-N 15-E
PRINE	4	1	Choctaw 8-N 17-E
PRINGLE	15	9	Choctaw 6-N 17-E
" "	20	4	Choctaw 5-N 17-E
" "	21	2	Choctaw 5-N 18-E
" "	16	1	CHOCTAW 6-N 18-E
" "	11	1	CHOCTAW 7-N 18-E
PRIVETT	18	1	Choctaw 5-N 15-E
PRYER	2	1	Choctaw 8-N 15-E
PUGH	10	3	Choctaw 7-N 17-E
RAGSDALE	3	3	Choctaw 8-N 16-E
" "	14	2	Choctaw 6-N 16-E
" "	4	1	Choctaw 8-N 17-E
RAGSDALL	14	3	Choctaw 6-N 16-E
RAINER	20	1	Choctaw 5-N 17-E
RAMSAY	9	3	Choctaw 7-N 16-E
RAMSEY	11	1	Choctaw 7-N 18-E
" "	3	1	Choctaw 8-N 16-E
" "	4	1	Choctaw 8-N 17-E
RANDALL	14	1	Choctaw 6-N 16-E
RASBERRY	1	1	Choctaw 8-N 14-E
RATCLIFF	3	11	Choctaw 8-N 16-E
" "	4	3	Choctaw 8-N 17-E
RATSBURG	4	1	Choctaw 8-N 17-E
RAWLINGS	16	1	Choctaw 6-N 18-E
RAWSON	15	2	Choctaw 6-N 17-E
" "	4	2	Choctaw 8-N 17-E
" "	10	1	Choctaw 7-N 17-E
RAY	8	1	Choctaw 7-N 15-E
REA	11	1	Choctaw 7-N 18-E
REASON	8	2	Choctaw 7-N 15-E
REAVES	13	1	Choctaw 6-N 15-E
REECE	17	1	Choctaw 5-N 14-E
REED	16	2	Choctaw 6-N 18-E
" "	13	1	Choctaw 6-N 15-E
REESE	18	3	Choctaw 5-N 15-E
REEVES	13	3	Choctaw 6-N 15-E
REID	9	2	Choctaw 7-N 16-E
REIVES	13	2	Choctaw 6-N 15-E
REW	18	1	Choctaw 5-N 15-E
REYNOLDS	4	4	Choctaw 8-N 17-E
" "	5	4	Choctaw 8-N 18-E
" "	21	1	Choctaw 5-N 18-E
RHODES	13	2	Choctaw 6-N 15-E
" "	16	1	Choctaw 6-N 18-E
RICE	11	1	Choctaw 7-N 18-E
RICH	17	3	CHOCTAW 5-N 14-E
RICHARDSON	2	9	Choctaw 8-N 15-E
" "	9	7	Choctaw 7-N 16-E
" "	3	2	Choctaw 8-N 16-E
RICHE	1	1	Choctaw 8-N 14-E

Surname/Township Index

Surname	Map Group	Parcels of Land	Meridian/Township/Range
RICHEY	15	1	Choctaw 6-N 17-E
RILEY	17	1	Choctaw 5-N 14-E
" "	19	1	Choctaw 5-N 16-E
RIVERS	3	1	Choctaw 8-N 16-E
ROBBINS	19	2	Choctaw 5-N 16-E
ROBERSON	21	1	Choctaw 5-N 18-E
ROBERTS	14	1	Choctaw 6-N 16-E
" "	9	1	Choctaw 7-N 16-E
" "	4	1	Choctaw 8-N 17-E
ROBERTSON	21	1	Choctaw 5-N 18-E
" "	7	1	Choctaw 7-N 14-E
ROBINSON	21	1	Choctaw 5-N 18-E
ROBISON	21	3	Choctaw 5-N 18-E
ROCHEL	4	2	Choctaw 8-N 17-E
RODGERS	15	2	Choctaw 6-N 17-E
" "	8	2	Choctaw 7-N 15-E
" "	16	1	Choctaw 6-N 18-E
ROGERS	2	3	Choctaw 8-N 15-E
" "	10	2	Choctaw 7-N 17-E
" "	11	2	Choctaw 7-N 18-E
" "	8	1	Choctaw 7-N 15-E
" "	3	1	Choctaw 8-N 16-E
" "	5	1	Choctaw 8-N 18-E
RUNNELS	16	2	Choctaw 6-N 18-E
RUSH	12	1	Choctaw 6-N 14-E
RUSHING	14	6	Choctaw 6-N 16-E
" "	15	5	Choctaw 6-N 17-E
" "	10	5	Choctaw 7-N 17-E
" "	8	4	Choctaw 7-N 15-E
" "	9	4	Choctaw 7-N 16-E
" "	1	2	Choctaw 8-N 14-E
" "	18	1	Choctaw 5-N 15-E
" "	19	1	Choctaw 5-N 16-E
" "	13	1	Choctaw 6-N 15-E
" "	2	1	Choctaw 8-N 15-E
RUSSELL	13	14	Choctaw 6-N 15-E
" "	10	8	Choctaw 7-N 17-E
" "	7	2	Choctaw 7-N 14-E
" "	9	2	Choctaw 7-N 16-E
" "	15	1	Choctaw 6-N 17-E
" "	16	1	Choctaw 6-N 18-E
" "	8	1	Choctaw 7-N 15-E
RUSSING	19	1	Choctaw 5-N 16-E
" "	14	1	Choctaw 6-N 16-E
RUTLEDGE	14	2	Choctaw 6-N 16-E
SADLER	1	1	Choctaw 8-N 14-E
SALTER	3	1	Choctaw 8-N 16-E
SANDERFORD	2	2	Choctaw 8-N 15-E
SANDIFORD	2	2	Choctaw 8-N 15-E
SANDYFORD	8	3	Choctaw 7-N 15-E
SANFORD	14	1	Choctaw 6-N 16-E
SAPP	21	1	Choctaw 5-N 18-E
SATCHER	16	1	Choctaw 6-N 18-E
SAVELL	12	1	Choctaw 6-N 14-E
" "	7	1	Choctaw 7-N 14-E
SCARBOROUGH	14	7	Choctaw 6-N 16-E
" "	19	3	Choctaw 5-N 16-E
" "	10	2	Choctaw 7-N 17-E
" "	21	1	CHOCTAW 5-N 18-E
SCARBROUGH	18	1	Choctaw 5-N 15-E

Family Maps of Lauderdale County, Mississippi

Surname	Map Group	Parcels of Land	Meridian/Township/Range
SCARBROUGH (Cont'd)	19	1	Choctaw 5-N 16-E
" "	15	1	Choctaw 6-N 17-E
" "	16	1	Choctaw 6-N 18-E
SCHANROCK	14	1	Choctaw 6-N 16-E
SCOTT	18	2	Choctaw 5-N 15-E
" "	14	2	Choctaw 6-N 16-E
" "	10	1	Choctaw 7-N 17-E
SCRUGGS	18	1	Choctaw 5-N 15-E
SCRUGS	8	1	Choctaw 7-N 15-E
SEALE	5	3	Choctaw 8-N 18-E
SECREST	10	1	Choctaw 7-N 17-E
SEGARS	4	1	Choctaw 8-N 17-E
SELLERS	3	2	Choctaw 8-N 16-E
SEMMES	13	3	Choctaw 6-N 15-E
SHACKELFORD	14	2	Choctaw 6-N 16-E
" "	13	1	Choctaw 6-N 15-E
SHAMBERGER	16	1	Choctaw 6-N 18-E
SHAMBURGER	16	6	Choctaw 6-N 18-E
SHANBURGER	16	1	Choctaw 6-N 18-E
SHANNON	21	1	Choctaw 5-N 18-E
SHAW	13	7	Choctaw 6-N 15-E
" "	14	1	Choctaw 6-N 16-E
SHELBY	5	1	Choctaw 8-N 18-E
SHOWS	10	1	Choctaw 7-N 17-E
" "	11	1	Choctaw 7-N 18-E
SHUMATE	2	4	Choctaw 8-N 15-E
" "	3	3	Choctaw 8-N 16-E
SHURDEN	18	2	Choctaw 5-N 15-E
SIGRIS	4	1	Choctaw 8-N 17-E
SIKES	9	4	Choctaw 7-N 16-E
" "	13	1	Choctaw 6-N 15-E
" "	4	1	Choctaw 8-N 17-E
SILLIMAN	5	1	Choctaw 8-N 18-E
SIMMONS	5	8	Choctaw 8-N 18-E
" "	4	2	Choctaw 8-N 17-E
SIMPSON	12	1	Choctaw 6-N 14-E
SIMS	15	3	Choctaw 6-N 17-E
" "	19	1	Choctaw 5-N 16-E
" "	14	1	Choctaw 6-N 16-E
SKELTON	1	2	Choctaw 8-N 14-E
SKINNER	4	3	Choctaw 8-N 17-E
SKIPPER	19	3	Choctaw 5-N 16-E
SMITH	16	10	CHOCTAW 6-N 18-E
" "	21	9	Choctaw 5-N 18-E
" "	10	7	CHOCTAW 7-N 17-E
" "	17	6	CHOCTAW 5-N 14-E
" "	11	6	Choctaw 7-N 18-E
" "	14	3	Choctaw 6-N 16-E
" "	9	3	Choctaw 7-N 16-E
" "	19	2	Choctaw 5-N 16-E
" "	8	2	Choctaw 7-N 15-E
" "	3	2	Choctaw 8-N 16-E
" "	18	1	Choctaw 5-N 15-E
" "	5	1	Choctaw 8-N 18-E
SMITHWICK	5	3	Choctaw 8-N 18-E
SMITT	3	1	Choctaw 8-N 16-E
SNOWDEN	2	10	Choctaw 8-N 15-E
" "	12	2	Choctaw 6-N 14-E
SPEED	17	2	Choctaw 5-N 14-E
" "	18	1	Choctaw 5-N 15-E

Surname/Township Index

Surname	Map Group	Parcels of Land	Meridian/Township/Range
SPINKS	16	26	Choctaw 6-N 18-E
" "	15	2	Choctaw 6-N 17-E
" "	11	2	Choctaw 7-N 18-E
STAFFORD	13	7	Choctaw 6-N 15-E
" "	19	3	Choctaw 5-N 16-E
" "	14	1	Choctaw 6-N 16-E
STALLINGS	12	1	Choctaw 6-N 14-E
STANSEL	9	1	Choctaw 7-N 16-E
STANTON	5	4	Choctaw 8-N 18-E
" "	9	1	Choctaw 7-N 16-E
STEDIVANT	12	1	Choctaw 6-N 14-E
STEED	3	2	Choctaw 8-N 16-E
STEEDLEY	3	1	Choctaw 8-N 16-E
STEELE	8	1	Choctaw 7-N 15-E
STEINWINDER	18	2	Choctaw 5-N 15-E
" "	19	1	Choctaw 5-N 16-E
STEPHENS	12	2	Choctaw 6-N 14-E
" "	3	2	Choctaw 8-N 16-E
STEPHENSON	3	2	Choctaw 8-N 16-E
" "	8	1	Choctaw 7-N 15-E
STEWART	11	1	Choctaw 7-N 18-E
STINSON	19	4	Choctaw 5-N 16-E
" "	17	1	Choctaw 5-N 14-E
STOKES	8	3	Choctaw 7-N 15-E
" "	9	2	Choctaw 7-N 16-E
STONE	14	4	Choctaw 6-N 16-E
" "	19	2	Choctaw 5-N 16-E
STRAIT	5	16	Choctaw 8-N 18-E
" "	4	2	Choctaw 8-N 17-E
" "	6	2	Choctaw 8-N 19-E
STRANG	4	1	Choctaw 8-N 17-E
STRANGE	16	1	Choctaw 6-N 18-E
" "	1	1	Choctaw 8-N 14-E
STRINGER	8	2	Choctaw 7-N 15-E
STROUD	14	7	Choctaw 6-N 16-E
STUCKEY	17	3	Choctaw 5-N 14-E
SUGGS	15	1	Choctaw 6-N 17-E
" "	16	1	Choctaw 6-N 18-E
SULLIVAN	17	1	Choctaw 5-N 14-E
SUMMERS	12	1	Choctaw 6-N 14-E
SUTTLE	10	2	Choctaw 7-N 17-E
" "	17	1	Choctaw 5-N 14-E
" "	16	1	Choctaw 6-N 18-E
SUTTLES	1	4	Choctaw 8-N 14-E
SWAIN	5	3	Choctaw 8-N 18-E
SWIFT	4	3	Choctaw 8-N 17-E
" "	5	2	Choctaw 8-N 18-E
SWILLEY	14	4	Choctaw 6-N 16-E
" "	12	1	Choctaw 6-N 14-E
SWILLY	18	1	Choctaw 5-N 15-E
TABB	18	13	Choctaw 5-N 15-E
" "	13	11	Choctaw 6-N 15-E
" "	9	5	Choctaw 7-N 16-E
TALBERT	3	2	Choctaw 8-N 16-E
TAPPAN	9	5	Choctaw 7-N 16-E
" "	5	5	Choctaw 8-N 18-E
" "	2	3	Choctaw 8-N 15-E
" "	3	3	Choctaw 8-N 16-E
" "	8	2	Choctaw 7-N 15-E
TARRANT	3	1	Choctaw 8-N 16-E

Surname	Map Group	Parcels of Land	Meridian/Township/Range
TARTT	4	1	Choctaw 8-N 17-E
" "	5	1	Choctaw 8-N 18-E
TATUM	16	1	Choctaw 6-N 18-E
TAYLOR	19	7	Choctaw 5-N 16-E
" "	1	5	Choctaw 8-N 14-E
" "	13	3	Choctaw 6-N 15-E
" "	18	1	Choctaw 5-N 15-E
" "	16	1	Choctaw 6-N 18-E
" "	11	1	Choctaw 7-N 18-E
TEMPLETON	11	1	Choctaw 7-N 18-E
TERRAL	16	1	Choctaw 6-N 18-E
TERRELL	15	3	Choctaw 6-N 17-E
THOMAS	13	3	Choctaw 6-N 15-E
" "	18	1	Choctaw 5-N 15-E
" "	3	1	Choctaw 8-N 16-E
" "	4	1	Choctaw 8-N 17-E
THOMPSON	2	6	Choctaw 8-N 15-E
" "	17	4	Choctaw 5-N 14-E
" "	21	3	Choctaw 5-N 18-E
" "	18	2	Choctaw 5-N 15-E
" "	14	2	Choctaw 6-N 16-E
" "	19	1	Choctaw 5-N 16-E
" "	15	1	Choctaw 6-N 17-E
" "	5	1	Choctaw 8-N 18-E
THOMSON	16	1	Choctaw 6-N 18-E
THORNTON	8	2	Choctaw 7-N 15-E
" "	9	1	Choctaw 7-N 16-E
" "	11	1	Choctaw 7-N 18-E
TILL	10	1	Choctaw 7-N 17-E
TILLMAN	12	7	Choctaw 6-N 14-E
" "	17	1	Choctaw 5-N 14-E
TIMBREL	14	1	Choctaw 6-N 16-E
TIMMS	14	2	Choctaw 6-N 16-E
TIMS	14	1	Choctaw 6-N 16-E
TINNIN	9	5	Choctaw 7-N 16-E
TODD	1	3	Choctaw 8-N 14-E
TOLSON	8	3	Choctaw 7-N 15-E
TOOL	16	3	Choctaw 6-N 18-E
TOOMEY	16	1	Choctaw 6-N 18-E
TORRANS	9	1	Choctaw 7-N 16-E
TOUCHSTONE	11	1	Choctaw 7-N 18-E
TOURELL	19	1	Choctaw 5-N 16-E
TOWNSEND	8	2	Choctaw 7-N 15-E
TRAWICK	3	4	Choctaw 8-N 16-E
" "	2	3	Choctaw 8-N 15-E
TRUSSEL	1	1	Choctaw 8-N 14-E
TRUSSELL	1	49	Choctaw 8-N 14-E
" "	7	3	Choctaw 7-N 14-E
" "	2	2	Choctaw 8-N 15-E
TRUST	18	1	Choctaw 5-N 15-E
TUCKER	10	2	Choctaw 7-N 17-E
" "	4	1	Choctaw 8-N 17-E
TUGGLE	1	1	Choctaw 8-N 14-E
TURNER	18	2	Choctaw 5-N 15-E
" "	13	1	Choctaw 6-N 15-E
TUTON	5	6	Choctaw 8-N 18-E
" "	4	1	Choctaw 8-N 17-E
TUTT	4	12	Choctaw 8-N 17-E
" "	8	5	Choctaw 7-N 15-E
" "	1	5	Choctaw 8-N 14-E

Surname/Township Index

Surname	Map Group	Parcels of Land	Meridian/Township/Range
TUTT (Cont'd)	2	4	Choctaw 8-N 15-E
" "	9	2	Choctaw 7-N 16-E
" "	5	2	Choctaw 8-N 18-E
TWILLEY	12	1	Choctaw 6-N 14-E
ULMER	13	1	Choctaw 6-N 15-E
ULRIC	4	1	Choctaw 8-N 17-E
ULRICK	9	3	Choctaw 7-N 16-E
UPCHURCH	21	2	Choctaw 5-N 18-E
URY	10	1	Choctaw 7-N 17-E
VANCE	13	3	Choctaw 6-N 15-E
" "	9	1	Choctaw 7-N 16-E
VAUGHAN	17	2	Choctaw 5-N 14-E
VAUGHN	3	4	Choctaw 8-N 16-E
" "	17	3	Choctaw 5-N 14-E
" "	8	3	Choctaw 7-N 15-E
" "	18	2	Choctaw 5-N 15-E
VINSON	3	3	Choctaw 8-N 16-E
WABINGTON	14	1	Choctaw 6-N 16-E
" "	9	1	Choctaw 7-N 16-E
WADE	12	1	Choctaw 6-N 14-E
WADKINS	16	2	Choctaw 6-N 18-E
WAGGENER	3	1	Choctaw 8-N 16-E
WAGGONER	3	1	Choctaw 8-N 16-E
WAGONER	3	2	Choctaw 8-N 16-E
WAITS	14	1	Choctaw 6-N 16-E
" "	15	1	Choctaw 6-N 17-E
WALKER	11	14	Choctaw 7-N 18-E
" "	18	5	Choctaw 5-N 15-E
" "	5	5	Choctaw 8-N 18-E
" "	2	4	Choctaw 8-N 15-E
" "	12	3	Choctaw 6-N 14-E
" "	10	3	Choctaw 7-N 17-E
" "	17	2	Choctaw 5-N 14-E
" "	19	2	Choctaw 5-N 16-E
" "	8	2	Choctaw 7-N 15-E
" "	15	1	Choctaw 6-N 17-E
" "	16	1	Choctaw 6-N 18-E
WALLACE	18	16	Choctaw 5-N 15-E
" "	21	4	Choctaw 5-N 18-E
" "	12	1	Choctaw 6-N 14-E
WALTERER	4	1	Choctaw 8-N 17-E
WALTERS	4	2	Choctaw 8-N 17-E
WALTHALL	2	8	Choctaw 8-N 15-E
" "	14	2	Choctaw 6-N 16-E
" "	3	2	Choctaw 8-N 16-E
" "	13	1	Choctaw 6-N 15-E
WARBINGTON	9	3	Choctaw 7-N 16-E
" "	14	1	Choctaw 6-N 16-E
" "	15	1	Choctaw 6-N 17-E
" "	10	1	Choctaw 7-N 17-E
WARD	8	5	Choctaw 7-N 15-E
" "	16	2	Choctaw 6-N 18-E
" "	17	1	Choctaw 5-N 14-E
WARLINGTON	10	1	Choctaw 7-N 17-E
WARREN	5	6	Choctaw 8-N 18-E
" "	17	4	Choctaw 5-N 14-E
" "	1	4	Choctaw 8-N 14-E
WASHINGTON	18	2	Choctaw 5-N 15-E
" "	13	1	Choctaw 6-N 15-E
WATERS	4	2	Choctaw 8-N 17-E

Surname	Map Group	Parcels of Land	Meridian/Township/Range
WATES	15	1	Choctaw 6-N 17-E
WATKINS	16	3	Choctaw 6-N 18-E
" "	4	2	Choctaw 8-N 17-E
WATSON	13	1	Choctaw 6-N 15-E
" "	8	1	Choctaw 7-N 15-E
" "	11	1	Choctaw 7-N 18-E
" "	5	1	Choctaw 8-N 18-E
WATTERS	2	35	Choctaw 8-N 15-E
" "	1	17	Choctaw 8-N 14-E
WATTS	5	4	Choctaw 8-N 18-E
" "	15	1	Choctaw 6-N 17-E
" "	4	1	Choctaw 8-N 17-E
WEATHERFORD	19	1	Choctaw 5-N 16-E
WEATHERS	19	1	Choctaw 5-N 16-E
WEBB	18	1	Choctaw 5-N 15-E
WEBSTER	13	1	Choctaw 6-N 15-E
WEIR	1	2	Choctaw 8-N 14-E
WEISSINGER	10	1	Choctaw 7-N 17-E
WELCH	16	3	Choctaw 6-N 18-E
" "	21	1	Choctaw 5-N 18-E
" "	13	1	Choctaw 6-N 15-E
" "	15	1	Choctaw 6-N 17-E
WELLBORN	18	4	Choctaw 5-N 15-E
WELLS	7	1	Choctaw 7-N 14-E
" "	4	1	Choctaw 8-N 17-E
" "	5	1	Choctaw 8-N 18-E
WELSH	9	1	Choctaw 7-N 16-E
WEST	18	3	Choctaw 5-N 15-E
" "	17	1	Choctaw 5-N 14-E
" "	14	1	Choctaw 6-N 16-E
WESTBROOK	5	1	Choctaw 8-N 18-E
WHEAT	20	2	Choctaw 5-N 17-E
WHITE	4	32	Choctaw 8-N 17-E
" "	10	8	Choctaw 7-N 17-E
" "	5	7	Choctaw 8-N 18-E
" "	19	5	Choctaw 5-N 16-E
" "	14	4	CHOCTAW 6-N 16-E
" "	20	2	Choctaw 5-N 17-E
" "	15	2	Choctaw 6-N 17-E
" "	3	2	Choctaw 8-N 16-E
WHITEHEAD	18	3	Choctaw 5-N 15-E
" "	19	3	Choctaw 5-N 16-E
" "	13	2	Choctaw 6-N 15-E
" "	15	2	Choctaw 6-N 17-E
" "	9	1	Choctaw 7-N 16-E
WHITFIELD	4	4	Choctaw 8-N 17-E
WHITIKER	1	1	Choctaw 8-N 14-E
WHITLOCK	18	4	Choctaw 5-N 15-E
" "	17	2	Choctaw 5-N 14-E
WHITSETT	3	7	Choctaw 8-N 16-E
" "	5	2	Choctaw 8-N 18-E
" "	4	1	Choctaw 8-N 17-E
WIER	2	2	Choctaw 8-N 15-E
WIGGINS	16	2	Choctaw 6-N 18-E
WILKERSON	3	1	Choctaw 8-N 16-E
WILKINSON	21	4	Choctaw 5-N 18-E
" "	16	3	Choctaw 6-N 18-E
" "	5	3	Choctaw 8-N 18-E
" "	4	2	Choctaw 8-N 17-E
WILLARD	14	1	Choctaw 6-N 16-E

Surname/Township Index

Surname	Map Group	Parcels of Land	Meridian/Township/Range
WILLIAMS	14	4	Choctaw 6-N 16-E
" "	18	2	Choctaw 5-N 15-E
" "	19	2	Choctaw 5-N 16-E
" "	15	2	Choctaw 6-N 17-E
" "	3	2	Choctaw 8-N 16-E
" "	21	1	Choctaw 5-N 18-E
" "	13	1	Choctaw 6-N 15-E
" "	8	1	Choctaw 7-N 15-E
" "	10	1	Choctaw 7-N 17-E
WILLIAMSON	1	3	Choctaw 8-N 14-E
" "	17	1	Choctaw 5-N 14-E
" "	9	1	Choctaw 7-N 16-E
" "	10	1	Choctaw 7-N 17-E
WILLIS	20	1	Choctaw 5-N 17-E
WILSON	17	2	Choctaw 5-N 14-E
" "	12	2	Choctaw 6-N 14-E
" "	13	2	Choctaw 6-N 15-E
" "	14	2	Choctaw 6-N 16-E
" "	9	2	Choctaw 7-N 16-E
" "	15	1	Choctaw 6-N 17-E
" "	8	1	Choctaw 7-N 15-E
" "	11	1	Choctaw 7-N 18-E
WINDHAM	1	2	Choctaw 8-N 14-E
WINGATE	19	3	Choctaw 5-N 16-E
WINN	4	1	Choctaw 8-N 17-E
WINNINGHAM	4	3	Choctaw 8-N 17-E
" "	3	2	Choctaw 8-N 16-E
WITHERSPOON	11	2	Choctaw 7-N 18-E
WITT	1	2	Choctaw 8-N 14-E
WOLF	13	2	Choctaw 6-N 15-E
WOMACK	8	2	Choctaw 7-N 15-E
WOOD	14	1	Choctaw 6-N 16-E
" "	9	1	Choctaw 7-N 16-E
" "	11	1	Choctaw 7-N 18-E
WOODS	19	1	Choctaw 5-N 16-E
WOODWARD	18	1	Choctaw 5-N 15-E
WOOTAN	2	1	Choctaw 8-N 15-E
WOOTON	1	1	Choctaw 8-N 14-E
WORBINGTON	14	1	Choctaw 6-N 16-E
" "	9	1	Choctaw 7-N 16-E
" "	10	1	Choctaw 7-N 17-E
" "	1	1	Choctaw 8-N 14-E
WORMACK	8	1	Choctaw 7-N 15-E
WRIGHT	11	7	Choctaw 7-N 18-E
" "	17	2	Choctaw 5-N 14-E
" "	5	2	Choctaw 8-N 18-E
" "	14	1	Choctaw 6-N 16-E
" "	3	1	Choctaw 8-N 16-E
YARBOROUGH	5	2	Choctaw 8-N 18-E
YARBROUGH	19	2	Choctaw 5-N 16-E
" "	12	1	Choctaw 6-N 14-E
YARRELL	10	1	Choctaw 7-N 17-E
" "	4	1	Choctaw 8-N 17-E
YATES	10	2	Choctaw 7-N 17-E
YEAGER	17	2	Choctaw 5-N 14-E
" "	10	1	Choctaw 7-N 17-E
YOES	10	1	Choctaw 7-N 17-E
YOUNG	14	2	Choctaw 6-N 16-E
" "	18	1	Choctaw 5-N 15-E
" "	15	1	Choctaw 6-N 17-E

Surname	Map Group	Parcels of Land	Meridian/Township/Range
YOUNG (Cont'd)	**8**	1	Choctaw 7-N 15-E

– Part II –

Township Map Groups

Map Group 1: Index to Land Patents
Township 8-North Range 14-East (Choctaw)

After you locate an individual in this Index, take note of the Section and Section Part then proceed to the Land Patent map on the pages immediately following. You should have no difficulty locating the corresponding parcel of land.

The "For More Info" Column will lead you to more information about the underlying Patents. See the *Legend* at right, and the "How to Use this Book" chapter, for more information.

LEGEND
"For More Info . . . " column

- **A** = Authority (Legislative Act, See Appendix "A")
- **B** = Block or Lot (location in Section unknown)
- **C** = Cancelled Patent
- **F** = Fractional Section
- **G** = Group (Multi-Patentee Patent, see Appendix "C")
- **V** = Overlaps another Parcel
- **R** = Re-Issued (Parcel patented more than once)

(A & G items require you to look in the Appendixes referred to above. All other Letter-designations followed by a number require you to locate line-items in this index that possess the ID number found after the letter).

ID	Individual in Patent	Sec.	Sec. Part	Date Issued	Other Counties	For More Info . . .
110	ALFORD, Julius	2	1	1844-09-10		A1 G6
2	BAILES, Alfred	1	1	1841-02-27		A1
3	BAIRD, Allen C	31	SWSW	1841-02-27		A1
100	BEEMON, John W	17	S½SW	1895-02-21		A4
101	" "	17	W½SE	1895-02-21		A4
30	BOWEN, Horatio	28	N½	1841-02-27		A1 G24
127	BRECK, Samuel	15	W½	1841-02-27		A1 G30
128	" "	15	W½NE	1841-02-27		A1 G30
129	" "	22	N½	1841-02-27		A1 G30
205	CALLUM, Willis	3	16	1841-02-27		A1
206	" "	3	17	1841-02-27		A1
151	COOKSEY, Tabitha	30	E½SE	1859-10-01		A1
127	CRUFT, Edward	15	W½	1841-02-27		A1 G30
128	" "	15	W½NE	1841-02-27		A1 G30
129	" "	22	N½	1841-02-27		A1 G30
153	DANIEL, Thompson M	6	10	1860-05-01		A1
154	" "	6	11	1860-05-01		A1
155	" "	6	7	1860-05-01		A1
156	" "	6	9	1860-05-01		A1
78	DAVIS, James T	17	NWNW	1905-03-30		A4 G61
78	DAVIS, Mattie	17	NWNW	1905-03-30		A4 G61
33	DICKSON, James F	9	NESW	1859-10-01		A1
34	" "	9	SENW	1859-10-01		A1
157	DOBBS, Tyrie D	24	NESW	1854-03-15		A1
158	" "	24	SENW	1854-03-15		A1
89	DUNKLEY, John G	6	15	1895-05-11		A4
90	" "	6	16	1895-05-11		A4
31	ETHRIDGE, Isaac	3	1	1859-10-01		A1
32	" "	3	2	1859-10-01		A1
29	EVERETT, Horace	30	SENE	1855-03-15		A1
15	FOSTER, Crofford	12	SWSW	1859-10-01		A1
4	GIBSON, Allen	10	NWSW	1859-10-01		A1
5	" "	10	SWNW	1859-10-01		A1
123	GIBSON, Rosier W	6	1	1894-04-10		A4
124	" "	6	2	1894-04-10		A4
125	" "	6	3	1894-04-10		A4
126	" "	6	6	1894-04-10		A4
1	GREEN, Abner	7	E½NE	1841-02-27		A1
160	HERRINGTON, Vincent J	14	NWSE	1859-10-01		A1
92	HILL, John R	4	12	1859-10-01		A1
93	" "	4	5	1859-10-01		A1
111	HOUSE, Marcus L	4	16	1856-04-01		A1
114	HOUSE, Marquis D	4	10	1859-10-01		A1
115	" "	4	9	1859-10-01		A1
180	HOUSE, Willie J	1	12	1841-02-27		A1
181	" "	1	13	1841-02-27		A1

Township 8-N Range 14-E (Choctaw) - Map Group 1

ID	Individual in Patent	Sec.	Sec. Part	Date Issued Other Counties	For More Info . . .
182	HOUSE, Willie J (Cont'd)	1	20	1841-02-27	A1
184	"	"	2	10	1841-02-27	A1
185	"	"	2	11	1841-02-27	A1
187	"	"	2	13	1841-02-27	A1
188	"	"	2	14	1841-02-27	A1
189	"	"	2	15	1841-02-27	A1
190	"	"	2	16	1841-02-27	A1
192	"	"	2	20	1841-02-27	A1
197	"	"	2	9	1841-02-27	A1
186	"	"	2	12	1854-03-15	A1
193	"	"	2	3	1854-03-15	A1
194	"	"	2	4	1854-03-15	A1
196	"	"	2	6	1854-03-15	A1
183	"	"	12	NWSE	1856-04-01	A1
191	"	"	2	2	1856-04-01	A1
195	"	"	2	5	1856-04-01	A1
198	"	"	3	10	1859-10-01	A1
199	"	"	4	20	1859-10-01	A1
200	"	"	8	NESE	1859-10-01	A1
201	"	"	8	NESW	1859-10-01	A1
202	"	"	8	NWSE	1859-10-01	A1
203	"	"	8	SESE	1859-10-01	A1
204	"	"	9	NWSW	1859-10-01	A1
102	JAMES, Joseph A	19	NESE	1904-07-13	A4
103	"	"	19	S½SE	1904-07-13	A4
104	"	"	19	SESW	1904-07-13	A4
130	KING, Seaborn	3	5	1859-10-01	A1
131	"	"	3	6	1859-10-01	A1
35	LITCHFIELD, James	3	7	1860-05-01	A1
11	MARTIN, Benjamin	14	SESE	1859-10-01	A1
159	MARTIN, Valentine	35	NENE	1841-02-27	A1
118	MASON, Obadiah T	9	SESE	1860-05-01	A1
38	MAY, James M	9	NWNE	1854-03-15	A1
39	"	"	9	SENE	1854-03-15	A1
36	"	"	4	19	1855-03-15	A1
37	"	"	9	NENE	1860-05-01	A1
117	MAY, Nathaniel T	4	18	1856-04-01	A1
91	MAYATT, John	17	NESE	1890-12-20	A1
8	MEEK, Amzi	8	E½NW	1841-02-27	A1 G154
119	MIEATT, Peter	12	SWSE	1859-10-01	A1
168	MILLS, Willard C	4	17	1851/03/15	A2
164	"	"	12	NWSW	1854-03-15	A1
165	"	"	12	SESW	1854-03-15	A1
166	"	"	24	NESE	1854-03-15	A1
167	"	"	24	SESE	1856-04-01	A1
169	"	"	5	1	1856-04-01	A1
177	MONTAGUE, William V	4	3	1859-10-01	A1
178	"	"	4	4	1859-10-01	A1
179	"	"	4	6	1859-10-01	A1
6	MOORE, Allen	30	NWNW	1854-03-15	A1
127	MORROW, Alexander P	15	W½	1841-02-27	A1 G30
128	"	"	15	W½NE	1841-02-27	A1 G30
129	"	"	22	N½	1841-02-27	A1 G30
12	PARKS, Charity	18	SENE	1848-09-01	A1
161	PARKS, Wayman A	18	SWSE	1859-10-01	A1
162	"	"	19	E½NW	1859-10-01	A1
163	"	"	19	SWNW	1859-10-01	A1
173	PARKS, William J	17	NESW	1854-03-15	A1
174	"	"	18	NESE	1854-03-15	A1
175	"	"	18	SWNE	1854-03-15	A1
10	PAYNE, Ben	5	8	1899-07-15	A4
150	PEARL, Sylvester	26	E½SE	1841-02-27	A1 G160
149	"	"	25	W½SW	1860-08-15	A1 G160
17	POWELL, Francis M	3	8	1899-04-01	A4
18	"	"	3	9	1899-04-01	A4
75	POWELL, James R	5	6	1896-10-31	A4
76	"	"	5	7	1896-10-31	A4
108	RASBERRY, Joseph	36	SWNW	1854-03-15	A1
77	RICHE, James S	19	N½NE	1902-01-17	A4
13	RUSHING, Charles E	17	NWSW	1860-10-01	A1
14	"	"	17	SWNW	1860-10-01	A1
84	SADLER, John A	6	8	1856-04-01	A1
23	SKELTON, George H	4	7	1859-10-01	A1

53

… Family Maps of Lauderdale County, Mississippi

ID	Individual in Patent	Sec.	Sec. Part	Date Issued	Other Counties	For More Info . . .
24	SKELTON, George H (Cont'd)	4	8	1859-10-01		A1
16	STRANGE, Ernest T	19	SENE	1916-07-25		A4
85	SUTTLES, John D	6	12	1898-08-27		A4
86	" "	6	13	1898-08-27		A4
87	" "	6	4	1898-08-27		A4
88	" "	6	5	1898-08-27		A4
120	TAYLOR, Preston	23	E½SW	1841-02-27		A1
121	" "	36	NWNW	1841-02-27		A1
122	" "	7	SWNW	1841-02-27		A1 G176
30	TAYLOR, Swepson	28	N½	1841-02-27		A1 G24
207	TAYLOR, Zachariah	10	NWNW	1859-10-01		A1
26	TODD, Harriet S	19	NESW	1902-07-03		A4
27	" "	19	NWSE	1902-07-03		A4
28	" "	19	SWNE	1902-07-03		A4
8	TRUSSEL, Matthew	8	E½NW	1841-02-27		A1 G154
9	TRUSSELL, Andrew J	31	W½NW	1841-02-27		A1
79	TRUSSELL, James	33	W½SE	1841-02-27		A1 G177
80	" "	7	W½SW	1841-02-27		A1 G178
42	TRUSSELL, James M	17	NE	1841-02-27		A1
58	" "	30	W½SE	1841-02-27		A1
59	" "	30	W½SW	1841-02-27		A1
69	" "	7	SE	1841-02-27		A1
122	" "	7	SWNW	1841-02-27		A1 G176
70	" "	7	W½NE	1841-02-27		A1
71	" "	8	E½NE	1841-02-27		A1
73	" "	8	W½NE	1841-02-27		A1
48	" "	23	SWNW	1848-09-01		A1
40	" "	10	NENE	1854-03-15		A1
41	" "	14	SWSE	1854-03-15		A1
44	" "	17	SENW	1854-03-15		A1
45	" "	17	SESE	1854-03-15		A1
46	" "	18	NENE	1854-03-15		A1
50	" "	26	NESW	1854-03-15		A1
51	" "	3	11	1854-03-15		A1
52	" "	3	12	1854-03-15		A1
53	" "	3	13	1854-03-15		A1
54	" "	3	14	1854-03-15		A1
55	" "	3	15	1854-03-15		A1
56	" "	3	19	1854-03-15		A1
57	" "	3	20	1854-03-15		A1
60	" "	34	E½SE	1854-03-15		A1
61	" "	36	NE	1854-03-15		A1
62	" "	36	NWSW	1854-03-15		A1
63	" "	5	5	1854-03-15		A1
64	" "	6	14	1854-03-15		A1
67	" "	6	19	1854-03-15		A1
49	" "	24	SWNW	1855-06-15		A1
43	" "	17	NENW	1859-10-01		A1
47	" "	20	SENW	1859-10-01		A1
74	" "	9	SWSW	1859-10-01		A1
68	" "	7	NWNW	1860-05-01		A1
72	" "	8	NWNW	1860-05-01		A1
65	" "	6	17	1897-05-05		A1
66	" "	6	18	1897-05-05		A1
150	TRUSSELL, John	26	E½SE	1841-02-27		A1 G160
96	" "	26	W½SE	1841-02-27		A1
79	" "	33	W½SE	1841-02-27		A1 G177
98	" "	34	W½NE	1841-02-27		A1
99	" "	34	W½SE	1841-02-27		A1
95	" "	24	SWSE	1854-03-15		A1
97	" "	34	S½SW	1859-10-01		A1
149	" "	25	W½SW	1860-08-15		A1 G160
116	TRUSSELL, Matthew	33	SESE	1841-02-27		A1
171	TRUSSELL, William C	18	NWNE	1841-02-27		A1
172	" "	30	SESW	1841-02-27		A1
80	" "	7	W½SW	1841-02-27		A1 G178
176	TUGGLE, William L	4	2	1859-10-01		A1
19	TUTT, Gabriel H	1	4	1841-02-27		A1
20	" "	1	5	1841-02-27		A1
110	" "	2	1	1844-09-10		A1 G6
112	TUTT, Marina	24	SESW	1841-02-27		A1
113	" "	24	W½SW	1841-02-27		A1
109	WARREN, Joseph	7	E½SW	1841-02-27		A1

Township 8-N Range 14-E (Choctaw) - Map Group 1

ID	Individual in Patent	Sec.	Sec. Part	Date Issued	Other Counties	For More Info . . .
105	WARREN, Joseph M	18	NWNW	1841-02-27		A1
106	" "	18	SENW	1841-02-27		A1
107	" "	7	E½NW	1841-02-27		A1
132	WATTERS, Stacy B	1	10	1841-02-27		A1
137	" "	1	2	1841-02-27		A1
138	" "	1	3	1841-02-27		A1
139	" "	1	6	1841-02-27		A1
140	" "	1	7	1841-02-27		A1
143	" "	12	E½NE	1841-02-27		A1
144	" "	12	E½SE	1841-02-27		A1
145	" "	13	E½SE	1841-02-27		A1
148	" "	24	NE	1841-02-27		A1
133	" "	1	15	1841-12-15		A1
134	" "	1	16	1841-12-15		A1
135	" "	1	17	1841-12-15		A1
136	" "	1	18	1841-12-15		A1
141	" "	1	8	1841-12-15		A1
142	" "	1	9	1841-12-15		A1
146	" "	15	E½NE	1841-12-15		A1
147	" "	15	NESE	1841-12-15		A1
150	WEIR, Robert	26	E½SE	1841-02-27		A1 G160
149	" "	25	W½SW	1860-08-15		A1 G160
94	WHITIKER, John S	24	NWNW	1854-03-15		A1
7	WILLIAMSON, Allen	10	SWSE	1859-10-01		A1
82	WILLIAMSON, Jessie T	9	SESW	1892-04-29		A4
83	" "	9	SWSE	1892-04-29		A4
25	WINDHAM, George W	18	S½W½NW	1841-02-27		A1
152	WINDHAM, Thomas J	18	N½E½NW	1841-02-27		A1
21	WITT, Gabriel H	2	7	1841-02-27		A1
22	" "	2	8	1841-02-27		A1
81	WOOTON, Jesse	27	SENW	1841-02-27		A1
170	WORBINGTON, William B	1	11	1841-02-27		A1

Family Maps of Lauderdale County, Mississippi

Patent Map

T8-N R14-E
Choctaw Meridian

Map Group 1

Township Statistics

Parcels Mapped	:	207
Number of Patents	:	138
Number of Individuals	:	81
Patentees Identified	:	77
Number of Surnames	:	61
Multi-Patentee Parcels	:	12
Oldest Patent Date	:	2/27/1841
Most Recent Patent	:	7/25/1916
Block/Lot Parcels	:	91
Parcels Re - Issued	:	0
Parcels that Overlap	:	0
Cities and Towns	:	2
Cemeteries	:	7

Lots-Sec. 6
1 GIBSON, Rosier W   1894
2 GIBSON, Rosier W   1894
3 GIBSON, Rosier W   1894
4 SUTTLES, John D   1898
5 SUTTLES, John D   1898
6 GIBSON, Rosier W   1894
7 DANIEL, Thompson M 1860
8 SADLER, John A   1856
9 DANIEL, Thompson M 1860
10 DANIEL, Thompson M 1860
11 DANIEL, Thompson M 1860
12 SUTTLES, John D   1898
13 SUTTLES, John D   1898
14 TRUSSELL, James M   1854
15 DUNKLEY, John G   1895
16 DUNKLEY, John G   1895
17 TRUSSELL, James M   1897
18 TRUSSELL, James M   1897
19 TRUSSELL, James M   1854

Lots-Sec. 5
1 MILLS, Willard C   1856
5 TRUSSELL, James M   1854
6 POWELL, James R   1896
7 POWELL, James R   1896
8 PAYNE, Ben   1899

Lots-Sec. 4
2 TUGGLE, William L   1859
3 MONTAGUE, William V 1859
4 MONTAGUE, William V 1859
5 HILL, John R   1859
6 MONTAGUE, William V 1859
7 SKELTON, George H   1859
8 SKELTON, George H   1859
9 HOUSE, Marquis D   1859
10 HOUSE, Marquis D   1859
12 HILL, John R   1859
16 HOUSE, Marcus L   1856
17 MILLS, Willard C   1851
18 MAY, Nathaniel T   1856
19 MAY, James M   1855
20 HOUSE, Willie J   1859

Copyright 2006 Boyd IT, Inc. All Rights Reserved

56

Township 8-N Range 14-E (Choctaw) - Map Group 1

Lots-Sec. 3
1	ETHRIDGE, Isaac	1859
2	ETHRIDGE, Isaac	1859
5	KING, Seaborn	1859
6	KING, Seaborn	1859
7	LITCHFIELD, James	1860
8	POWELL, Francis M	1899
9	POWELL, Francis M	1899
10	HOUSE, Willie J	1859
11	TRUSSELL, James M	1854
12	TRUSSELL, James M	1854
13	TRUSSELL, James M	1854
14	TRUSSELL, James M	1854
15	TRUSSELL, James M	1854
16	CALLUM, Willis	1841
17	CALLUM, Willis	1841
19	TRUSSELL, James M	1854
20	TRUSSELL, James M	1854

Lots-Sec. 2
1	ALFORD, Julius [6]	1844
2	HOUSE, Willie J	1856
3	HOUSE, Willie J	1854
4	HOUSE, Willie J	1854
5	HOUSE, Willie J	1856
6	HOUSE, Willie J	1854
7	WITT, Gabriel H	1841
8	WITT, Gabriel H	1841
9	HOUSE, Willie J	1841
10	HOUSE, Willie J	1841
11	HOUSE, Willie J	1841
12	HOUSE, Willie J	1854
13	HOUSE, Willie J	1841
14	HOUSE, Willie J	1841
15	HOUSE, Willie J	1841
16	HOUSE, Willie J	1841
20	HOUSE, Willie J	1841

Lots-Sec. 1
1	BAILES, Alfred	1841
2	WATTERS, Stacy B	1841
3	WATTERS, Stacy B	1841
4	TUTT, Gabriel H	1841
5	TUTT, Gabriel H	1841
6	WATTERS, Stacy B	1841
7	WATTERS, Stacy B	1841
8	WATTERS, Stacy B	1841
9	WATTERS, Stacy B	1841
10	WATTERS, Stacy B	1841
11	WORBINGTON, William	1841
12	HOUSE, Willie J	1841
13	HOUSE, Willie J	1841
15	WATTERS, Stacy B	1841
16	WATTERS, Stacy B	1841
17	WATTERS, Stacy B	1841
18	WATTERS, Stacy B	1841
20	HOUSE, Willie J	1841

Helpful Hints

1. This Map's INDEX can be found on the preceding pages.

2. Refer to Map "C" to see where this Township lies within Lauderdale County, Mississippi.

3. Numbers within square brackets [ ] denote a multi-patentee land parcel (multi-owner). Refer to Appendix "C" for a full list of members in this group.

4. Areas that look to be crowded with Patentees usually indicate multiple sales of the same parcel (Re-issues) or Overlapping parcels. See this Township's Index for an explanation of these and other circumstances that might explain "odd" groupings of Patentees on this map.

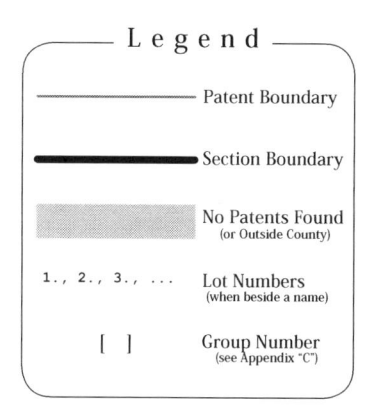

Copyright 2006 Boyd IT, Inc. All Rights Reserved

Legend

— Patent Boundary
— Section Boundary
(shaded) No Patents Found (or Outside County)
1., 2., 3., ... Lot Numbers (when beside a name)
[ ] Group Number (see Appendix "C")

Scale: Section = 1 mile X 1 mile (generally, with some exceptions)

Section 10: TAYLOR Zachariah 1859; GIBSON Allen 1859; GIBSON Allen 1859; TRUSSELL James M 1854; WILLIAMSON Allen 1859

Section 11

Section 12: WATTERS Stacy B 1841; MILLS Willard C 1854; HOUSE Willie J 1856; FOSTER Crofford 1859; MILLS Willard C 1854; MIEATT Peter 1859; WATTERS Stacy B 1841

Section 15: BRECK [30] Samuel 1841; BRECK [30] Samuel 1841; WATTERS Stacy B 1841; WATTERS Stacy B 1841

Section 14: HERRINGTON Vincent J 1859; TRUSSELL James M 1854; MARTIN Benjamin 1859

Section 13: WATTERS Stacy B 1841

Section 22: BRECK [30] Samuel 1841

Section 23: TRUSSELL James M 1848; TAYLOR Preston 1841

Section 24: WHITIKER John S 1854; TRUSSELL James M 1855; DOBBS Tyrie D 1854; WATTERS Stacy B 1841; TUTT Marina 1841; DOBBS Tyrie D 1854; MILLS Willard C 1854; TUTT Marina 1841; TRUSSELL John 1854; MILLS Willard C 1856

Section 27: WOOTON Jesse 1841

Section 26: TRUSSELL James M 1854; TRUSSELL John 1841; PEARL [160] Sylvester 1860; PEARL [160] Sylvester 1841

Section 25

Section 34: TRUSSELL John 1841; TRUSSELL John 1841; TRUSSELL John 1859; TRUSSELL James M 1854

Section 35: MARTIN Valentine 1841; TAYLOR Preston 1841; RASBERRY Joseph 1854; TRUSSELL James M 1854

Section 36: TRUSSELL James M 1854

57

Family Maps of Lauderdale County, Mississippi

Road Map
T8-N R14-E Choctaw Meridian
Map Group 1

Cities & Towns
Collinsville
Martin

Cemeteries
Collinsville Cemetery
Collinsville Church Cemetery
Hamrick Cemetery
Pine Forest Cemetery
Pine Grove Cemetery
Pleasant Ridge Cemetery
Trussel Cemetery

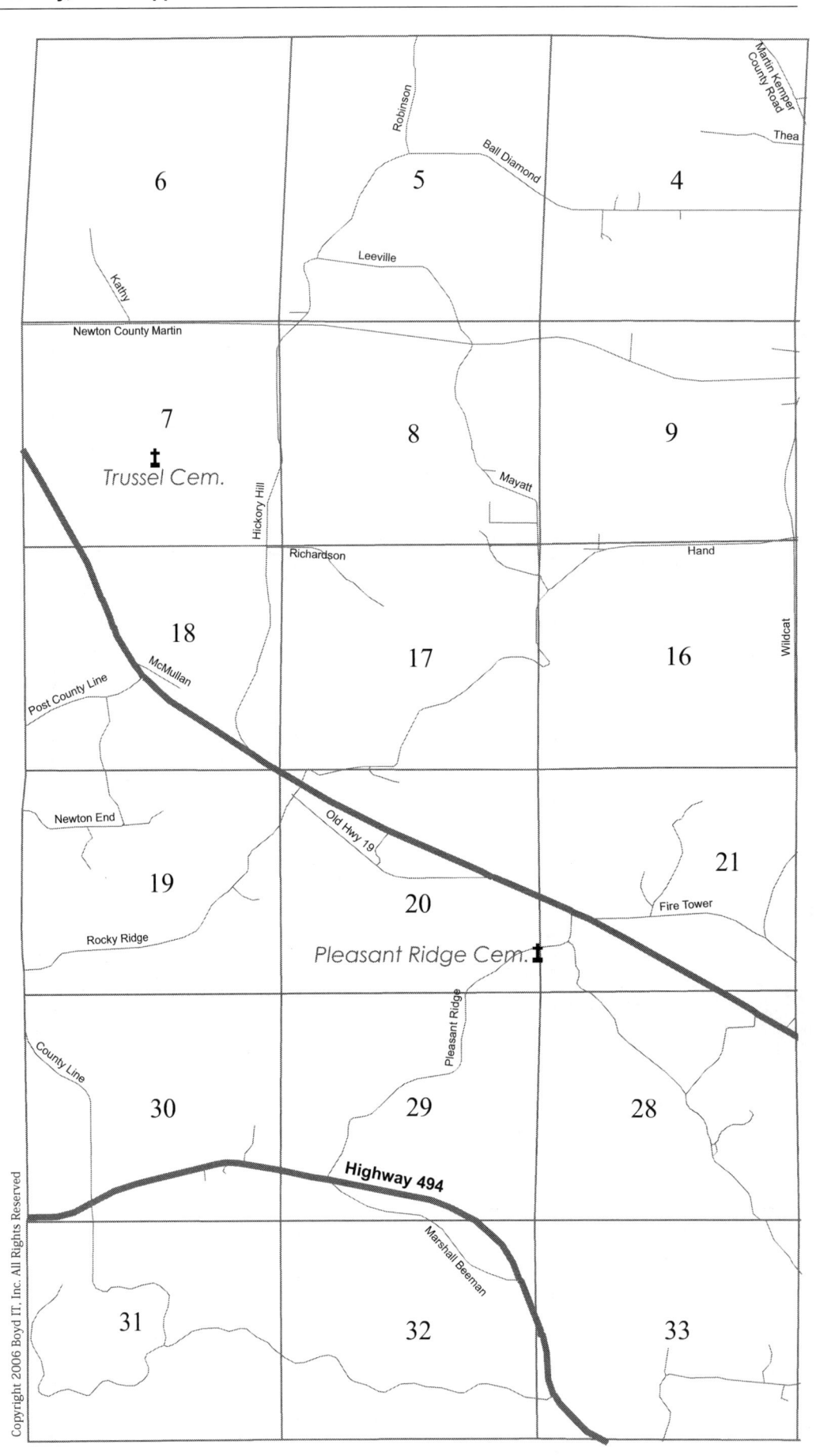

Township 8-N Range 14-E (Choctaw)   -   Map Group 1

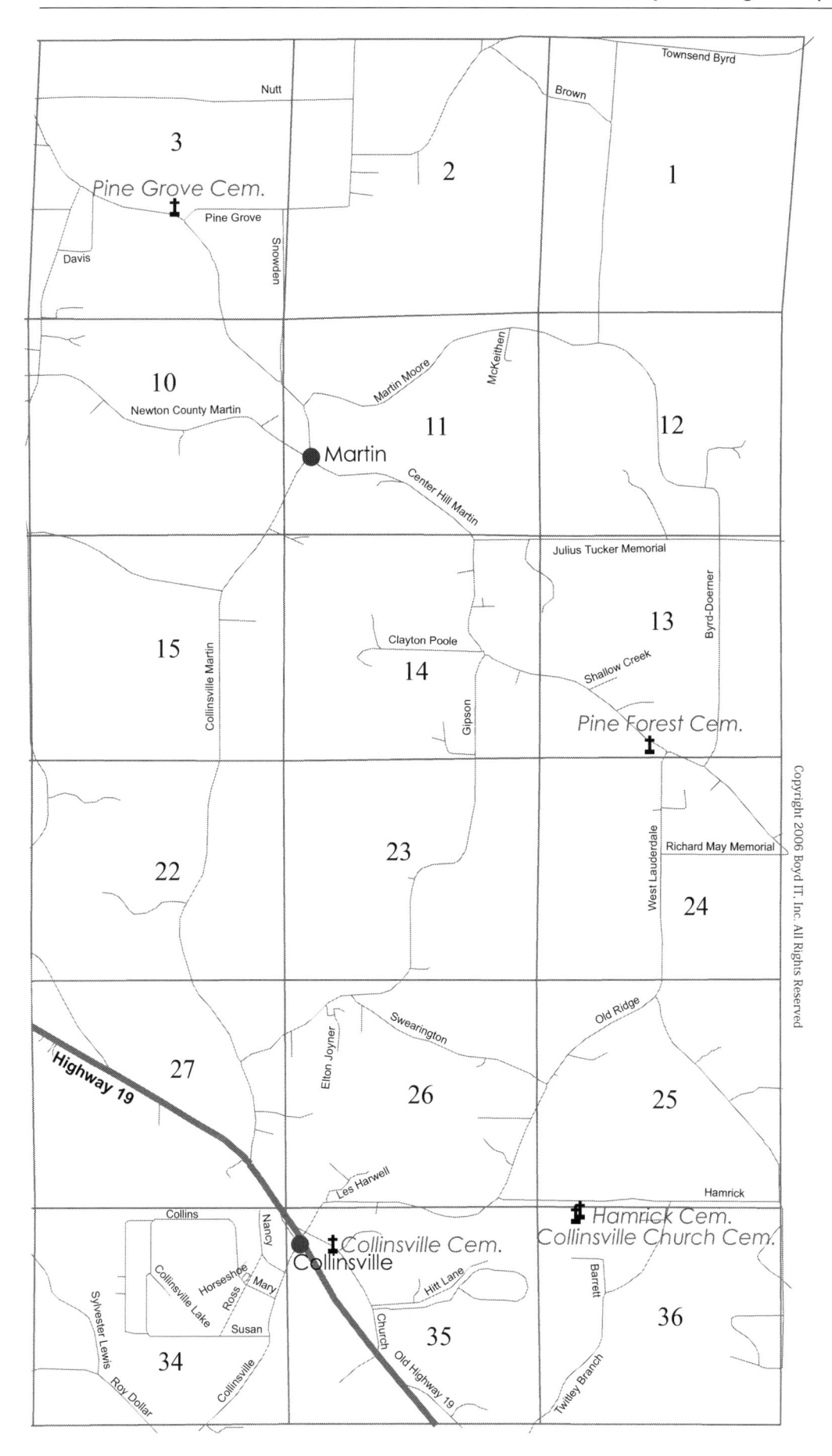

Helpful Hints

1. This road map has a number of uses, but primarily it is to help you: a) find the present location of land owned by your ancestors (at least the general area), b) find cemeteries and city-centers, and c) estimate the route/roads used by Census-takers & tax-assessors.

2. If you plan to travel to Lauderdale County to locate cemeteries or land parcels, please pick up a modern travel map for the area before you do. Mapping old land parcels on modern maps is not as exact a science as you might think. Just the slightest variations in public land survey coordinates, estimates of parcel boundaries, or road-map deviations can greatly alter a map's representation of how a road either does or doesn't cross a particular parcel of land.

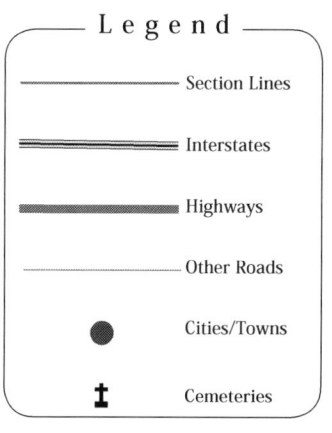

Legend

- Section Lines
- Interstates
- Highways
- Other Roads
- ● Cities/Towns
- ☦ Cemeteries

Scale: Section = 1 mile X 1 mile
(generally, with some exceptions)

Copyright 2006 Boyd IT, Inc. All Rights Reserved

59

Family Maps of Lauderdale County, Mississippi

Historical Map
T8-N R14-E
Choctaw Meridian
Map Group 1

Cities & Towns
Collinsville
Martin

Cemeteries
Collinsville Cemetery
Collinsville Church Cemetery
Hamrick Cemetery
Pine Forest Cemetery
Pine Grove Cemetery
Pleasant Ridge Cemetery
Trussel Cemetery

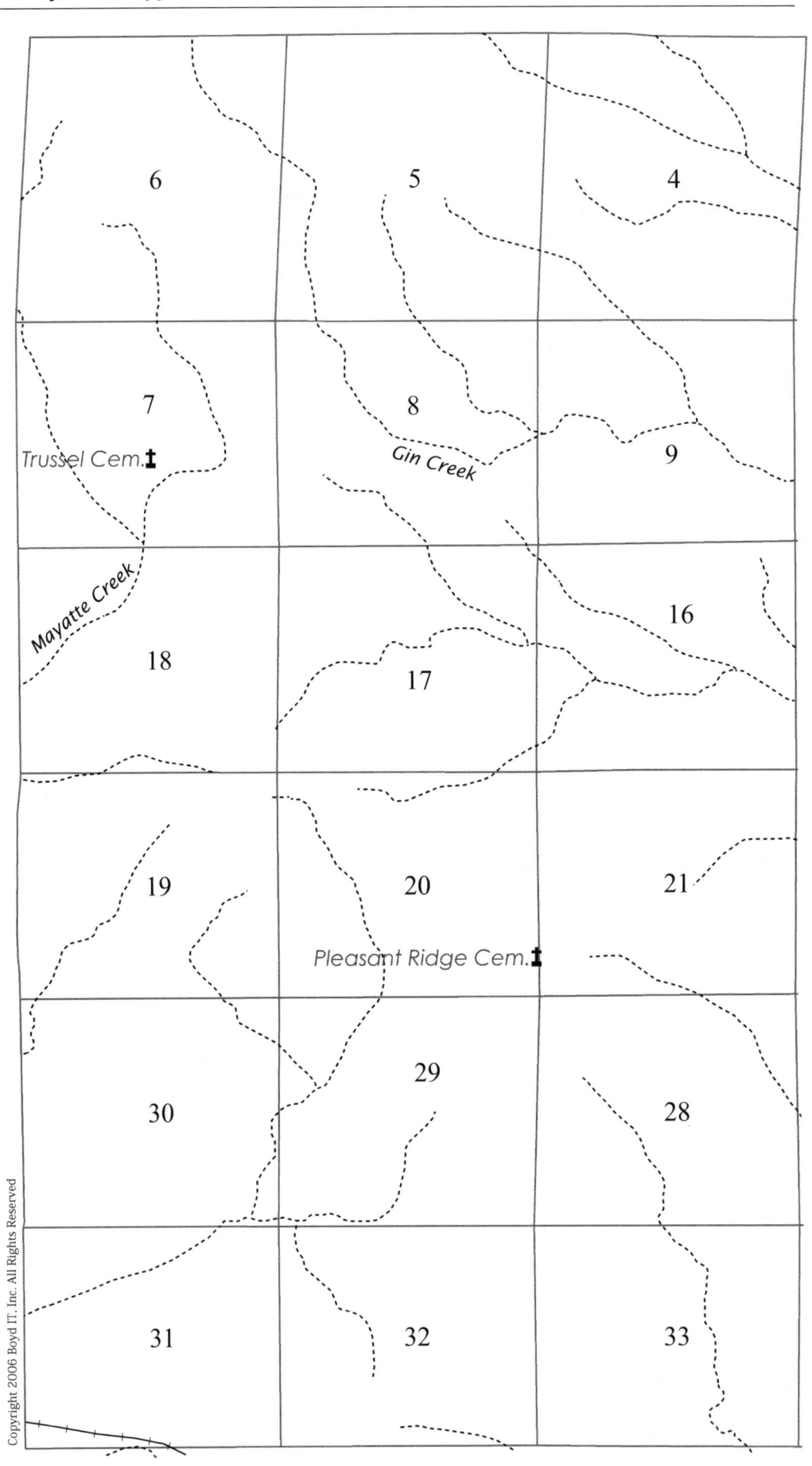

Township 8-N Range 14-E (Choctaw) - Map Group 1

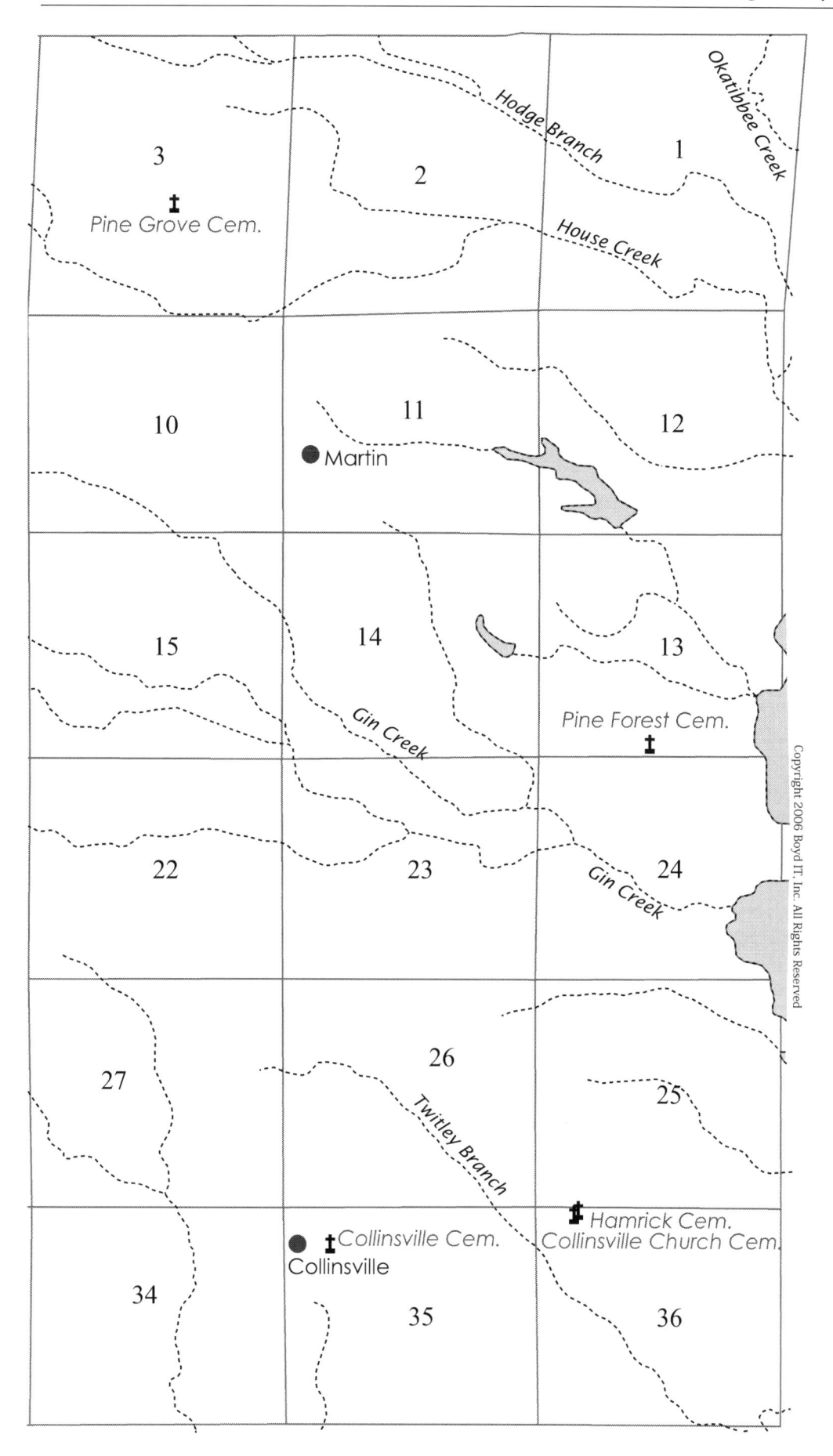

Helpful Hints

1. This Map takes a different look at the same Congressional Township displayed in the preceding two maps. It presents features that can help you better envision the historical development of the area: a) Water-bodies (lakes & ponds), b) Water-courses (rivers, streams, etc.), c) Railroads, d) City/town center-points (where they were oftentimes located when first settled), and e) Cemeteries.

2. Using this "Historical" map in tandem with this Township's Patent Map and Road Map, may lead you to some interesting discoveries. You will often find roads, towns, cemeteries, and waterways are named after nearby landowners: sometimes those names will be the ones you are researching. See how many of these research gems you can find here in Lauderdale County.

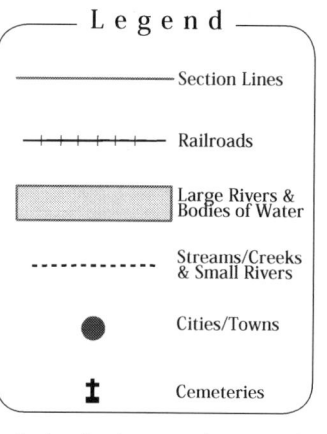

Legend

- Section Lines
- Railroads
- Large Rivers & Bodies of Water
- Streams/Creeks & Small Rivers
- Cities/Towns
- Cemeteries

Scale: Section = 1 mile X 1 mile
(there are some exceptions)

Features shown on map:
- Okatibbee Creek
- Hodge Branch
- House Creek
- Pine Grove Cem. (Section 3)
- Martin (Section 11)
- Gin Creek
- Pine Forest Cem. (Section 13)
- Twitley Branch
- Collinsville (Section 34/35)
- Collinsville Cem.
- Hamrick Cem.
- Collinsville Church Cem.

Sections: 1, 2, 3, 10, 11, 12, 13, 14, 15, 22, 23, 24, 25, 26, 27, 34, 35, 36

Copyright 2006 Boyd IT, Inc. All Rights Reserved

61

Map Group 2: Index to Land Patents
Township 8-North Range 15-East (Choctaw)

After you locate an individual in this Index, take note of the Section and Section Part then proceed to the Land Patent map on the pages immediately following. You should have no difficulty locating the corresponding parcel of land.

The "For More Info" Column will lead you to more information about the underlying Patents. See the *Legend* at right, and the "How to Use this Book" chapter, for more information.

LEGEND
"For More Info . . ." column

- **A** = Authority (Legislative Act, See Appendix "A")
- **B** = Block or Lot (location in Section unknown)
- **C** = Cancelled Patent
- **F** = Fractional Section
- **G** = Group (Multi-Patentee Patent, see Appendix "C")
- **V** = Overlaps another Parcel
- **R** = Re-Issued (Parcel patented more than once)

(A & G items require you to look in the Appendixes referred to above. All other Letter-designations followed by a number require you to locate line-items in this index that possess the ID number found after the letter).

ID	Individual in Patent	Sec.	Sec. Part	Date Issued	Other Counties	For More Info . . .
225	ADAIR, Armell F	3	10	1841-02-27		A1
226	"	3	15	1841-02-27		A1
227	"	3	16	1841-02-27		A1
228	"	3	9	1841-02-27		A1
229	"	4	14	1841-02-27		A1
230	ADAIR, Benjamin	4	4	1841-02-27		A1
314	ALFORD, Julius	14	E½SE	1841-02-27		A1
315	"	23	W½NE	1841-02-27		A1
316	"	31	E½NE	1841-02-27		A1
317	"	31	E½NW	1841-02-27		A1
339	ANDERSON, Richard	26	W½SW	1841-02-27		A1 G12
340	"	27	E½	1841-02-27		A1 G12
341	"	33	E½NW	1841-02-27		A1 G12
342	"	33	SW	1841-02-27		A1 G12
343	"	33	W½NE	1841-02-27		A1 G12
344	"	34	E½SE	1841-02-27		A1 G12
345	"	34	NE	1841-02-27		A1 G12
346	"	35	SW	1841-02-27		A1 G12
404	AVARA, Thomas	21	W½SW	1850-12-05		A1
405	AVARA, Thomas G	21	NENE	1848-09-01		A1
208	AVERY, Albert	28	NWSE	1854-03-15		A1
209	BAILES, Alfred	32	N½E½NE	1841-02-27		A1
210	"	6	4	1841-02-27		A1
211	"	6	5	1841-02-27		A1
319	BAINES, Littleberry	33	SE	1841-02-27		A1
349	BAINES, Richmond	24	E½SE	1841-02-27		A1
266	BAINS, Henry	33	E½NE	1841-02-27		A1
284	BROOKE, James R	1	15	1850-12-05		A1
253	BROWN, David M	28	NWNE	1854-03-15		A1
270	BROWN, James	21	E½SW	1841-02-27		A1
271	"	21	SENE	1841-02-27		A1
272	"	21	W½NE	1841-02-27		A1
274	"	35	W½SE	1841-02-27		A1
273	"	28	NENE	1848-09-01		A1
296	BROWN, John	36	SWNE	1859-10-01		A1
320	BROWN, Martha	22	W½NW	1841-02-27		A1
321	"	22	W½SW	1841-02-27		A1
322	"	25	W½SE	1841-02-27		A1
323	"	36	E½NW	1841-02-27		A1
329	BROWN, Randolph	36	SESW	1854-03-15		A1
293	BUTLER, Jesse T	10	W½SW	1841-02-27		A1
294	"	21	E½NW	1841-02-27		A1
295	"	21	SE	1841-02-27		A1
324	BUTLER, Moses H	29	W½SE	1841-02-27		A1
277	CASTLES, James	34	SWNW	1841-02-27		A1
236	COLEMAN, Charles H	2	11	1841-02-27		A1

Township 8-N Range 15-E (Choctaw) - Map Group 2

ID	Individual in Patent	Sec.	Sec. Part	Date Issued	Other Counties	For More Info...
237	COLEMAN, Charles H (Cont'd)	2	12	1841-02-27		A1
238	" "	2	13	1841-02-27		A1
239	" "	2	20	1841-02-27		A1
240	" "	2	3	1841-02-27		A1
241	" "	2	4	1841-02-27		A1
242	" "	2	5	1841-02-27		A1
243	" "	2	6	1841-02-27		A1
244	" "	3	1	1841-02-27		A1
245	" "	3	2	1841-02-27		A1
246	" "	3	7	1841-02-27		A1
247	" "	3	8	1841-02-27		A1
418	COOPER, William B	10	E½SE	1841-02-27		A1
351	CRAIG, Robert	13	NW	1841-02-27		A1
352	" "	13	SW	1841-02-27		A1
353	" "	14	SW	1841-02-27		A1
354	" "	14	W½NW	1841-02-27		A1
355	" "	15	E½NE	1841-02-27		A1
356	" "	15	E½SE	1841-02-27		A1
357	" "	23	NW	1841-02-27		A1
358	" "	23	SE	1841-02-27		A1
359	" "	24	NW	1841-02-27		A1
360	" "	24	W½SW	1841-02-27		A1
361	" "	34	SW	1841-02-27		A1
362	" "	34	W½SE	1841-02-27		A1
278	CRAWFORD, James	25	E½SE	1841-02-27		A1
279	" "	25	N½	1841-02-27		A1
280	" "	25	SW	1841-02-27		A1
289	CRAWLEY, James T	28	SWNW	1841-02-27		A1
234	DAVIS, Bluford G	2	2	1859-10-01		A1
257	DAVIS, Frederick A	29	E½SE	1841-05-27		A1 G60
258	" "	29	NE	1841-05-27		A1 G60
297	DAVIS, John	3	13	1841-02-27		A1
298	" "	3	14	1841-02-27		A1
299	" "	3	19	1841-02-27		A1
300	" "	3	20	1841-02-27		A1
363	DAWKINS, Samuel	9	NE	1841-02-27		A1
364	" "	9	SE	1841-02-27		A1
264	DENTON, Harvil	24	E½SW	1841-02-27		A1
265	" "	24	W½SE	1841-02-27		A1
267	DORMAN, Henry M	10	SENE	1859-10-01		A1
368	ETHRIDGE, Solomon	10	W½NE	1859-10-01		A1
339	GRANTLAND, Seaton	26	W½SW	1841-02-27		A1 G12
340	" "	27	E½	1841-02-27		A1 G12
341	" "	33	E½NW	1841-02-27		A1 G12
342	" "	33	SW	1841-02-27		A1 G12
343	" "	33	W½NE	1841-02-27		A1 G12
344	" "	34	E½SE	1841-02-27		A1 G12
345	" "	34	NE	1841-02-27		A1 G12
346	" "	35	SW	1841-02-27		A1 G12
259	GRAVES, George K	22	SENW	1854-03-15		A1
350	GRAVES, Robert B	28	SWSE	1849-12-01		A1
263	HOLLAND, Harfery W	29	NENW	1841-02-27		A1
406	HOLLAND, Thomas	15	SW	1841-02-27		A1
407	HOLLAND, Thomas M	28	NWSW	1859-10-01		A1
365	HUBBARD, Samuel	23	E½NE	1841-02-27		A1 G123
366	" "	31	W½NE	1841-02-27		A1 G123
367	" "	31	W½SE	1841-02-27		A1 G123
215	HUTTAN, Aquilla D	17	E½SE	1841-02-27		A1
216	" "	9	SW	1841-02-27		A1
217	" "	9	W½NW	1841-02-27		A1
421	HUTTAN, William J	4	12	1841-02-27		A1
422	" "	4	5	1841-02-27		A1
423	" "	5	1	1841-02-27		A1
424	" "	5	10	1841-02-27		A1
425	" "	5	11	1841-02-27		A1
426	" "	5	16	1841-02-27		A1
427	" "	5	17	1841-02-27		A1
428	" "	5	6	1841-02-27		A1
429	" "	5	7	1841-02-27		A1
430	" "	5	8	1841-02-27		A1
431	" "	5	9	1841-02-27		A1
218	HUTTON, Aquilla D	17	W½NE	1841-02-27		A1
302	HUTTON, John N	4	1	1841-02-27		A1

ID	Individual in Patent	Sec.	Sec. Part	Date Issued	Other Counties	For More Info . . .
303	HUTTON, John N (Cont'd)	4	15	1841-02-27		A1
304	" "	4	16	1841-02-27		A1
305	" "	4	17	1841-02-27		A1
306	" "	4	18	1841-02-27		A1
307	" "	4	8	1841-02-27		A1
308	" "	4	9	1841-02-27		A1
432	HUTTON, William J	4	13	1841-02-27		A1
433	" "	5	15	1841-02-27		A1
434	" "	5	18	1841-02-27		A1
420	JOLLY, William E	12	SWSE	1848-09-01		A1
281	KNOX, James	28	NWNW	1841-02-27		A1
365	LEWIS, Rufus G	23	E½NE	1841-02-27		A1 G123
366	" "	31	W½NE	1841-02-27		A1 G123
367	" "	31	W½SE	1841-02-27		A1 G123
301	LLOYD, John E	12	E½SE	1841-02-27		A1 G136
233	LONG, Benjamin S	35	NW	1841-02-27		A1
275	MCCRORY, James C	17	E½NE	1841-02-27		A1
276	" "	8	E½SE	1841-02-27		A1
348	MCLAMORE, Richard S	12	NWSE	1855-06-15		A1
313	MILES, Joshua	18	W½NE	1854-03-15		A1
416	MILLS, Willard C	12	NWNW	1854-03-15		A1
417	" "	12	SENW	1854-03-15		A1
415	" "	10	NENE	1859-10-01		A1
301	MORRIS, John H	12	E½SE	1841-02-27		A1 G136
254	MOSLEY, Elisha	12	E½NE	1854-03-15		A1
255	" "	12	NWNE	1854-03-15		A1
412	MOSLEY, Wiley	36	E½NE	1841-02-27		A1
413	" "	36	NESW	1841-02-27		A1
414	" "	36	NWNE	1841-02-27		A1
411	MURPHY, Van L	28	SESE	1854-03-15		A1
256	NASH, Ezekiel	10	NW	1841-02-27		A1 G155
256	NASH, Orsamus L	10	NW	1841-02-27		A1 G155
325	" "	11	SE	1841-02-27		A1 G156
326	" "	14	NE	1841-02-27		A1 G156
328	" "	21	W½NW	1841-02-27		A1 G156
257	" "	29	E½SE	1841-05-27		A1 G60
258	" "	29	NE	1841-05-27		A1 G60
327	" "	20	E½	1844-09-10		A1 G156
231	NEIGHBOURS, Benjamin	28	NESE	1849-12-01		A1
232	" "	28	SWNE	1849-12-01		A1
252	PRICE, Daniel M	12	NESW	1850-12-05		A1
435	PRIER, William	32	W½SE	1841-02-27		A1
436	PRYER, William	32	E½SE	1844-09-10		A1
330	RICHARDSON, Ransom	22	E½SW	1841-02-27		A1
331	" "	22	SESE	1841-02-27		A1
332	" "	22	W½SE	1841-02-27		A1
333	" "	23	SW	1841-02-27		A1
335	" "	26	NESE	1841-02-27		A1
336	" "	26	NW	1841-02-27		A1
338	" "	35	W½NE	1841-02-27		A1
334	" "	26	E½SW	1844-09-10		A1
337	" "	26	W½SE	1844-09-10		A1
310	ROGERS, John W	8	E½SW	1841-02-27		A1
311	" "	8	NE	1841-02-27		A1
312	" "	8	W½SE	1841-02-27		A1
235	RUSHING, Charles E	32	NWNW	1854-03-15		A1
260	SANDERFORD, Gray	22	NESE	1854-03-15		A1
318	SANDERFORD, Kemp	11	W½SW	1841-02-27		A1
262	SANDIFORD, Gray	24	NE	1841-02-27		A1
261	" "	13	SE	1844-09-10		A1
248	SHUMATE, Daniel D	1	1	1841-02-27		A1
251	" "	1	8	1841-02-27		A1
249	" "	1	2	1844-09-10		A1
250	" "	1	7	1844-09-10		A1
285	SNOWDEN, James	4	10	1841-02-27		A1
286	" "	4	11	1841-02-27		A1
287	" "	4	19	1841-02-27		A1
288	" "	9	E½NW	1841-02-27		A1
292	SNOWDEN, Jared	32	W½NE	1854-03-15		A1
309	SNOWDEN, John	4	20	1841-02-27		A1
408	SNOWDEN, Thomas V	20	E½SW	1854-03-15		A1
409	" "	20	NWSW	1854-03-15		A1
410	" "	20	SWSW	1854-03-15		A1

Township 8-N Range 15-E (Choctaw) - Map Group 2

ID	Individual in Patent	Sec.	Sec. Part	Date Issued	Other Counties	For More Info...
419	SNOWDEN, William B	20	NW	1854-03-15		A1
365	TAPPAN, John	23	E½NE	1841-02-27		A1 G123
366	" "	31	W½NE	1841-02-27		A1 G123
367	" "	31	W½SE	1841-02-27		A1 G123
219	THOMPSON, Archibald	3	11	1841-02-27		A1
220	" "	3	12	1841-02-27		A1
221	" "	3	3	1841-02-27		A1
222	" "	3	4	1841-02-27		A1
223	" "	3	5	1841-02-27		A1
224	" "	3	6	1841-02-27		A1
212	TRAWICK, Allen	10	E½SW	1841-02-27		A1
213	" "	10	W½SE	1841-02-27		A1
214	" "	15	NW	1841-02-27		A1
282	TRUSSELL, James M	29	SENW	1850-12-05		A1
283	" "	30	S½	1854-03-15		A1
269	TUTT, James B	19	E½NW	1841-02-27		A1
347	TUTT, Richard B	32	SENE	1841-02-27		A1
438	TUTT, Wilson G	33	NWNW	1841-02-27		A1
437	" "	28	SWSW	1845-01-21		A1
325	WALKER, Robert L	11	SE	1841-02-27		A1 G156
326	" "	14	NE	1841-02-27		A1 G156
328	" "	21	W½NW	1841-02-27		A1 G156
327	" "	20	E½	1844-09-10		A1 G156
339	WALTHALL, Madison	26	W½SW	1841-02-27		A1 G12
340	" "	27	E½	1841-02-27		A1 G12
341	" "	33	E½NW	1841-02-27		A1 G12
342	" "	33	SW	1841-02-27		A1 G12
343	" "	33	W½NE	1841-02-27		A1 G12
344	" "	34	E½SE	1841-02-27		A1 G12
345	" "	34	NE	1841-02-27		A1 G12
346	" "	35	SW	1841-02-27		A1 G12
369	WATTERS, Stacy B	18	E½SW	1841-02-27		A1
370	" "	18	NW	1841-02-27		A1
371	" "	18	W½SE	1841-02-27		A1
372	" "	19	E½SW	1841-02-27		A1
373	" "	19	SE	1841-02-27		A1
374	" "	19	W½NE	1841-02-27		A1
375	" "	19	W½NW	1841-02-27		A1
376	" "	30	NE	1841-02-27		A1
377	" "	30	W½NW	1841-02-27		A1
382	" "	5	3	1841-02-27		A1
383	" "	5	4	1841-02-27		A1
384	" "	5	5	1841-02-27		A1
388	" "	6	12	1841-02-27		A1
389	" "	6	13	1841-02-27		A1
390	" "	6	14	1841-02-27		A1
391	" "	6	15	1841-02-27		A1
392	" "	6	16	1841-02-27		A1
393	" "	6	17	1841-02-27		A1
394	" "	6	18	1841-02-27		A1
395	" "	6	19	1841-02-27		A1
397	" "	6	20	1841-02-27		A1
403	" "	7		1841-02-27		A1
378	" "	4	2	1841-12-15		A1
379	" "	4	3	1841-12-15		A1
380	" "	4	6	1841-12-15		A1
381	" "	4	7	1841-12-15		A1
385	" "	6	1	1841-12-15		A1
386	" "	6	10	1841-12-15		A1
387	" "	6	11	1841-12-15		A1
396	" "	6	2	1841-12-15		A1
398	" "	6	3	1841-12-15		A1
399	" "	6	6	1841-12-15		A1
400	" "	6	7	1841-12-15		A1
401	" "	6	8	1841-12-15		A1
402	" "	6	9	1841-12-15		A1
290	WIER, James	1	16	1841-02-27		A1
291	" "	1	9	1841-02-27		A1
268	WOOTAN, Israel F	12	SWNW	1859-10-01		A1

Family Maps of Lauderdale County, Mississippi

Patent Map

T8-N R15-E
Choctaw Meridian

Map Group 2

Township Statistics

Parcels Mapped	:	231
Number of Patents	:	148
Number of Individuals	:	87
Patentees Identified	:	81
Number of Surnames	:	62
Multi-Patentee Parcels	:	19
Oldest Patent Date	:	2/27/1841
Most Recent Patent	:	10/1/1859
Block/Lot Parcels	:	88
Parcels Re - Issued	:	0
Parcels that Overlap	:	0
Cities and Towns	:	3
Cemeteries	:	4

Lots-Sec. 6
1 WATTERS, Stacy B 1841
2 WATTERS, Stacy B 1841
3 WATTERS, Stacy B 1841
4 BAILES, Alfred 1841
5 BAILES, Alfred 1841
6 WATTERS, Stacy B 1841
7 WATTERS, Stacy B 1841
8 WATTERS, Stacy B 1841
9 WATTERS, Stacy B 1841
10 WATTERS, Stacy B 1841
11 WATTERS, Stacy B 1841
12 WATTERS, Stacy B 1841
13 WATTERS, Stacy B 1841
14 WATTERS, Stacy B 1841
15 WATTERS, Stacy B 1841
16 WATTERS, Stacy B 1841
17 WATTERS, Stacy B 1841
18 WATTERS, Stacy B 1841
19 WATTERS, Stacy B 1841
20 WATTERS, Stacy B 1841

Lots-Sec. 5
1 HUTTAN, William J 1841
3 WATTERS, Stacy B 1841
4 WATTERS, Stacy B 1841
5 WATTERS, Stacy B 1841
6 HUTTAN, William J 1841
7 HUTTAN, William J 1841
8 HUTTAN, William J 1841
9 HUTTAN, William J 1841
10 HUTTAN, William J 1841
11 HUTTAN, William J 1841
15 HUTTON, William J 1841
16 HUTTAN, William J 1841
17 HUTTAN, William J 1841
18 HUTTAN, William J 1841

Lots-Sec. 4
1 HUTTON, John N 1841
2 WATTERS, Stacy B 1841
3 WATTERS, Stacy B 1841
4 ADAIR, Benjamin 1841
5 HUTTAN, William J 1841
6 WATTERS, Stacy B 1841
7 WATTERS, Stacy B 1841
8 HUTTON, John N 1841
9 HUTTON, John N 1841
10 SNOWDEN, James 1841
11 SNOWDEN, James 1841
12 HUTTAN, William J 1841
13 HUTTON, William J 1841
14 ADAIR, Armell F 1841
15 HUTTON, John N 1841
16 HUTTON, John N 1841
17 HUTTON, John N 1841
18 HUTTON, John N 1841
19 SNOWDEN, James 1841
20 SNOWDEN, John 1841

66

Township 8-N Range 15-E (Choctaw) - Map Group 2

Lots-Sec. 3
1	COLEMAN, Charles H	1841
2	COLEMAN, Charles H	1841
3	THOMPSON, Archibald	1841
4	THOMPSON, Archibald	1841
5	THOMPSON, Archibald	1841
6	THOMPSON, Archibald	1841
7	COLEMAN, Charles H	1841
8	COLEMAN, Charles H	1841
9	ADAIR, Armell F	1841
10	ADAIR, Armell F	1841
11	THOMPSON, Archibald	1841
12	THOMPSON, Archibald	1841
13	DAVIS, John	1841
14	DAVIS, John	1841
15	ADAIR, Armell F	1841
16	ADAIR, Armell F	1841
19	DAVIS, John	1841
20	DAVIS, John	1841

Lots-Sec. 2
2	DAVIS, Bluford G	1859
3	COLEMAN, Charles H	1841
4	COLEMAN, Charles H	1841
5	COLEMAN, Charles H	1841
6	COLEMAN, Charles H	1841
11	COLEMAN, Charles H	1841
12	COLEMAN, Charles H	1841
13	COLEMAN, Charles H	1841
20	COLEMAN, Charles H	1841

Lots-Sec. 1
1	SHUMATE, Daniel D	1841
2	SHUMATE, Daniel D	1844
7	SHUMATE, Daniel D	1841
8	SHUMATE, Daniel D	1841
9	WIER, James	1841
15	BROOKE, James R	1850
16	WIER, James	1841

Helpful Hints

1. This Map's INDEX can be found on the preceding pages.

2. Refer to Map "C" to see where this Township lies within Lauderdale County, Mississippi.

3. Numbers within square brackets [ ] denote a multi-patentee land parcel (multi-owner). Refer to Appendix "C" for a full list of members in this group.

4. Areas that look to be crowded with Patentees usually indicate multiple sales of the same parcel (Re-issues) or Overlapping parcels. See this Township's Index for an explanation of these and other circumstances that might explain "odd" groupings of Patentees on this map.

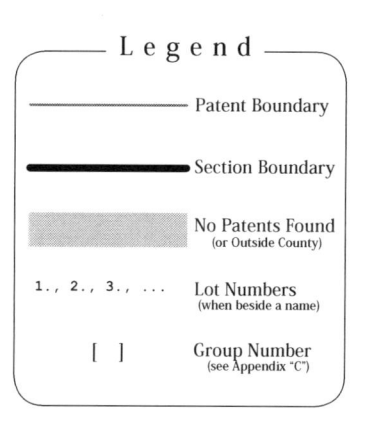

Copyright 2006 Boyd IT, Inc. All Rights Reserved

Legend

— Patent Boundary
— Section Boundary
▓ No Patents Found (or Outside County)
1., 2., 3., ... Lot Numbers (when beside a name)
[ ] Group Number (see Appendix "C")

Scale: Section = 1 mile X 1 mile (generally, with some exceptions)

67

Family Maps of Lauderdale County, Mississippi

Road Map
T8-N R15-E
Choctaw Meridian
Map Group 2

Cities & Towns
Center Hill
Obadiah
Shucktown

Cemeteries
Arkadelphia Cemetery
Fellowship Cemetery
Gum Log Cemetery
Mount Carmel Cemetery

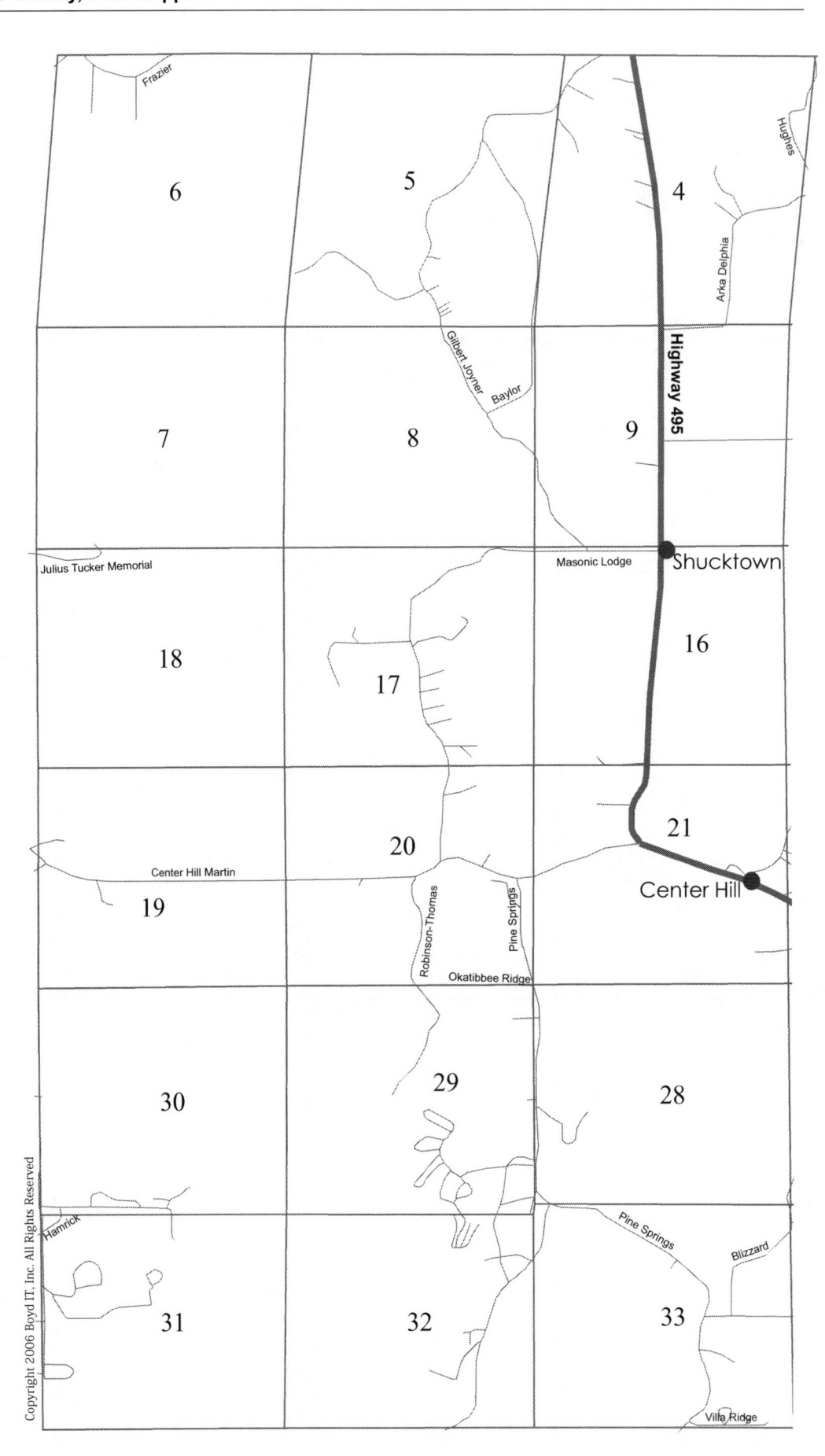

Township 8-N Range 15-E (Choctaw) - Map Group 2

Helpful Hints

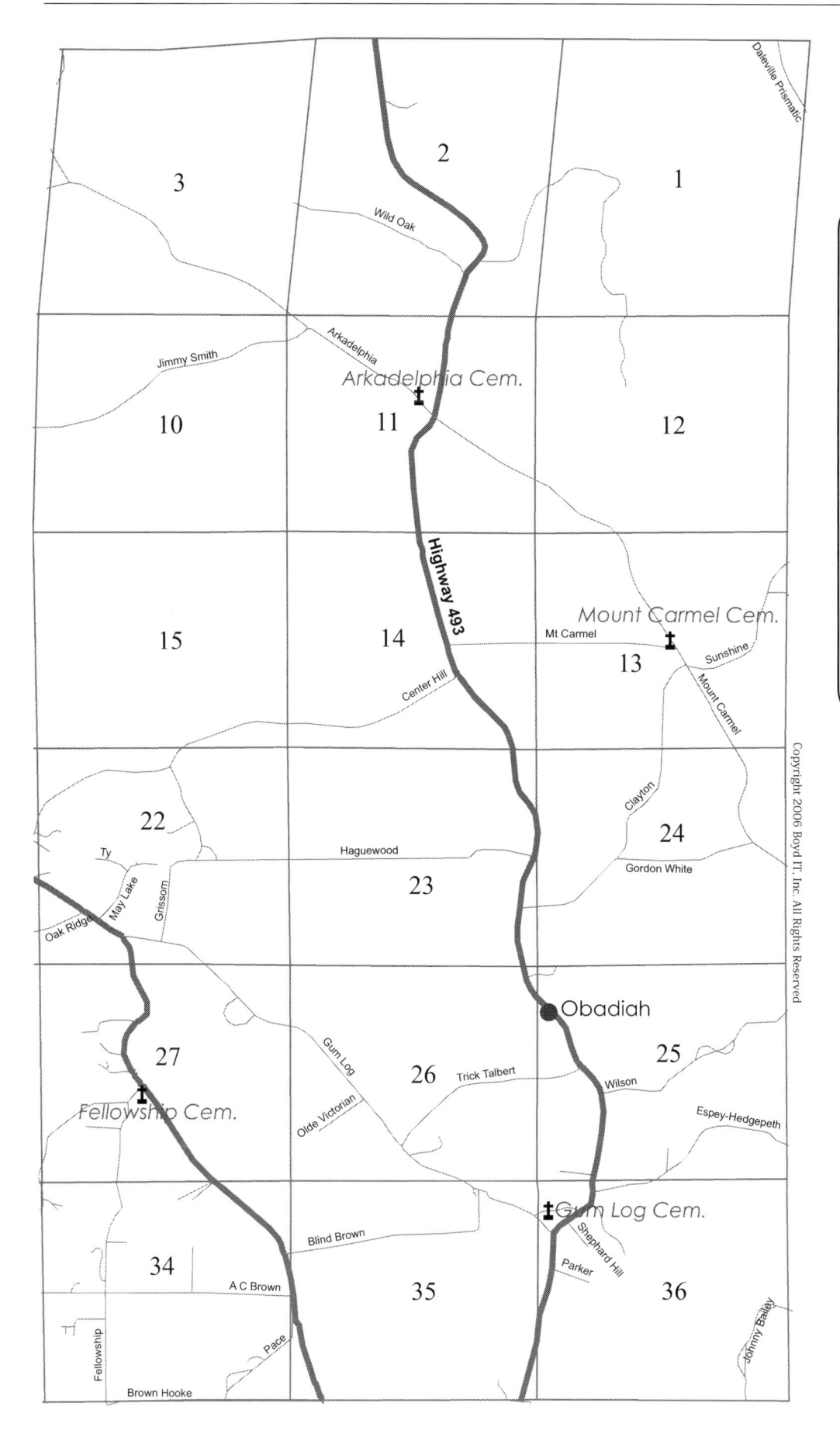

1. This road map has a number of uses, but primarily it is to help you: a) find the present location of land owned by your ancestors (at least the general area), b) find cemeteries and city-centers, and c) estimate the route/roads used by Census-takers & tax-assessors.

2. If you plan to travel to Lauderdale County to locate cemeteries or land parcels, please pick up a modern travel map for the area before you do. Mapping old land parcels on modern maps is not as exact a science as you might think. Just the slightest variations in public land survey coordinates, estimates of parcel boundaries, or road-map deviations can greatly alter a map's representation of how a road either does or doesn't cross a particular parcel of land.

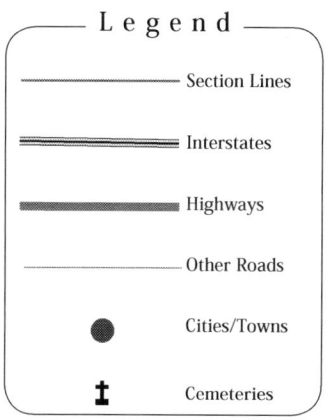

Copyright 2006 Boyd IT, Inc. All Rights Reserved

Legend

- Section Lines
- Interstates
- Highways
- Other Roads
- ● Cities/Towns
- ☦ Cemeteries

Scale: Section = 1 mile X 1 mile
(generally, with some exceptions)

69

Family Maps of Lauderdale County, Mississippi

Historical Map
T8-N R15-E
Choctaw Meridian
Map Group 2

Cities & Towns
Center Hill
Obadiah
Shucktown

Cemeteries
Arkadelphia Cemetery
Fellowship Cemetery
Gum Log Cemetery
Mount Carmel Cemetery

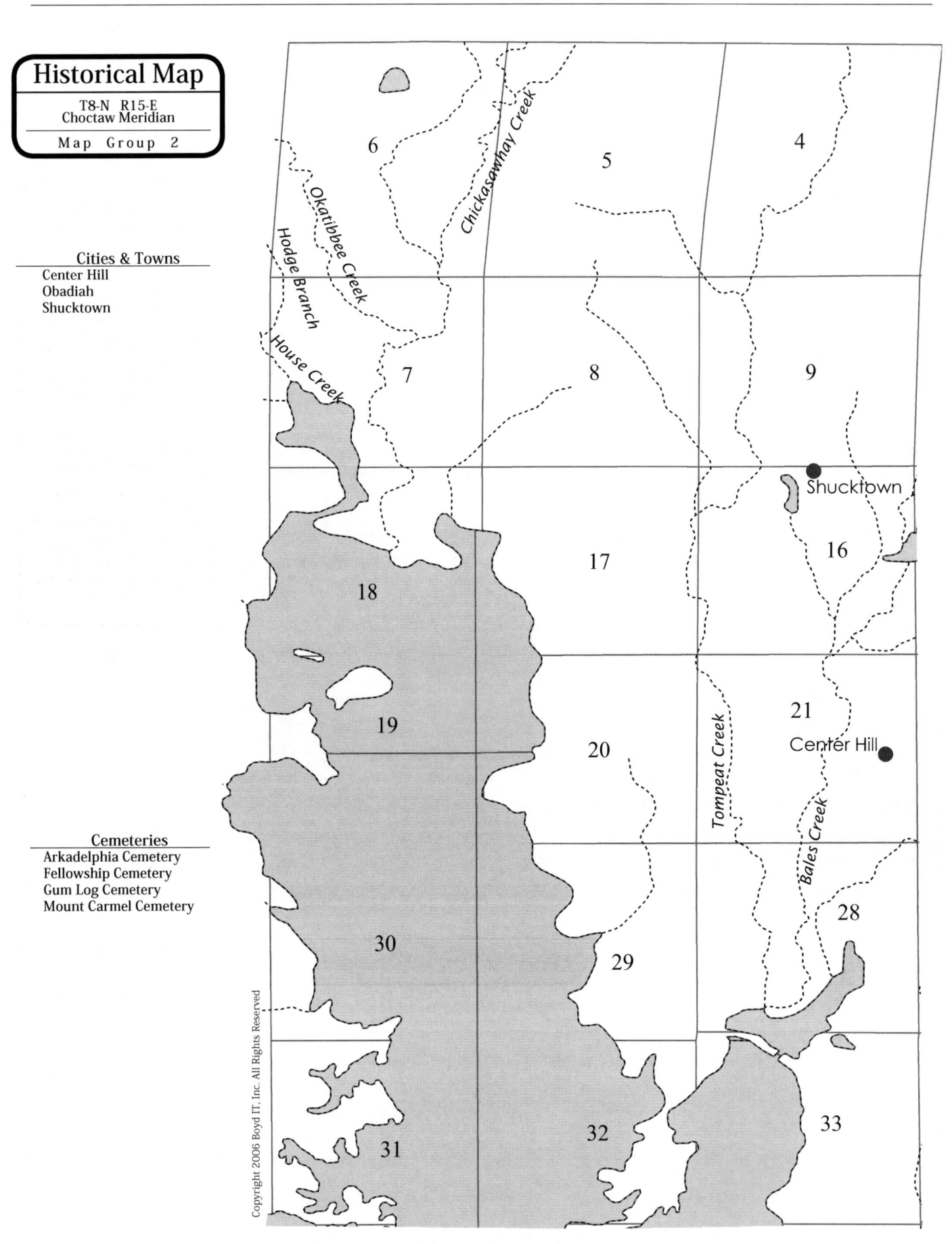

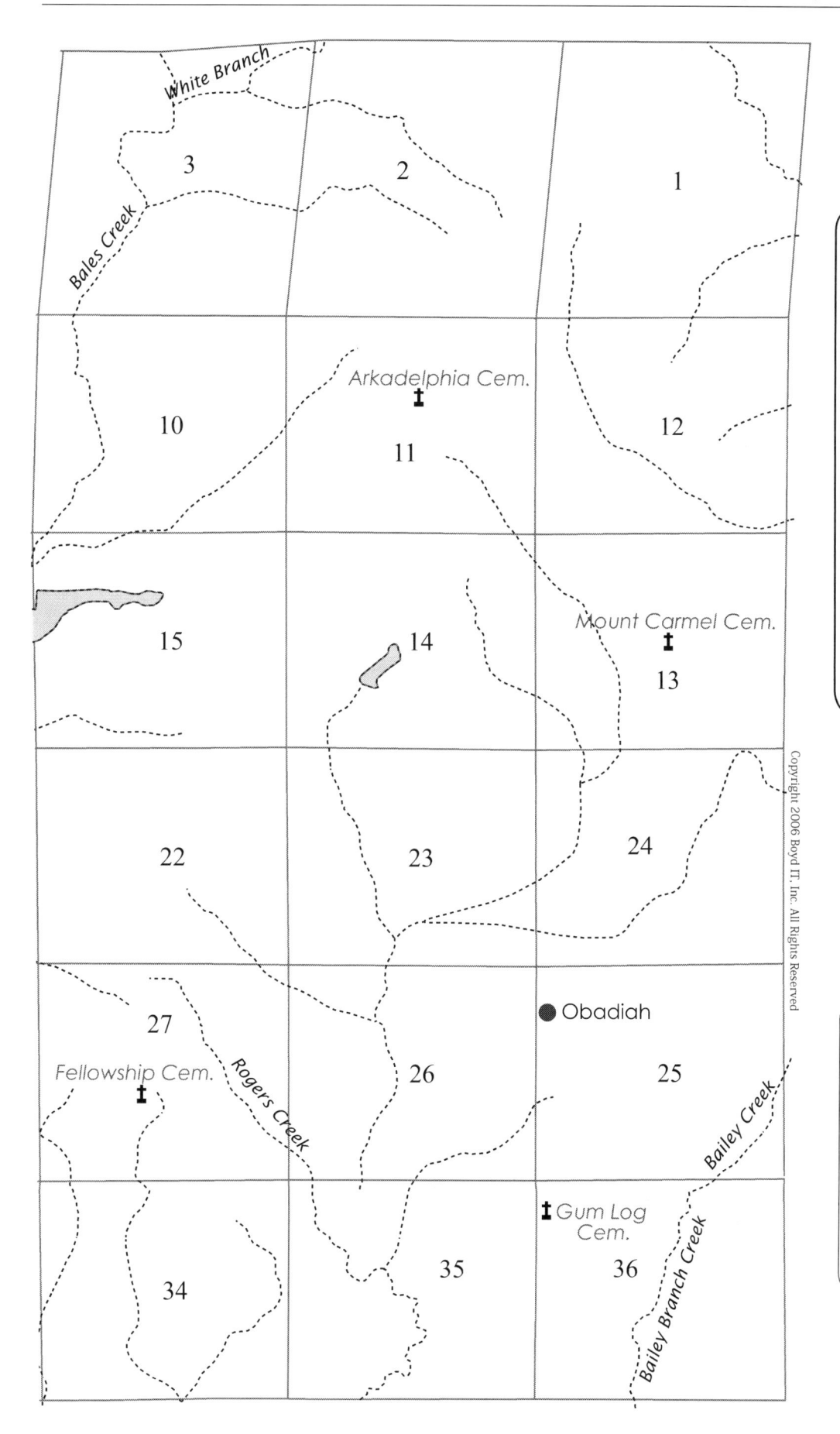

Township 8-N Range 15-E (Choctaw) — Map Group 2

Helpful Hints

1. This Map takes a different look at the same Congressional Township displayed in the preceding two maps. It presents features that can help you better envision the historical development of the area: a) Water-bodies (lakes & ponds), b) Water-courses (rivers, streams, etc.), c) Railroads, d) City/town center-points (where they were oftentimes located when first settled), and e) Cemeteries.

2. Using this "Historical" map in tandem with this Township's Patent Map and Road Map, may lead you to some interesting discoveries. You will often find roads, towns, cemeteries, and waterways are named after nearby landowners: sometimes those names will be the ones you are researching. See how many of these research gems you can find here in Lauderdale County.

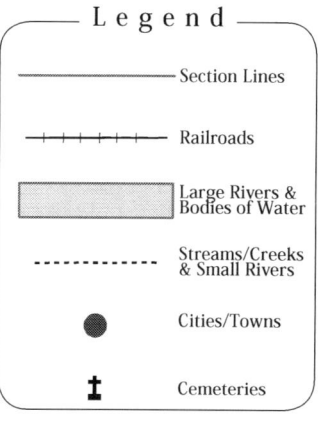

Legend

- Section Lines
- Railroads
- Large Rivers & Bodies of Water
- Streams/Creeks & Small Rivers
- ● Cities/Towns
- ✝ Cemeteries

Scale: Section = 1 mile X 1 mile
(there are some exceptions)

Family Maps of Lauderdale County, Mississippi

Map Group 3: Index to Land Patents
Township 8-North Range 16-East (Choctaw)

After you locate an individual in this Index, take note of the Section and Section Part then proceed to the Land Patent map on the pages immediately following. You should have no difficulty locating the corresponding parcel of land.

The "For More Info" Column will lead you to more information about the underlying Patents. See the *Legend* at right, and the "How to Use this Book" chapter, for more information.

```
                          LEGEND
              "For More Info . . . " column
    A = Authority (Legislative Act, See Appendix "A")
    B = Block or Lot (location in Section unknown)
    C = Cancelled Patent
    F = Fractional Section
    G = Group  (Multi-Patentee Patent, see Appendix "C")
    V = Overlaps another Parcel
    R = Re-Issued (Parcel patented more than once)

    (A & G items require you to look in the Appendixes referred
    to above. All other Letter-designations followed by a number
    require you to locate line-items in this index that possess
    the ID number found after the letter).
```

ID	Individual in Patent	Sec.	Sec. Part	Date Issued	Other Counties	For More Info . . .
447	ADAIR, Armell F	24	SWSW	1841-02-27		A1
448	" "	9	SESW	1841-02-27		A1
614	ALEXANDER, Lindsey	32	N½NE	1841-02-27		A1
474	ALVICE, Elijah	8	S½W½NW	1841-02-27		A1
475	" "	8	W½SW	1841-02-27		A1
636	ANDERSON, Richard	31	SE	1841-02-27		A1 G12
637	" "	32	NW	1841-02-27		A1 G12
486	ANDREWS, Ezekiel	3	4	1841-02-27		A1
487	" "	3	5	1841-02-27		A1
687	ANDREWS, Warren	31	N½	1841-02-27		A1
688	" "	31	SW	1841-02-27		A1
686	ARMSTRONG, Tom	1	17	1895-05-11		A4
570	BAILIFF, John	13	E½SW	1841-02-27		A1
571	" "	13	W½SW	1841-02-27		A1
572	" "	14	E½SW	1841-02-27		A1
573	" "	14	W½SW	1841-02-27		A1
574	" "	15	S½SE	1841-02-27		A1 V511
575	" "	22	E½NW	1841-02-27		A1
576	" "	22	W½NW	1841-02-27		A1
577	" "	24	N½W½NW	1841-02-27		A1 G13
624	BARNES, Maxy	15	N½E½SE	1841-02-27		A1
692	BARNETT, William	7	E½SE	1841-02-27		A1
532	BLANKS, James L	27	NESW	1850-12-05		A1
533	" "	27	SESW	1860-05-01		A1
625	BOSTON, Moody	27	W½NW	1897-11-01		A4
482	BROWN, Epps R	23	SWSW	1841-02-27		A1
483	" "	24	E½SW	1841-02-27		A1
485	" "	26	NW	1841-02-27		A1
484	" "	26	NESW	1845-01-21		A1
489	BROWN, Fairchild B	3	8	1897-02-15		A4
490	" "	3	9	1897-02-15		A4
693	BROWN, William	1	11	1906-06-21		A4
694	" "	1	12	1906-06-21		A4
703	BROWN, William J	29	SWSW	1860-05-01		A1
442	BULLARD, Alexander	32	S½W½NE	1841-02-27		A1
639	BURRESS, Robert N	29	NESE	1860-10-01		A1
640	BURRIS, Robert N	28	SWNW	1850-12-05		A1
547	CAMPBELL, James R	23	N½W½SW	1841-02-27		A1
453	CARPENTER, Benjamin	19	N½W½NW	1841-02-27		A1
644	CARPENTER, Samuel	19	S½W½NW	1841-02-27		A1
690	CARTER, William B	23	E½NE	1841-02-27		A1
668	CATES, Sion	11	NESE	1841-02-27		A1
548	COATES, James R	12	S½SE	1841-02-27		A1
549	" "	14	E½SE	1841-02-27		A1
550	" "	14	W½NE	1841-02-27		A1
592	COATES, John W	14	SENE	1841-02-27		A1

72

Township 8-N Range 16-E (Choctaw) - Map Group 3

ID	Individual in Patent	Sec.	Sec. Part	Date Issued	Other Counties	For More Info . . .
696	COATES, William	11	NWNE	1841-02-27		A1
697	" "	2	14	1841-02-27		A1
698	" "	2	19	1841-02-27		A1
699	" "	24	E½NW	1841-02-27		A1
700	" "	27	NENW	1841-02-27		A1
701	" "	9	NESE	1841-02-27		A1
467	COATS, Daniel	2	9	1841-02-27		A1
691	COATS, William B	24	S½W½NW	1841-02-27		A1
441	COLE, Albert	15	W½NE	1906-05-01		A4
597	COLE, Joseph	27	SWSW	1905-10-19		A4
618	COLE, Margaret A	1	15	1884-12-30		A4 G46
630	COLE, Peter H	4	9	1854-03-15		A1
643	COLE, Samos	27	S½SE	1891-08-19		A4
618	COLE, Stephen	1	15	1884-12-30		A4 G46
704	COLE, William L	20	NESE	1849-12-01		A1
716	COLE, Wily	15	NENE	1892-04-29		A4
590	COLEMAN, John H	12	NESE	1849-12-01		A1
669	COTES, Sion	12	NWSW	1841-02-27		A1
561	CRAIL, Joel D	12	NENE	1848-09-01		A1
610	CRANE, Lewis	32	E½SE	1841-02-27		A1
611	" "	32	S½E½NE	1841-02-27		A1
612	" "	32	W½SE	1841-02-27		A1
613	" "	33	SW	1841-02-27		A1
494	CROCKETT, George	19	E½SE	1890-06-25		A4
514	DALE, James	22	SE	1843-02-28		A1 G57
601	DALE, Joseph P	21	NWNE	1841-02-27		A1
645	DALE, Samuel	15	NW	1841-02-27		A1
646	" "	15	NWSW	1841-02-27		A1
647	" "	15	S½W½SW	1841-02-27		A1
648	" "	21	SWNE	1841-02-27		A1
649	" "	22	E½SW	1841-02-27		A1
650	" "	22	S½E½NE	1841-02-27		A1
651	" "	22	W½NE	1841-02-27		A1
514	" "	22	SE	1843-02-28		A1 G57
488	DANIEL, Ezekiel	3	18	1848-09-01		A1
558	DANIEL, James W	8	E½SW	1841-02-27		A1
587	DAVIS, John	3	10	1841-02-27		A1
588	" "	3	7	1841-02-27		A1
652	DAWKINS, Samuel	6	14	1841-02-27		A1
653	" "	6	16	1841-02-27		A1
654	" "	6	17	1841-02-27		A1
655	" "	6	19	1841-02-27		A1
656	" "	7	E½NE	1841-02-27		A1
657	" "	8	N½W½NW	1841-02-27		A1
658	DAWKINS, Samuel F	5	12	1841-02-27		A1
659	" "	5	5	1841-02-27		A1
591	DENTON, John R	18	S½NE	1859-10-01		A1
685	DOBBS, Thomas N	4	10	1854-03-15		A1
695	DOBBS, William C	20	SWSE	1849-12-01		A1
577	DOWD, John K	24	N½W½NW	1841-02-27		A1 G13
491	DUKE, Frederick	13	NWNE	1841-02-27		A1
471	EASOM, Edward	1	16	1841-02-27		A1
589	ENGLISH, John	28	SWNE	1850-12-05		A1
498	FINLEY, Harrison M	18	NWNE	1841-02-27		A1
499	" "	7	E½SW	1841-02-27		A1
500	" "	7	W½SE	1841-02-27		A1 G74
706	GABEL, William P	18	NWNW	1859-10-01		A1
515	GILLESPIE, James F	3	11	1841-02-27		A1
516	" "	3	15	1841-02-27		A1
517	" "	3	16	1841-02-27		A1
518	" "	3	6	1841-02-27		A1
519	" "	4	1	1844-09-10		A1
520	" "	4	8	1844-09-10		A1
702	GILLESPIE, William	2	18	1841-02-27		A1
609	GLASS, Levi J	11	W½SW	1897-11-01		A4
641	GORDON, Russell	1	2	1904-12-31		A4
642	" "	1	3	1904-12-31		A4
667	GORMAN, Simpson	4	20	1841-02-27		A1
497	GRANT, Green W	33	SE	1841-02-27		A1
680	GRANT, Thomas D	13	NENE	1841-02-27		A1
636	GRANTLAND, Seaton	31	SE	1841-02-27		A1 G12
637	" "	32	NW	1841-02-27		A1 G12
452	GREEN, Benjamin B	35	E½NE	1895-06-28		A4

Family Maps of Lauderdale County, Mississippi

ID	Individual in Patent	Sec.	Sec. Part	Date Issued	Other Counties	For More Info...
598	GRIGSBY, Joseph	21	E½SW	1846-08-10		A1 G92
459	HAMLET, Charles R	8	E½NW	1841-02-27		A1
460	"	8	NE	1841-02-27		A1
461	"	8	NESE	1841-02-27		A1
505	HARRIS, Hilton	29	E½NE	1905-03-30		A4
709	HARRIS, William T	29	NWNE	1892-04-29		A4
568	HAWKINS, John B	35	SESW	1841-02-27		A1
569	"	35	SWSE	1841-02-27		A1
559	HENDERSON, Jesse B	22	NENE	1841-02-27		A1
560	HENDERSON, Jessee B	9	SESE	1841-02-27		A1
681	HIGHTOWER, Thomas	14	E½NW	1841-02-27		A1
683	"	15	E½SW	1841-02-27		A1
684	"	23	SWNE	1841-02-27		A1
682	"	14	W½SE	1844-09-10		A1
537	HILL, James P	23	E½SE	1841-02-27		A1
538	"	24	NWSW	1841-02-27		A1
539	"	26	E½NE	1841-02-27		A1
540	"	26	NWNE	1841-02-27		A1 V635
541	"	26	S½W½NE	1841-02-27		A1 V635
542	"	26	SESW	1849-12-01		A1
544	"	27	NWNE	1849-12-01		A1
546	"	27	SENW	1849-12-01		A1
543	"	27	NENE	1850-12-05		A1
545	"	27	NWSW	1850-12-05		A1
628	HOLLADAY, Owen F	6	4	1854-03-15		A1
660	HUBBARD, Samuel	6	1	1841-02-27		A1 G122
661	"	6	8	1841-02-27		A1 G122
662	"	6	9	1841-02-27		A1 G122
527	HUGHES, James F	9	NW	1841-02-27		A1 V529
528	"	9	W½SW	1841-02-27		A1
521	"	5	13	1844-09-10		A1
522	"	5	14	1844-09-10		A1
523	"	5	15	1844-09-10		A1
524	"	5	18	1844-09-10		A1
525	"	5	19	1844-09-10		A1
526	"	5	20	1844-09-10		A1
455	HUSSEY, Bless	1	5	1904-12-31		A4
472	HUZZA, Edward	23	E½SW	1841-02-27		A1
473	"	34	SW	1841-02-27		A1
440	JOHNSON, Abraham B	1	SWSW	1906-06-16		A4
615	JOHNSON, Luther E	13	NESE	1854-03-15		A1
616	"	13	SENE	1854-03-15		A1
617	"	13	SWNE	1856-04-01		A1
456	JOLLY, Bradley	7	E½NW	1841-02-27		A1
476	JONES, Elijah C	21	E½NE	1841-02-27		A1
478	"	22	W½SW	1841-02-27		A1
477	"	21	SWSW	1849-12-01		A1
501	KITERAL, Henry M	27	NESE	1849-12-01		A1
626	KITERAL, Nathaniel	36	NWNW	1849-12-01		A1
479	KITRELL, Elisha B	35	SENW	1885-05-25		A4
480	"	35	SWNE	1885-05-25		A4
619	KITRELL, Margaret J	35	SWSW	1883-09-15		A4
627	KITRELL, Nathaniel	13	S½NW	1841-02-27		A1
502	KITTRELL, Henry M	27	NWSE	1851/03/15		A2
711	KITTRELL, William W	1	13	1897-05-07		A4
712	"	1	14	1897-05-07		A4
713	"	1	18	1897-05-07		A4
714	"	1	19	1897-05-07		A4
534	LACKEY, James N	35	E½SE	1890-02-21		A4
535	"	35	NESW	1890-02-21		A4
536	"	35	NWSE	1890-02-21		A4
670	LACY, Stephen	6	10	1841-02-27		A1
671	"	6	12	1841-02-27		A1
672	"	6	15	1841-02-27		A1
673	"	6	18	1841-02-27		A1
674	"	6	2	1841-02-27		A1
677	"	6	3	1841-02-27		A1 G130
678	"	6	6	1841-02-27		A1 G130
675	"	6	7	1841-02-27		A1
676	"	7	W½NE	1841-02-27		A1
660	LEWIS, Moses	6	1	1841-02-27		A1 G122
661	"	6	8	1841-02-27		A1 G122
662	"	6	9	1841-02-27		A1 G122

Township 8-N Range 16-E (Choctaw) - Map Group 3

ID	Individual in Patent	Sec.	Sec. Part	Date Issued	Other Counties	For More Info . . .
660	LEWIS, Rufus G	6	1	1841-02-27		A1 G122
661	" "	6	8	1841-02-27		A1 G122
662	" "	6	9	1841-02-27		A1 G122
705	LOVE, William	18	S½SE	1859-10-01		A1
600	MANGUM, Joseph	18	NENE	1841-02-27		A1
621	MARTIN, Mary	29	SESW	1884-12-30		A4
506	MCCARTY, Houston	23	NENW	1895-02-21		A4
507	" "	23	NWNE	1895-02-21		A4
503	MCGEE, Henry W	11	SESW	1897-11-01		A4
562	MCKELLAR, John A	5	10	1841-02-27		A1
563	" "	5	2	1841-02-27		A1
564	" "	5	3	1841-02-27		A1
565	" "	5	7	1841-02-27		A1
566	" "	5	8	1841-02-27		A1
567	" "	5	9	1841-02-27		A1
446	MILLER, Annie	27	SWNE	1905-08-26		A4
689	MILLS, Willard C	4	18	1854-03-15		A1
638	MOSELEY, Robert	4	14	1854-03-15		A1
504	MURPHY, Hezekiah R	13	NENW	1841-02-27		A1
606	OBRINK, Julia A	35	NWSW	1895-01-17		A4 G157
607	" "	35	SWNW	1895-01-17		A4 G157
606	OBRINK, William J	35	NWSW	1895-01-17		A4 G157
607	" "	35	SWNW	1895-01-17		A4 G157
508	ODAM, Jacob	1	1	1841-02-27		A1
509	" "	1	8	1841-02-27		A1
715	OWENS, Willis	21	S½SE	1841-02-27		A1
598	" "	21	E½SW	1846-08-10		A1 G92
593	PARKER, John W	26	NWSW	1849-12-01		A1
594	" "	27	SENE	1854-03-15		A1
595	" "	35	NWNE	1854-03-15		A1
596	" "	35	NWNW	1854-03-15		A1
707	PARKER, William	23	NWNW	1854-03-15		A1
629	PARTIN, Patrick H	9	E½NE	1886-03-20		A4
602	PHILLIPS, Joseph	8	S½E½SE	1841-02-27		A1
677	RAGSDALE, Samuel	6	3	1841-02-27		A1 G130
678	" "	6	6	1841-02-27		A1 G130
500	" "	7	W½SE	1841-02-27		A1 G74
511	RAMSEY, James B	15	SWSE	1850-12-05		A1 V574
457	RATCLIFF, Burdett E	11	NESW	1904-12-31		A4
492	RATCLIFF, Gabriel A	1	6	1849-12-01		A1
493	" "	1	7	1849-12-01		A1
531	RATCLIFF, James H	11	NWSE	1875-07-01		A4
553	RATCLIFF, James S	13	NWNW	1841-02-27		A1
554	" "	13	SESE	1841-02-27		A1
555	" "	13	W½SE	1841-02-27		A1
556	" "	14	NENE	1841-02-27		A1
551	" "	11	SESE	1848-09-01		A1
552	" "	11	SWSE	1849-12-01		A1
620	RATCLIFF, Martha J	11	NW	1898-08-27		A4
466	RICHARDSON, Daniel B	34	W½SE	1841-02-27		A1
635	RICHARDSON, Ransom	26	W½NE	1859-10-01		A1 V540, 541
708	RIVERS, William	8	W½SE	1841-02-27		A1
481	ROGERS, Elizabeth	11	SWNE	1856-04-01		A1
439	SALTER, Aaron L	4	13	1859-10-01		A1
578	SELLERS, John C	11	E½NE	1841-02-27		A1
579	" "	12	SWNW	1841-02-27		A1
468	SHUMATE, Daniel D	6	11	1841-02-27		A1
469	" "	6	5	1841-02-27		A1
470	" "	7	W½SW	1841-02-27		A1
464	SMITH, Craven	6	13	1841-02-27		A1
465	" "	6	20	1841-02-27		A1
557	SMITT, James	29	SESE	1913-09-29		A1
495	STEED, George	4	2	1841-02-27		A1
496	" "	4	7	1841-02-27		A1
454	STEEDLEY, Benjamin F	35	NENW	1913-10-27		A4
622	STEPHENS, Matthew	1	10	1841-02-27		A1
623	" "	1	9	1841-02-27		A1
462	STEPHENSON, Charles W	5	16	1844-09-10		A1
463	" "	5	17	1844-09-10		A1
449	TALBERT, Arthur	18	NENW	1859-10-01		A1
450	" "	18	SENW	1859-10-01		A1
660	TAPPAN, John	6	1	1841-02-27		A1 G122
661	" "	6	8	1841-02-27		A1 G122

75

ID	Individual in Patent	Sec.	Sec. Part	Date Issued	Other Counties	For More Info . . .
662	TAPPAN, John (Cont'd)	6	9	1841-02-27		A1 G122
608	TARRANT, Larkin Y	15	SENE	1848-09-01		A1
710	THOMAS, William	23	S½NW	1841-02-27		A1
443	TRAWICK, Allen	20	SW	1841-02-27		A1
444	" "	29	NW	1841-02-27		A1
510	TRAWICK, James A	20	NWSE	1841-02-27		A1
679	TRAWICK, Stephen T	29	NWSW	1896-04-28		A4
631	VAUGHN, Pinckney	21	N½SE	1841-02-27		A1
632	" "	21	NW	1841-02-27		A1
634	" "	36	W½NE	1841-02-27		A1
633	" "	21	NWSW	1850-12-05		A1
603	VINSON, Joseph W	33	E½NE	1892-04-29		A4
604	" "	33	SENW	1892-04-29		A4
605	" "	33	SWNE	1892-04-29		A4
663	WAGGENER, Seth	29	SWNE	1860-10-01		A1
664	WAGGONER, Seth	29	NESW	1859-10-01		A1
665	WAGONER, Seth	29	NWSE	1850-12-05		A1
666	" "	29	SWSE	1854-03-15		A1
636	WALTHALL, Madison	31	SE	1841-02-27		A1 G12
637	" "	32	NW	1841-02-27		A1 G12
458	WHITE, Charles Edward	19	SWSE	1919-01-07		A4
599	WHITE, Joseph Lee	33	W½NW	1913-11-19		A1
580	WHITSETT, John C	2	13	1841-02-27		A1
581	" "	2	20	1841-02-27		A1
582	" "	3	13	1841-02-27		A1
583	" "	3	14	1841-02-27		A1
584	" "	3	19	1841-02-27		A1
585	" "	3	20	1841-02-27		A1
586	" "	4	3	1841-02-27		A1
445	WILKERSON, Allen	12	NESW	1841-02-27		A1
512	WILLIAMS, James D	28	NENW	1841-02-27		A1
513	" "	28	NWNE	1841-02-27		A1
529	WINNINGHAM, James G	9	NENW	1841-02-27		A1 V527
530	" "	9	W½SE	1841-02-27		A1
451	WRIGHT, Asa	23	W½SE	1841-02-27		A1

Family Maps of Lauderdale County, Mississippi

Patent Map

T8-N R16-E
Choctaw Meridian

Map Group 3

Township Statistics

Parcels Mapped	:	278
Number of Patents	:	218
Number of Individuals	:	155
Patentees Identified	:	149
Number of Surnames	:	115
Multi-Patentee Parcels	:	14
Oldest Patent Date	:	2/27/1841
Most Recent Patent	:	1/7/1919
Block/Lot Parcels	:	86
Parcels Re-Issued	:	0
Parcels that Overlap	:	7
Cities and Towns	:	2
Cemeteries	:	6

Lots-Sec. 6
1. HUBBARD, Samuel [122] 1841
2. LACY, Stephen 1841
3. LACY, Stephen [130] 1841
4. HOLLADAY, Owen F 1854
5. SHUMATE, Daniel D 1841
6. LACY, Stephen [130] 1841
7. LACY, Stephen 1841
8. HUBBARD, Samuel [122] 1841
9. HUBBARD, Samuel [122] 1841
10. LACY, Stephen 1841
11. SHUMATE, Daniel D 1841
12. LACY, Stephen 1841
13. SMITH, Craven 1841
14. DAWKINS, Samuel 1841
15. LACY, Stephen 1841
16. DAWKINS, Samuel 1841
17. DAWKINS, Samuel 1841
18. LACY, Stephen 1841
19. DAWKINS, Samuel 1841
20. SMITH, Craven 1841

Lots-Sec. 5
2. MCKELLAR, John A 1841
3. MCKELLAR, John A 1841
5. DAWKINS, Samuel F 1841
7. MCKELLAR, John A 1841
8. MCKELLAR, John A 1841
9. MCKELLAR, John A 1841
10. MCKELLAR, John A 1841
12. DAWKINS, Samuel F 1841
13. HUGHES, James F 1844
14. HUGHES, James F 1844
15. HUGHES, James F 1844
16. STEPHENSON, Charles 1844
17. STEPHENSON, Charles 1844
18. HUGHES, James F 1844
19. HUGHES, James F 1844
20. HUGHES, James F 1844

Lots-Sec. 4
1. GILLESPIE, James F 1844
2. STEED, George 1841
3. WHITSETT, John C 1841
7. STEED, George 1841
8. GILLESPIE, James F 1844
9. COLE, Peter H 1854
10. DOBBS, Thomas N 1854
13. SALTER, Aaron L 1859
14. MOSELEY, Robert 1854
18. MILLS, Willard C 1854
20. GORMAN, Simpson 1841

Copyright 2006 Boyd IT, Inc. All Rights Reserved

Township 8-N Range 16-E (Choctaw) — Map Group 3

Lots-Sec. 3
4	ANDREWS, Ezekiel	1841
5	ANDREWS, Ezekiel	1841
6	GILLESPIE, James F	1841
7	DAVIS, John	1841
8	BROWN, Fairchild B	1897
9	BROWN, Fairchild B	1897
10	DAVIS, John	1841
11	GILLESPIE, James F	1841
13	WHITSETT, John C	1841
14	WHITSETT, John C	1841
15	GILLESPIE, James F	1841
16	GILLESPIE, James F	1841
18	DANIEL, Ezekiel	1848
19	WHITSETT, John C	1841
20	WHITSETT, John C	1841

Lots-Sec. 2
9	COATS, Daniel	1841
13	WHITSETT, John C	1841
14	COATES, William	1841
18	GILLESPIE, William	1841
19	COATES, William	1841
20	WHITSETT, John C	1841

Lots-Sec. 1
1	ODAM, Jacob	1841
2	GORDON, Russell	1904
3	GORDON, Russell	1904
5	HUSSEY, Bless	1904
6	RATCLIFF, Gabriel A	1849
7	RATCLIFF, Gabriel A	1849
8	ODAM, Jacob	1841
9	STEPHENS, Matthew	1841
10	STEPHENS, Matthew	1841
11	BROWN, William	1906
12	BROWN, William	1906
13	KITTRELL, William W	1897
14	KITTRELL, William W	1897
15	COLE, Margaret A[46]	1884
16	EASOM, Edward	1841
17	ARMSTRONG, Tom	1895
18	KITTRELL, William W	1897
19	KITTRELL, William W	1897

Helpful Hints

1. This Map's INDEX can be found on the preceding pages.
2. Refer to Map "C" to see where this Township lies within Lauderdale County, Mississippi.
3. Numbers within square brackets [ ] denote a multi-patentee land parcel (multi-owner). Refer to Appendix "C" for a full list of members in this group.
4. Areas that look to be crowded with Patentees usually indicate multiple sales of the same parcel (Re-issues) or Overlapping parcels. See this Township's Index for an explanation of these and other circumstances that might explain "odd" groupings of Patentees on this map.

Section 1 area
JOHNSON, Abraham B 1906

Section 11
- COATES, William 1841
- RATCLIFF, Martha J 1898
- ROGERS, Elizabeth 1856
- SELLERS, John C 1841
- GLASS, Levi J 1897
- RATCLIFF, Burdett E 1904
- RATCLIFF, James H 1875
- CATES, Sion 1841
- MCGEE, Henry W 1897
- RATCLIFF, James S 1849
- RATCLIFF, James R 1848

Section 12
- CRAIL, Joel D 1848
- SELLERS, John C 1841
- COTES, Sion 1841
- WILKERSON, Allen 1841
- COLEMAN, John H 1849
- COATES, James R 1841

Section 13
- RATCLIFF, James R 1841
- RATCLIFF, James S 1841
- MURPHY, Hezekiah R 1841
- DUKE, Frederick 1841
- GRANT, Thomas D 1841
- COATES, John W 1841
- KITRELL, Nathaniel 1841
- JOHNSON, Luther E 1856
- JOHNSON, Luther E 1854
- BAILIFF, John 1841
- RATCLIFF, James S 1841
- JOHNSON, Luther E 1854
- BAILIFF, John 1841
- COATES, James R 1841
- BAILIFF, John 1841
- RATCLIFF, James S 1841

Section 14
- HIGHTOWER, Thomas 1841
- BAILIFF, John 1841
- HIGHTOWER, Thomas 1844
- BAILIFF, John 1841

Section 15
- DALE, Samuel 1841
- COLE, Albert 1906
- COLE, Wily 1892
- TARRANT, Larkin Y 1848
- DALE, Samuel 1841
- HIGHTOWER, Thomas 1841
- BARNES, Maxy 1841
- DALE, Samuel 1841
- RAMSEY, James B 1850
- BAILIFF, John 1841

Section 22
- BAILIFF, John 1841
- DALE, Samuel 1841
- HENDERSON, Jesse B 1841
- BAILIFF, John 1841
- DALE, Samuel 1841
- JONES, Elijah C 1841
- DALE [57], James 1843
- DALE, Samuel 1841

Section 23
- PARKER, William 1854
- MCCARTY, Houston 1895
- MCCARTY, Houston 1895
- HIGHTOWER, Thomas 1841
- CARTER, William B 1841
- COATS, William 1841
- COATES, William 1841
- THOMAS, William 1841
- CAMPBELL, James R 1841
- WRIGHT, Asa 1841
- HILL, James P 1841
- BROWN, Epps R 1841
- HUZZA, Edward 1841
- HILL, James P 1841
- ADAIR, Armell F 1841
- BROWN, Epps R 1841

Section 24
(mostly blank)

Section 25
(mostly blank)

Section 26
- RICHARDSON, Ransom 1859
- HILL, James P 1841
- BROWN, Epps R 1841
- HILL, James P 1841
- HILL, James P 1841
- PARKER, John W 1849
- BROWN, Epps R 1845
- HILL, James P 1849

Section 27
- BOSTON, Moody 1897
- COATES, William 1841
- HILL, James P 1849
- HILL, James P 1850
- HILL, James P 1849
- MILLER, Annie 1905
- PARKER, John W 1854
- HILL, James P 1850
- BLANKS, James L 1850
- KITTRELL, Henry M 1851
- KITERAL, Henry M 1849
- COLE, Joseph 1905
- BLANKS, James L 1860
- COLE, Samos 1891

Section 34
- HUZZA, Edward 1841
- RICHARDSON, Daniel B 1841

Section 35
- PARKER, John W 1854
- STEEDLEY, Benjamin F 1913
- PARKER, John W 1854
- KITERAL, Nathaniel 1849
- OBRINK [157], Julia A 1895
- KITRELL, Elisha B 1885
- KITRELL, Elisha B 1885
- GREEN, Benjamin B 1895
- OBRINK [157], Julia A 1895
- LACKEY, James N 1890
- LACKEY, James N 1890
- KITRELL, Margaret J 1883
- HAWKINS, John B 1841
- HAWKINS, John B 1841
- LACKEY, James N 1890

Section 36
- VAUGHN, Pinckney 1841

Copyright 2006 Boyd IT, Inc. All Rights Reserved

Legend
- Patent Boundary
- Section Boundary
- No Patents Found (or Outside County)
- 1., 2., 3., ... — Lot Numbers (when beside a name)
- [ ] — Group Number (see Appendix "C")

Scale: Section = 1 mile X 1 mile (generally, with some exceptions)

Family Maps of Lauderdale County, Mississippi

Road Map
T8-N R16-E
Choctaw Meridian

Map Group 3

Cities & Towns
Daleville
Lizelia

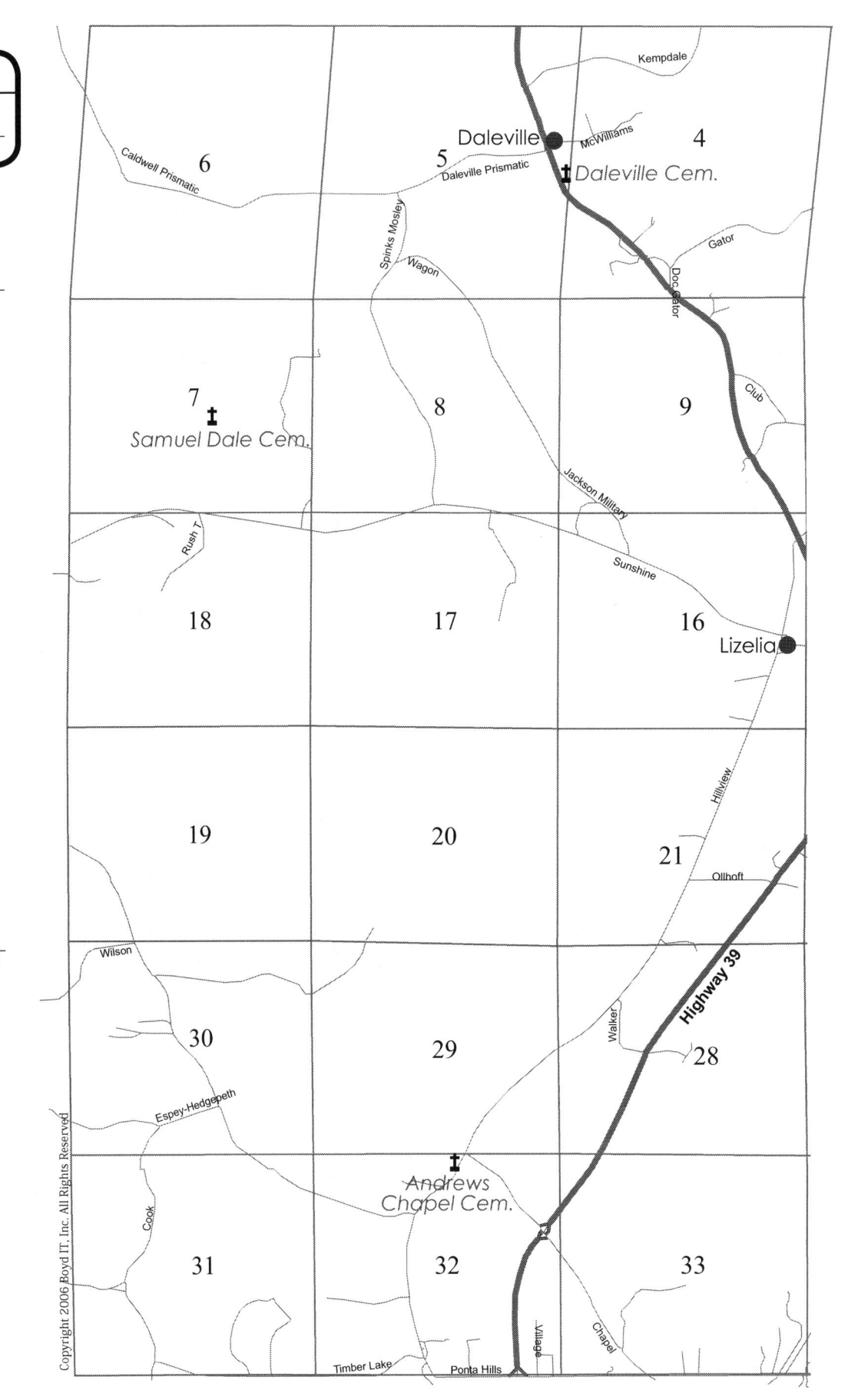

Cemeteries
Andrews Chapel Cemetery
Daleville Cemetery
Daleville Cemetery
Hickory Grove Cemetery
Moore Cemetery
Samuel Dale Cemetery

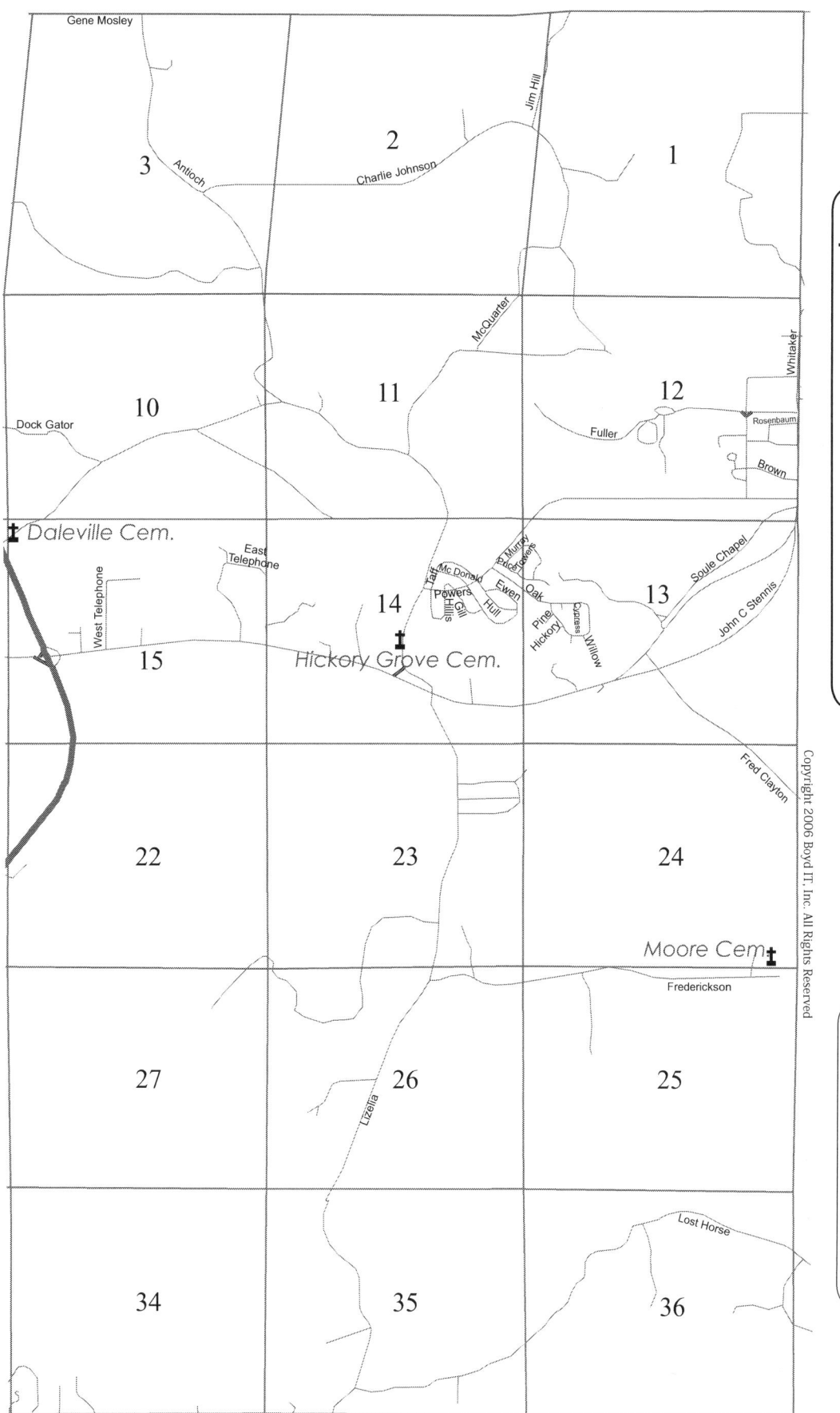

Township 8-N Range 16-E (Choctaw) - Map Group 3

Helpful Hints

1. This road map has a number of uses, but primarily it is to help you: a) find the present location of land owned by your ancestors (at least the general area), b) find cemeteries and city-centers, and c) estimate the route/roads used by Census-takers & tax-assessors.

2. If you plan to travel to Lauderdale County to locate cemeteries or land parcels, please pick up a modern travel map for the area before you do. Mapping old land parcels on modern maps is not as exact a science as you might think. Just the slightest variations in public land survey coordinates, estimates of parcel boundaries, or road-map deviations can greatly alter a map's representation of how a road either does or doesn't cross a particular parcel of land.

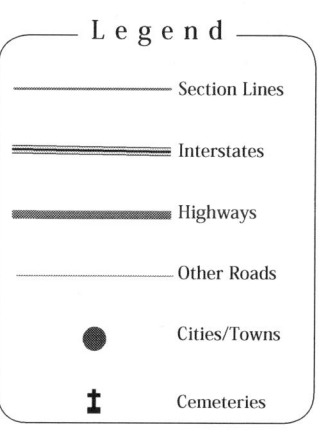

Legend

- Section Lines
- Interstates
- Highways
- Other Roads
- ● Cities/Towns
- ✝ Cemeteries

Scale: Section = 1 mile X 1 mile
(generally, with some exceptions)

Roads/labels visible on map: Gene Mosley, Antioch, Charlie Johnson, Jim Hill, McQuarter, Whitaker, Dock Gator, Fuller, Rosenbaum, Brown, Daleville Cem., East Telephone, West Telephone, Taff, McDonald, Powers, Gill, Hills, Hull, Murphy, Price, Loete, Ewen, Oak, Pine, Hickory, Cypress, Willow, Soule Chapel, John C Stennis, Hickory Grove Cem., Fred Clayton, Moore Cem., Frederickson, Lizelia, Lost Horse

Sections: 1, 2, 3, 10, 11, 12, 13, 14, 15, 22, 23, 24, 25, 26, 27, 34, 35, 36

Copyright 2006 Boyd IT, Inc. All Rights Reserved

81

Family Maps of Lauderdale County, Mississippi

Historical Map
T8-N R16-E
Choctaw Meridian
Map Group 3

Cities & Towns
Daleville
Lizelia

Cemeteries
Andrews Chapel Cemetery
Daleville Cemetery
Daleville Cemetery
Hickory Grove Cemetery
Moore Cemetery
Samuel Dale Cemetery

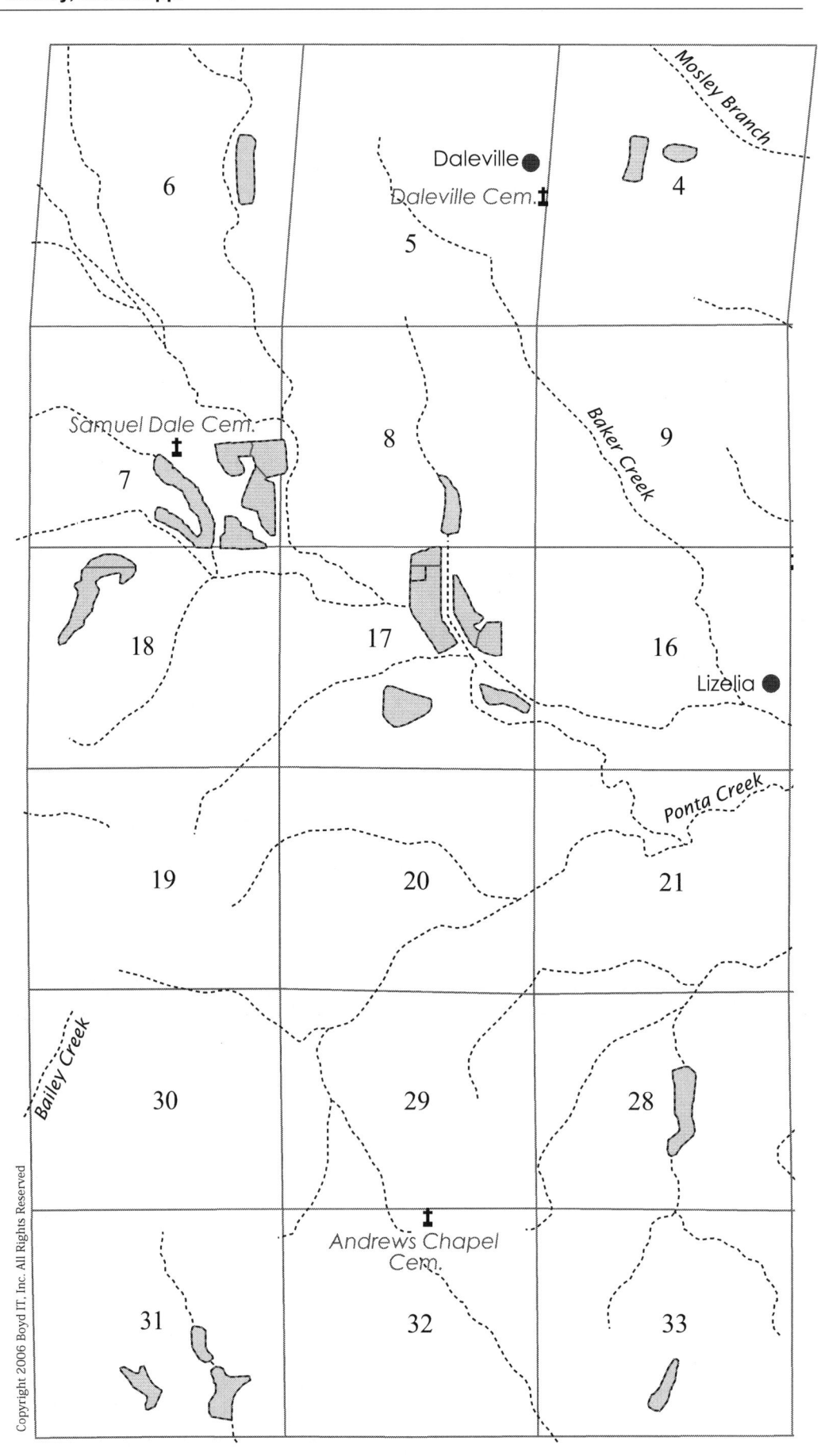

Township 8-N Range 16-E (Choctaw)   -   Map Group 3

Helpful Hints

1. This Map takes a different look at the same Congressional Township displayed in the preceding two maps. It presents features that can help you better envision the historical development of the area: a) Water-bodies (lakes & ponds), b) Water-courses (rivers, streams, etc.), c) Railroads, d) City/town center-points (where they were oftentimes located when first settled), and e) Cemeteries.

2. Using this "Historical" map in tandem with this Township's Patent Map and Road Map, may lead you to some interesting discoveries. You will often find roads, towns, cemeteries, and waterways are named after nearby landowners: sometimes those names will be the ones you are researching. See how many of these research gems you can find here in Lauderdale County.

Sections: 3, 2, 1, 10, 11, 12, 15, 14, 13, 22, 23, 24, 27, 26, 25, 34, 35, 36

Features:
- Wright Creek
- Daleville Cem.
- Hickory Grove Cem.
- Ponta Creek
- Moore Cem.
- Lost Horse Creek
- Wildhorse Creek

Copyright 2006 Boyd IT, Inc. All Rights Reserved

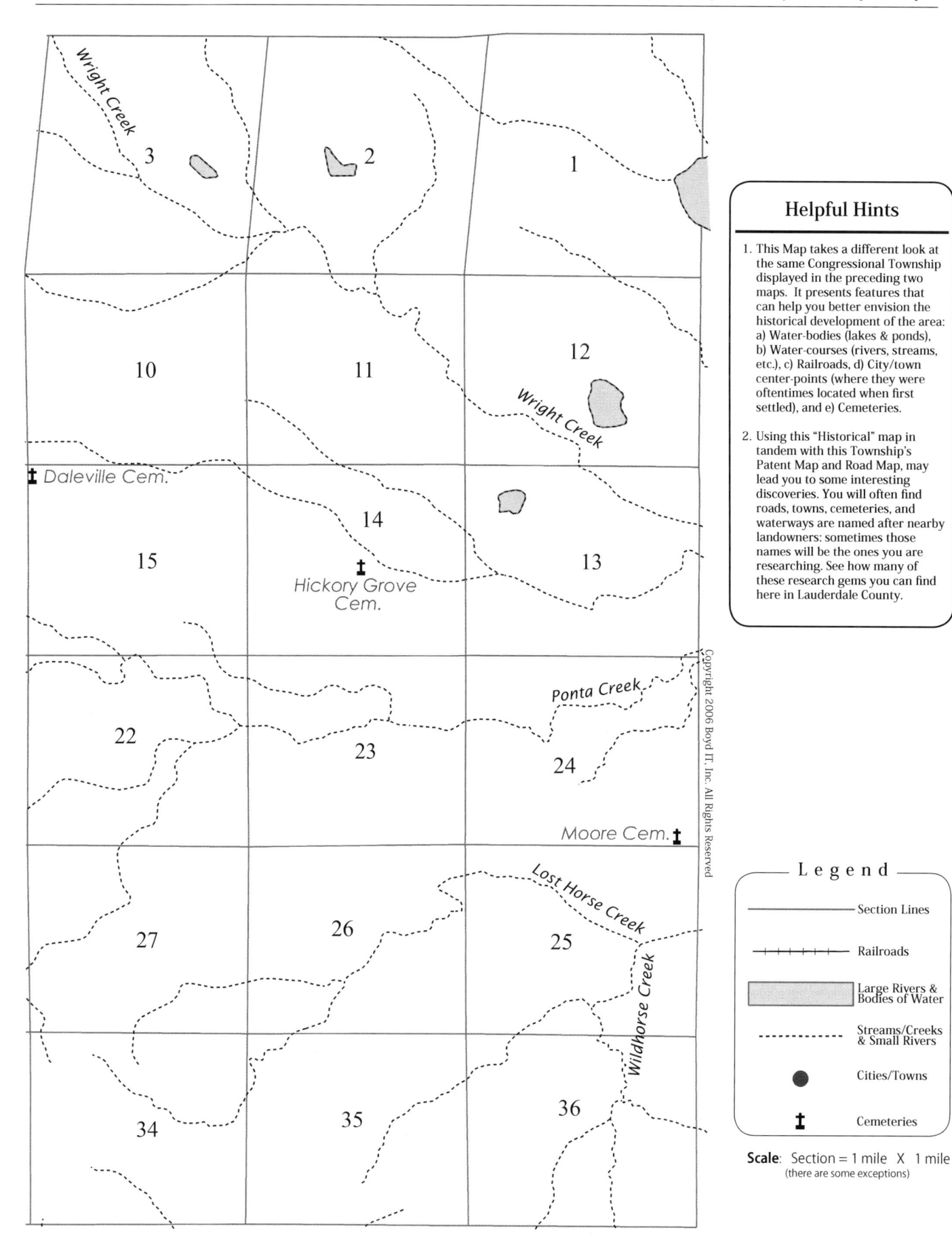

Legend
- Section Lines
- Railroads
- Large Rivers & Bodies of Water
- Streams/Creeks & Small Rivers
- Cities/Towns
- Cemeteries

Scale: Section = 1 mile X 1 mile
(there are some exceptions)

83

Map Group 4: Index to Land Patents
Township 8-North Range 17-East (Choctaw)

After you locate an individual in this Index, take note of the Section and Section Part then proceed to the Land Patent map on the pages immediately following. You should have no difficulty locating the corresponding parcel of land.

The "For More Info" Column will lead you to more information about the underlying Patents. See the *Legend* at right, and the "How to Use this Book" chapter, for more information.

```
                        LEGEND
              "For More Info . . . " column
A = Authority (Legislative Act, See Appendix "A")
B = Block or Lot (location in Section unknown)
C = Cancelled Patent
F = Fractional Section
G = Group (Multi-Patentee Patent, see Appendix "C")
V = Overlaps another Parcel
R = Re-Issued (Parcel patented more than once)

(A & G items require you to look in the Appendixes referred
to above. All other Letter-designations followed by a number
require you to locate line-items in this index that possess
the ID number found after the letter).
```

ID	Individual in Patent	Sec.	Sec. Part	Date Issued	Other Counties	For More Info . . .
864	ALLEN, John H	35	NESE	1860-05-01		A1
993	ALLEN, William	36	SW	1841-02-27		A1
994	" "	36	W½SE	1841-02-27		A1
992	" "	35	SESE	1854-03-15		A1
780	ANDING, David	28	E½NW	1841-02-27		A1
857	BAREFOOT, John	4	2	1841-02-27		A1
859	" "	4	3	1841-02-27		A1 G17
860	" "	4	6	1841-02-27		A1 G17
858	" "	4	7	1841-02-27		A1
895	BARFOOT, Miles	4	13	1841-02-27		A1
718	BROWN, Alfred	27	N½SE	1859-10-01		A1
719	" "	27	SENE	1859-10-01		A1
720	" "	35	E½SW	1859-10-01		A1
721	" "	35	NWNE	1859-10-01		A1
844	BROWN, James W	31	NWNW	1860-05-01		A1
845	" "	31	SENW	1860-05-01		A1
909	BROWN, Robert	3	1	1848-09-01		A1
910	" "	3	8	1848-09-01		A1
894	BURNS, Michael	27	NWNE	1854-03-15		A1
917	BUSBY, Shepherd	33	SENE	1860-05-01		A1
918	" "	33	SESE	1860-05-01		A1
885	BUSH, Lewis B	23	E½NE	1841-02-27		A1 G34
886	" "	23	E½NW	1841-02-27		A1 G34
887	" "	23	W½NE	1841-02-27		A1 G34
888	" "	23	W½NW	1841-02-27		A1 G34
774	BUTCHEE, Daniel	21	NWNE	1854-03-15		A1
775	" "	21	SWNE	1854-03-15		A1
776	" "	29	NENW	1854-03-15		A1
777	" "	29	SWNW	1854-03-15		A1
778	" "	29	W½SW	1854-03-15		A1
861	BUTCHEE, John	6	14	1849-12-01		A1
782	CALHOUN, Duncan	11	NENE	1859-10-01		A1
783	" "	11	NWNE	1859-10-01		A1
1001	CHILDE, William H	23	W½SW	1859-10-01		A1
760	CLAYTON, Charles G	21	NWNW	1854-03-15		A1
799	CLAYTON, George W	29	SESW	1854-03-15		A1
798	" "	29	NESW	1856-04-01		A1
805	CLAYTON, Henry A	30	NENE	1841-02-27		A1
806	" "	32	SWSW	1841-02-27		A1
873	CLAYTON, John W	18	NE	1841-02-27		A1
879	" "	7	SWSE	1841-02-27		A1
875	" "	18	NWSE	1846-09-01		A1
874	" "	18	NW	1848-09-01		A1
876	" "	30	SENE	1848-09-01		A1
877	" "	7	NWSE	1848-09-01		A1
878	" "	7	SESE	1848-09-01		A1

Township 8-N Range 17-E (Choctaw) - Map Group 4

ID	Individual in Patent	Sec.	Sec. Part	Date Issued	Other Counties	For More Info . . .
880	CLAYTON, John W (Cont'd)	9	NW	1860-05-01		A1
881	" "	9	NWSW	1860-05-01		A1
999	CLAYTON, William	29	NWNW	1841-02-27		A1
987	CLAYTON, William A	18	E½SW	1846-09-01		A1
989	" "	21	SWNW	1846-09-01		A1
991	" "	29	SENW	1856-04-01		A1
990	" "	29	NWSE	1858-07-15		A1
988	" "	21	N½SW	1859-10-01		A1
862	COCKE, John	15	SESW	1859-10-01		A1
939	COOPWOOD, Thomas	2	11	1841-02-27		A1 G50
940	" "	2	14	1841-02-27		A1 G50
941	" "	2	19	1841-02-27		A1 G50
925	" "	3	10	1841-02-27		A1 G52
933	" "	3	11	1841-02-27		A1 G51
934	" "	3	12	1841-02-27		A1 G51
935	" "	3	13	1841-02-27		A1 G51
936	" "	3	14	1841-02-27		A1 G51
926	" "	3	15	1841-02-27		A1 G52
927	" "	3	16	1841-02-27		A1 G52
928	" "	3	17	1841-02-27		A1 G52
929	" "	3	18	1841-02-27		A1 G52
937	" "	3	19	1841-02-27		A1 G51
938	" "	3	20	1841-02-27		A1 G51
923	" "	3	4	1841-02-27		A1
924	" "	3	5	1841-02-27		A1
930	" "	3	9	1841-02-27		A1 G52
931	" "	4	1	1841-02-27		A1 G52
932	" "	4	8	1841-02-27		A1 G52
906	DARDEN, Reddick	25	NWSW	1854-03-15		A1
907	" "	25	SENW	1858-07-15		A1
791	DAVIS, Edwin B	11	SENE	1841-02-27		A1 G58
790	" "	12	E½SW	1841-02-27		A1 G59
792	" "	12	SWNW	1841-02-27		A1 G58
789	" "	12	W½SW	1841-02-27		A1
791	DAVIS, Thomas	11	SENE	1841-02-27		A1 G58
792	" "	12	SWNW	1841-02-27		A1 G58
942	" "	12	SE	1846-09-01		A1
981	DELK, Vincent	4	11	1841-02-27		A1
982	DELK, Vinson	2	20	1841-02-27		A1
841	DEMENT, James P	13	SWNW	1859-10-01		A1 R842
842	" "	13	SWNW	1896-12-01		A1 R841
770	EASTHAM, Crawford	6	4	1841-02-27		A1 G67
771	" "	6	5	1841-02-27		A1 G67
769	" "	7	NENE	1841-02-27		A1
772	EATHAM, Crawford	5	13	1841-02-27		A1 G68
773	" "	5	20	1841-02-27		A1 G68
863	FEDRICK, John	9	SWSE	1856-04-01		A1
978	FINDLEY, Tira B	8	SESW	1841-02-27		A1
757	FINLEY, Caswell W	17	E½NE	1841-02-27		A1
758	" "	17	W½NE	1841-02-27		A1
964	FLEWELLEN, Thomas	6	1	1840-02-20		A1 G77
965	" "	6	10	1840-02-20		A1 G77
966	" "	6	15	1840-02-20		A1 G77
967	" "	6	16	1840-02-20		A1 G77
968	" "	6	17	1840-02-20		A1 G77
969	" "	6	18	1840-02-20		A1 G77
970	" "	6	2	1840-02-20		A1 G77
971	" "	6	7	1840-02-20		A1 G77
972	" "	6	8	1840-02-20		A1 G77
973	" "	6	9	1840-02-20		A1 G77
946	" "	5	1	1841-02-27		A1 G77
947	" "	5	10	1841-02-27		A1 G77
948	" "	5	11	1841-02-27		A1 G77
949	" "	5	12	1841-02-27		A1 G77
950	" "	5	14	1841-02-27		A1 G77
951	" "	5	15	1841-02-27		A1 G77
952	" "	5	16	1841-02-27		A1 G77
953	" "	5	17	1841-02-27		A1 G77
954	" "	5	18	1841-02-27		A1 G77
955	" "	5	19	1841-02-27		A1 G77
956	" "	5	2	1841-02-27		A1 G77
957	" "	5	3	1841-02-27		A1 G77
958	" "	5	4	1841-02-27		A1 G77

Family Maps of Lauderdale County, Mississippi

ID	Individual in Patent	Sec.	Sec. Part	Date Issued	Other Counties	For More Info . . .
959	FLEWELLEN, Thomas (Cont'd)	5	5	1841-02-27		A1 G77
960	" "	5	6	1841-02-27		A1 G77
961	" "	5	7	1841-02-27		A1 G77
962	" "	5	8	1841-02-27		A1 G77
963	" "	5	9	1841-02-27		A1 G77
939	FOWLER, Samuel	2	11	1841-02-27		A1 G50
940	" "	2	14	1841-02-27		A1 G50
941	" "	2	19	1841-02-27		A1 G50
1000	FOWLER, William	33	SWSW	1841-02-27		A1
854	GILHAM, John A	14	E½SW	1841-02-27		A1
855	" "	14	W½SE	1841-02-27		A1
856	" "	14	W½SW	1841-02-27		A1
717	GLASCOCK, Alexander	25	NESW	1859-10-01		A1
820	GRAHAM, James M	27	E½SW	1851/03/15		A2
964	GRANT, David B	6	1	1840-02-20		A1 G77
965	" "	6	10	1840-02-20		A1 G77
966	" "	6	15	1840-02-20		A1 G77
967	" "	6	16	1840-02-20		A1 G77
968	" "	6	17	1840-02-20		A1 G77
969	" "	6	18	1840-02-20		A1 G77
970	" "	6	2	1840-02-20		A1 G77
971	" "	6	7	1840-02-20		A1 G77
972	" "	6	8	1840-02-20		A1 G77
973	" "	6	9	1840-02-20		A1 G77
946	" "	5	1	1841-02-27		A1 G77
947	" "	5	10	1841-02-27		A1 G77
948	" "	5	11	1841-02-27		A1 G77
949	" "	5	12	1841-02-27		A1 G77
950	" "	5	14	1841-02-27		A1 G77
951	" "	5	15	1841-02-27		A1 G77
952	" "	5	16	1841-02-27		A1 G77
953	" "	5	17	1841-02-27		A1 G77
954	" "	5	18	1841-02-27		A1 G77
955	" "	5	19	1841-02-27		A1 G77
956	" "	5	2	1841-02-27		A1 G77
957	" "	5	3	1841-02-27		A1 G77
958	" "	5	4	1841-02-27		A1 G77
959	" "	5	5	1841-02-27		A1 G77
960	" "	5	6	1841-02-27		A1 G77
961	" "	5	7	1841-02-27		A1 G77
962	" "	5	8	1841-02-27		A1 G77
963	" "	5	9	1841-02-27		A1 G77
800	GRANT, Green W	22	E½NE	1841-02-27		A1
801	" "	22	W½NE	1841-02-27		A1
802	" "	24	E½SE	1841-02-27		A1
803	" "	25	E½NE	1841-02-27		A1
847	GREEN, Jesse	24	SWSE	1851/03/15		A2
848	" "	33	SESW	1851/03/15		A2
849	" "	33	SWSE	1851/03/15		A2
736	GRIFFIN, Archibald M	13	E½NE	1841-02-27		A1
737	" "	13	W½NE	1841-02-27		A1
738	" "	17	E½NW	1841-02-27		A1
739	" "	17	W½NW	1841-02-27		A1
740	" "	23	E½SE	1841-02-27		A1
741	" "	23	W½SE	1841-02-27		A1
742	" "	25	W½NE	1841-02-27		A1
913	GRIFFITH, Samuel	27	SWSE	1859-10-01		A1 G91
912	GRIFFITH, Samuel A	27	SESE	1856-04-01		A1
729	HANNA, Andrew W	1	11	1841-02-27		A1
730	" "	1	12	1841-02-27		A1
731	" "	1	13	1841-02-27		A1
732	" "	1	14	1841-02-27		A1
733	" "	1	19	1841-02-27		A1
734	" "	1	20	1841-02-27		A1
735	" "	1	8	1844-09-10		A1
722	HANNAH, Andrew	2	10	1841-02-27		A1 G101
723	" "	2	15	1841-02-27		A1 G101
724	" "	2	18	1841-02-27		A1 G101
964	HARDY, Charles	6	1	1840-02-20		A1 G77
965	" "	6	10	1840-02-20		A1 G77
966	" "	6	15	1840-02-20		A1 G77
967	" "	6	16	1840-02-20		A1 G77
968	" "	6	17	1840-02-20		A1 G77

Township 8-N Range 17-E (Choctaw) - Map Group 4

ID	Individual in Patent	Sec.	Sec. Part	Date Issued	Other Counties	For More Info . . .
969	HARDY, Charles (Cont'd)	6	18	1840-02-20		A1 G77
970	"	"	6	2	1840-02-20	A1 G77
971	"	"	6	7	1840-02-20	A1 G77
972	"	"	6	8	1840-02-20	A1 G77
973	"	"	6	9	1840-02-20	A1 G77
946	"	"	5	1	1841-02-27	A1 G77
947	"	"	5	10	1841-02-27	A1 G77
948	"	"	5	11	1841-02-27	A1 G77
949	"	"	5	12	1841-02-27	A1 G77
950	"	"	5	14	1841-02-27	A1 G77
951	"	"	5	15	1841-02-27	A1 G77
952	"	"	5	16	1841-02-27	A1 G77
953	"	"	5	17	1841-02-27	A1 G77
954	"	"	5	18	1841-02-27	A1 G77
955	"	"	5	19	1841-02-27	A1 G77
956	"	"	5	2	1841-02-27	A1 G77
957	"	"	5	3	1841-02-27	A1 G77
958	"	"	5	4	1841-02-27	A1 G77
959	"	"	5	5	1841-02-27	A1 G77
960	"	"	5	6	1841-02-27	A1 G77
961	"	"	5	7	1841-02-27	A1 G77
962	"	"	5	8	1841-02-27	A1 G77
963	"	"	5	9	1841-02-27	A1 G77
743	HARLAN, Benjamin	21	E½NE	1841-02-27		A1
744	"	"	22	W½NW	1841-02-27	A1
745	"	"	9	NENE	1841-02-27	A1
746	"	"	9	W½NE	1841-02-27	A1
747	HARLIN, Benjamin	4	14	1844-09-10		A1
748	"	"	4	18	1844-09-10	A1
749	"	"	4	19	1844-09-10	A1
754	HENDRICK, Bernard G	33	NESW	1848-09-01		A1
750	"	"	22	SW	1851/03/15	A2
752	"	"	27	NWNW	1851/03/15	A2
753	"	"	27	NWSW	1851/03/15	A2
755	"	"	36	SESE	1851/03/15	A2
751	"	"	27	NENW	1856-04-01	A1
892	HOLDERNESS, Mckinney	4	4	1841-02-27		A1 G114
893	"	"	4	5	1841-02-27	A1 G114
1002	HUMPHRIES, William	26	NENW	1860-07-02		A1
1003	"	"	26	NWNE	1860-07-02	A1
784	HUSSEY, Edward G	17	SW	1846-09-01		A1
785	"	"	18	E½SE	1846-09-01	A1
786	"	"	32	SESE	1849-12-01	A1
865	JACKSON, John M	13	E½SE	1841-02-27		A1
867	JEMISON, John S	1	15	1841-02-27		A1
868	"	"	1	18	1841-02-27	A1
890	JOHNSON, Luther E	18	W½SW	1841-02-27		A1
919	JOHNSTON, Spencer S	14	SWNW	1841-02-27		A1
781	JONES, David W	33	NENW	1849-12-01		A1
882	JONES, John W	15	NESW	1850-12-05		A1
914	JONES, Seborn	10	NWNE	1849-12-01		A1
915	JONES, Sebrne	10	E½NW	1841-02-27		A1
916	"	"	10	SWNE	1841-02-27	A1
921	JONES, Starling	7	E½SW	1841-02-27		A1 V911
920	"	"	15	SWSW	1856-04-01	A1
853	KILLINGSWORTH, Jesse V	33	NESE	1860-05-01		A1
943	LAMB, Thomas E	28	SWSW	1849-12-01		A1
944	"	"	29	SESE	1849-12-01	A1
945	"	"	33	NWNW	1849-12-01	A1
911	LEWIS, Rufus G	7	SW	1841-02-27		A1 G134 V921
974	LEWIS, Thomas	28	W½NE	1841-02-27		A1
975	"	"	29	NE	1841-02-27	A1
1004	LUNSFORD, William	17	SE	1846-09-01		A1
821	MCCOWN, James	1	4	1841-02-27		A1
822	"	"	1	5	1841-02-27	A1
832	"	"	12	E½NE	1841-02-27	A1 G144
823	"	"	12	E½NW	1841-02-27	A1
833	"	"	12	W½NE	1841-02-27	A1 G144
824	"	"	15	E½NE	1841-02-27	A1
825	"	"	15	W½NE	1841-02-27	A1
722	"	"	2	10	1841-02-27	A1 G101
723	"	"	2	15	1841-02-27	A1 G101
826	"	"	2	16	1841-02-27	A1

87

Family Maps of Lauderdale County, Mississippi

ID	Individual in Patent	Sec.	Sec. Part	Date Issued	Other Counties	For More Info . . .
827	MCCOWN, James (Cont'd)	2	17	1841-02-27		A1
724	" "	2	18	1841-02-27		A1 G101
828	" "	2	9	1841-02-27		A1
829	" "	22	E½SE	1841-02-27		A1
834	" "	24	E½NW	1841-02-27		A1 G145
933	" "	3	11	1841-02-27		A1 G51
934	" "	3	12	1841-02-27		A1 G51
935	" "	3	13	1841-02-27		A1 G51
936	" "	3	14	1841-02-27		A1 G51
937	" "	3	19	1841-02-27		A1 G51
938	" "	3	20	1841-02-27		A1 G51
830	" "	3	3	1841-02-27		A1
831	" "	3	6	1841-02-27		A1
1005	MCDANIEL, William	8	SE	1841-02-27		A1 G149
891	MCELROY, Mary C	35	SWNE	1860-05-01		A1
756	MCGREW, Cabell	13	E½NW	1859-10-01		A1
819	MCLEAN, James C	21	E½NW	1848-12-01		A1
817	MCMILLAN, James A	24	NWSE	1849-12-01		A1
835	MCNEILL, James	11	NENW	1850-12-05		A1
836	" "	2	3	1850-12-05		A1
837	" "	2	4	1850-12-05		A1
838	" "	2	6	1850-12-05		A1
839	" "	2	7	1850-12-05		A1
832	MCPHAUL, Daniel	12	E½NE	1841-02-27		A1 G144
833	" "	12	W½NE	1841-02-27		A1 G144
1009	MCWHORTER, William T	6	19	1848-09-01		A1
1010	" "	7	E½NW	1848-09-01		A1
1011	" "	7	NESE	1848-09-01		A1
1012	" "	7	NWNW	1848-09-01		A1
1015	" "	8	W½SW	1848-09-01		A1
1013	" "	7	SENE	1848-12-01		A1
1014	" "	7	W½NE	1848-12-01		A1
1016	MILLER, William T	28	SWNW	1849-12-01		A1
1017	" "	29	NESE	1849-12-01		A1
983	MILLS, Willard C	13	NWNW	1860-10-01		A1
984	" "	27	NENE	1860-10-01		A1
985	" "	35	SENW	1860-10-01		A1
986	" "	9	SENE	1860-10-01		A1
790	MONTGOMERY, John D	12	E½SW	1841-02-27		A1 G59
761	MOORE, Charles	15	E½NW	1841-02-27		A1
762	" "	15	E½SE	1841-02-27		A1
763	" "	15	W½NW	1841-02-27		A1
764	" "	15	W½SE	1841-02-27		A1
793	MOORE, Elias	2	5	1850-12-05		A1
807	MOORE, Henry	11	SWSW	1911-05-03		A4
889	MOORE, Lewis	15	NWSW	1854-03-15		A1
902	MOORE, Nelson	4	10	1841-02-27		A1
908	MOORE, Richard	9	E½SE	1848-09-01		A1
903	NEW, Nicholas F	1	16	1841-02-27		A1
904	" "	1	17	1841-02-27		A1
905	" "	1	9	1841-02-27		A1
840	NOEL, James	30	W½NE	1846-09-01		A1
812	ODOM, Jacob	6	12	1841-02-27		A1
813	" "	6	13	1841-02-27		A1
814	" "	6	20	1841-02-27		A1
770	" "	6	4	1841-02-27		A1 G67
771	" "	6	5	1841-02-27		A1 G67
810	" "	2	12	1848-09-01		A1
815	" "	9	SESW	1848-09-01		A1
811	" "	3	2	1849-12-01		A1
728	OFALLIN, Andrew	25	W½NW	1848-12-01		A1
726	" "	24	SWSW	1850-12-05		A1
725	" "	24	NWSW	1851/03/15		A2
727	" "	25	NENW	1851/03/15		A2
1006	ONEAL, William	11	NWSW	1841-02-27		A1
794	PARKER, Frank	1	1	1906-10-24		A4 G159
794	PARKER, Nancy	1	1	1906-10-24		A4 G159
997	PISTOLE, William C	22	W½SE	1841-02-27		A1
998	" "	24	E½SW	1841-02-27		A1
779	PRINE, Daniel	7	SWNW	1841-02-27		A1
1005	RAGSDALE, Samuel	8	SE	1841-02-27		A1 G149
818	RAMSEY, James B	27	SWNW	1859-10-01		A1
795	RATCLIFF, Gabriel A	31	NENW	1860-05-01		A1

Township 8-N Range 17-E (Choctaw) - Map Group 4

ID	Individual in Patent	Sec.	Sec. Part	Date Issued	Other Counties	For More Info . . .
796	RATCLIFF, Gabriel A (Cont'd)	31	NESW	1860-05-01		A1
797	" "	31	SWNW	1860-05-01		A1
816	RATSBURG, Jacob	31	SENE	1859-10-01		A1
765	RAWSON, Charles	27	SWSW	1841-02-27		A1
766	"	34	NWNW	1841-02-27		A1
897	REYNOLDS, Moses	25	W½SE	1841-02-27		A1
898	"	36	E½NW	1841-02-27		A1
899	"	36	NE	1841-02-27		A1
896	"	25	E½SE	1922-08-07		A1
809	ROBERTS, Hiram W	26	NENE	1849-12-01		A1
772	ROCHEL, Etherington	5	13	1841-02-27		A1 G68
773	"	5	20	1841-02-27		A1 G68
1007	SEGARS, William	26	S½SE	1841-02-27		A1
1008	SIGRIS, William	4	12	1841-02-27		A1
843	SIKES, James	21	SE	1846-09-01		A1
787	SIMMONS, Edward W	9	NESW	1856-04-01		A1
788	"	9	NWSE	1860-05-01		A1
850	SKINNER, Jesse	25	S½SW	1841-02-27		A1
851	"	35	E½NE	1841-02-27		A1
852	"	36	W½NW	1841-02-27		A1
869	STRAIT, John	12	NWNW	1841-02-27		A1 G173
866	STRAIT, John N	35	W½NW	1858-07-15		A1
976	STRANG, Thomas	29	SWSE	1858-07-15		A1 G174
870	SWIFT, John	13	E½SW	1841-02-27		A1
871	"	13	W½SE	1841-02-27		A1
872	"	13	W½SW	1841-02-27		A1
922	TARTT, Thirsa	3	7	1850-12-02		A1
1018	THOMAS, William	31	W½NE	1856-04-01		A1
977	TUCKER, Thomas	32	NESW	1849-12-01		A1
759	TUTON, Charles A	1	2	1841-02-27		A1
925	TUTT, James B	3	10	1841-02-27		A1 G52
926	"	3	15	1841-02-27		A1 G52
927	"	3	16	1841-02-27		A1 G52
928	"	3	17	1841-02-27		A1 G52
929	"	3	18	1841-02-27		A1 G52
930	"	3	9	1841-02-27		A1 G52
931	"	4	1	1841-02-27		A1 G52
859	"	4	3	1841-02-27		A1 G17
892	"	4	4	1841-02-27		A1 G114
893	"	4	5	1841-02-27		A1 G114
860	"	4	6	1841-02-27		A1 G17
932	"	4	8	1841-02-27		A1 G52
976	ULRIC, John G	29	SWSE	1858-07-15		A1 G174
900	WALTERER, Nedom	28	SE	1846-09-01		A1
901	WALTERS, Needham	21	S½SW	1854-03-15		A1
1019	WALTERS, William	26	SENE	1850-12-05		A1
768	WATERS, Charles	33	NWNE	1849-12-01		A1
767	"	33	NENE	1856-04-01		A1
883	WATKINS, John	28	E½SW	1848-09-01		A1
884	"	28	NWSW	1848-09-01		A1
804	WATTS, Haden	2	13	1841-02-27		A1
869	WELLS, Samuel G	12	NWNW	1841-02-27		A1 G173
964	WHITE, James F	6	1	1840-02-20		A1 G77
965	"	6	10	1840-02-20		A1 G77
966	"	6	15	1840-02-20		A1 G77
967	"	6	16	1840-02-20		A1 G77
968	"	6	17	1840-02-20		A1 G77
969	"	6	18	1840-02-20		A1 G77
970	"	6	2	1840-02-20		A1 G77
971	"	6	7	1840-02-20		A1 G77
972	"	6	8	1840-02-20		A1 G77
973	"	6	9	1840-02-20		A1 G77
946	"	5	1	1841-02-27		A1 G77
947	"	5	10	1841-02-27		A1 G77
948	"	5	11	1841-02-27		A1 G77
949	"	5	12	1841-02-27		A1 G77
950	"	5	14	1841-02-27		A1 G77
951	"	5	15	1841-02-27		A1 G77
952	"	5	16	1841-02-27		A1 G77
953	"	5	17	1841-02-27		A1 G77
954	"	5	18	1841-02-27		A1 G77
955	"	5	19	1841-02-27		A1 G77
956	"	5	2	1841-02-27		A1 G77

Family Maps of Lauderdale County, Mississippi

ID	Individual in Patent	Sec.	Sec. Part	Date Issued	Other Counties	For More Info . . .
957	WHITE, James F (Cont'd)	5	3	1841-02-27		A1 G77
958	" "	5	4	1841-02-27		A1 G77
959	" "	5	5	1841-02-27		A1 G77
960	" "	5	6	1841-02-27		A1 G77
961	" "	5	7	1841-02-27		A1 G77
962	" "	5	8	1841-02-27		A1 G77
963	" "	5	9	1841-02-27		A1 G77
1020	WHITE, William	24	E½NE	1841-02-27		A1
834	" "	24	E½NW	1841-02-27		A1 G145
1021	" "	24	W½NE	1841-02-27		A1
1022	" "	24	W½NW	1841-02-27		A1
885	WHITFIELD, Boaz	23	E½NE	1841-02-27		A1 G34
886	" "	23	E½NW	1841-02-27		A1 G34
887	" "	23	W½NE	1841-02-27		A1 G34
888	" "	23	W½NW	1841-02-27		A1 G34
911	WHITSETT, John C	7	SW	1841-02-27		A1 G134 V921
995	WILKINSON, William B	11	NWNW	1850-12-05		A1
996	" "	11	W½SE	1859-10-01		A1
808	WINN, Hinchy	10	NESE	1850-12-02		A1
913	WINNINGHAM, James H	27	SWSE	1859-10-01		A1 G91
979	WINNINGHAM, Vandy V	31	E½SE	1859-10-01		A1 G181
980	" "	31	NWSE	1859-10-01		A1 G181
979	WINNINGHAM, William W	31	E½SE	1859-10-01		A1 G181
980	" "	31	NWSE	1859-10-01		A1 G181
846	YARRELL, Jared	23	E½SW	1846-09-01		A1

Family Maps of Lauderdale County, Mississippi

Patent Map

T8-N R17-E
Choctaw Meridian

Map Group 4

Township Statistics

Parcels Mapped	:	306
Number of Patents	:	217
Number of Individuals	:	136
Patentees Identified	:	129
Number of Surnames	:	106
Multi-Patentee Parcels	:	74
Oldest Patent Date	:	2/20/1840
Most Recent Patent	:	8/7/1922
Block/Lot Parcels	:	105
Parcels Re-Issued	:	1
Parcels that Overlap	:	2
Cities and Towns	:	2
Cemeteries	:	6

Lots-Sec. 6

1. FLEWELLEN, Thoma[77] 1840
2. FLEWELLEN, Thoma[77] 1840
4. EASTHAM, Crawfor[67] 1841
5. EASTHAM, Crawfor[67] 1841
7. FLEWELLEN, Thoma[77] 1840
8. FLEWELLEN, Thoma[77] 1840
9. FLEWELLEN, Thoma[77] 1840
10. FLEWELLEN, Thoma[77] 1840
12. ODOM, Jacob 1841
13. ODOM, Jacob 1841
14. BUTCHEE, John 1849
15. FLEWELLEN, Thoma[77] 1840
16. FLEWELLEN, Thoma[77] 1840
17. FLEWELLEN, Thoma[77] 1840
18. FLEWELLEN, Thoma[77] 1840
19. MCWHORTER, William T 1848
20. ODOM, Jacob 1841

Lots-Sec. 5

1. FLEWELLEN, Thoma[77] 1841
2. FLEWELLEN, Thoma[77] 1841
3. FLEWELLEN, Thoma[77] 1841
4. FLEWELLEN, Thoma[77] 1841
5. FLEWELLEN, Thoma[77] 1841
6. FLEWELLEN, Thoma[77] 1841
7. FLEWELLEN, Thoma[77] 1841
8. FLEWELLEN, Thoma[77] 1841
9. FLEWELLEN, Thoma[77] 1841
10. FLEWELLEN, Thoma[77] 1841
11. FLEWELLEN, Thoma[77] 1841
12. FLEWELLEN, Thoma[77] 1841
13. EATHAM, Crawford[68] 1841
14. FLEWELLEN, Thoma[77] 1841
15. FLEWELLEN, Thoma[77] 1841
16. FLEWELLEN, Thoma[77] 1841
17. FLEWELLEN, Thoma[77] 1841
18. FLEWELLEN, Thoma[77] 1841
19. FLEWELLEN, Thoma[77] 1841
20. EATHAM, Crawford[68] 1841

Lots-Sec. 4

1. COOPWOOD, Thomas[52] 1841
2. BAREFOOT, John 1841
3. BAREFOOT, John [17] 1841
4. HOLDERNESS, Mck[114] 1841
5. HOLDERNESS, Mck[114] 1841
6. BAREFOOT, John [17] 1841
7. BAREFOOT, John 1841
8. COOPWOOD, Thomas[52] 1841
10. MOORE, Nelson 1841
11. DELK, Vincent 1841
12. SIGRIS, William 1841
13. BARFOOT, Miles 1841
14. HARLIN, Benjamin 1844
18. HARLIN, Benjamin 1844
19. HARLIN, Benjamin 1844

Section 7:
- MCWHORTER, William T 1848
- MCWHORTER, William T 1848
- EASTHAM, Crawford 1841
- PRINE, Daniel 1841
- MCWHORTER, William T 1848
- MCWHORTER, William T 1848
- LEWIS [134], Rufus G 1841
- CLAYTON, John W 1848
- MCWHORTER, William T 1848
- JONES, Starling 1841
- CLAYTON, John W 1841
- CLAYTON, John W 1848

Section 8:
- FINDLEY, Tira B 1841
- MCDANIEL [149], William 1841

Section 9:
- HARLAN, Benjamin 1841
- HARLAN, Benjamin 1841
- CLAYTON, John W 1860
- MILLS, Willard C 1860
- CLAYTON, John W 1860
- SIMMONS, Edward W 1856
- SIMMONS, Edward W 1856
- ODOM, Jacob 1848
- FEDRICK, John 1856
- MOORE, Richard 1848

Section 18:
- CLAYTON, John W 1848
- CLAYTON, John W 1841
- JOHNSON, Luther E 1841
- CLAYTON, John W 1846
- CLAYTON, William A 1846
- HUSSEY, Edward G 1846

Section 17:
- GRIFFIN, Archibald M 1841
- FINLEY, Caswell W 1841
- GRIFFIN, Archibald M 1841
- FINLEY, Caswell W 1841
- HUSSEY, Edward G 1846
- LUNSFORD, William 1846

Section 16

Section 19

Section 20

Section 21:
- CLAYTON, Charles G 1854
- BUTCHEE, Daniel 1854
- CLAYTON, William A 1846
- MCLEAN, James C 1848
- BUTCHEE, Daniel 1854
- HARLAN, Benjamin 1841
- CLAYTON, William A 1859
- SIKES, James 1846
- WALTERS, Needham 1854

Section 30

Section 29:
- NOEL, James 1846
- CLAYTON, Henry A 1841
- CLAYTON, William 1841
- BUTCHEE, Daniel 1854
- LEWIS, Thomas 1841
- CLAYTON, John W 1848
- BUTCHEE, Daniel 1854
- CLAYTON, William A 1856
- BUTCHEE, Daniel 1854
- CLAYTON, George W 1856
- CLAYTON, William A 1858
- MILLER, William T 1849
- STRANG [174], Thomas 1858
- LAMB, Thomas E 1849
- CLAYTON, George W 1854

Section 28:
- LEWIS, Thomas 1841
- MILLER, William T 1849
- ANDING, David 1841
- WATKINS, John 1848
- LAMB, Thomas E 1849
- WATKINS, John 1848
- WALTERER, Nedom 1846

Section 31:
- BROWN, James W 1860
- RATCLIFF, Gabriel A 1860
- THOMAS, William 1856
- RATCLIFF, Gabriel A 1860
- BROWN, James W 1860
- RATSBURG, Jacob 1859
- RATCLIFF, Gabriel A 1860
- WINNINGHAM [181], Vandy V 1859
- WINNINGHAM [181], Vandy V 1859

Section 32:
- TUCKER, Thomas 1849
- CLAYTON, Henry A 1841
- HUSSEY, Edward G 1849

Section 33:
- LAMB, Thomas E 1849
- JONES, David W 1849
- WATERS, Charles 1849
- WATERS, Charles 1856
- BUSBY, Shepherd 1860
- HENDRICK, Bernard G 1848
- KILLINGSWORTH, Jesse V 1860
- FOWLER, William 1841
- GREEN, Jesse 1851
- GREEN, Jesse 1851
- BUSBY, Shepherd 1860

Copyright 2006 Boyd IT, Inc. All Rights Reserved

Township 8-N Range 17-E (Choctaw) - Map Group 4

Lots-Sec. 3
1. BROWN, Robert 1848
2. ODOM, Jacob 1849
3. MCCOWN, James 1841
4. COOPWOOD, Thomas 1841
5. COOPWOOD, Thomas 1841
6. MCCOWN, James 1841
7. TARTT, Thirsa 1850
8. BROWN, Robert 1848
9. COOPWOOD, Thomas[52]1841
10. COOPWOOD, Thomas[52]1841
11. COOPWOOD, Thomas[51]1841
12. COOPWOOD, Thomas[51]1841
13. COOPWOOD, Thomas[51]1841
14. COOPWOOD, Thomas[51]1841
15. COOPWOOD, Thomas[52]1841
16. COOPWOOD, Thomas[52]1841
17. COOPWOOD, Thomas[52]1841
18. COOPWOOD, Thomas[52]1841
19. COOPWOOD, Thomas[51]1841

Lots-Sec. 2
3. MCNEILL, James 1850
4. MCNEILL, James 1850
5. MOORE, Elias 1850
6. MCNEILL, James 1850
7. MCNEILL, James 1850
9. MCCOWN, James 1841
10. HANNAH, Andrew [101]1841
11. COOPWOOD, Thomas[50]1841
12. ODOM, Jacob 1848
13. WATTS, Haden 1841
14. COOPWOOD, Thomas[50]1841
15. HANNAH, Andrew [101]1841
16. MCCOWN, James 1841
17. MCCOWN, James 1841
18. HANNAH, Andrew [101]1841
19. COOPWOOD, Thomas[50]1841
20. DELK, Vinson 1841

Lots-Sec. 1
1. PARKER, Frank [159]1906
2. TUTON, Charles A 1841
4. MCCOWN, James 1841
5. MCCOWN, James 1841
8. HANNA, Andrew W 1844
9. NEW, Nicholas F 1841
11. HANNA, Andrew W 1841
12. HANNA, Andrew W 1841
13. HANNA, Andrew W 1841
14. HANNA, Andrew W 1841
15. JEMISON, John S 1841
16. NEW, Nicholas F 1841
17. NEW, Nicholas F 1841
18. JEMISON, John S 1841
19. HANNA, Andrew W 1841
20. HANNA, Andrew W 1841

Helpful Hints

1. This Map's INDEX can be found on the preceding pages.
2. Refer to Map "C" to see where this Township lies within Lauderdale County, Mississippi.
3. Numbers within square brackets [ ] denote a multi-patentee land parcel (multi-owner). Refer to Appendix "C" for a full list of members in this group.
4. Areas that look to be crowded with Patentees usually indicate multiple sales of the same parcel (Re-issues) or Overlapping parcels. See this Township's Index for an explanation of these and other circumstances that might explain "odd" groupings of Patentees on this map.

Copyright 2006 Boyd IT, Inc. All Rights Reserved

Legend

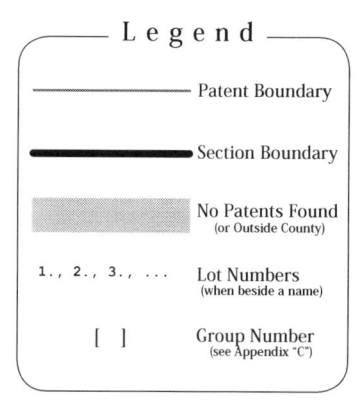

— Patent Boundary
— Section Boundary
▓ No Patents Found (or Outside County)
1., 2., 3., ... Lot Numbers (when beside a name)
[ ] Group Number (see Appendix "C")

Scale: Section = 1 mile X 1 mile (generally, with some exceptions)

93

Family Maps of Lauderdale County, Mississippi

Road Map
T8-N R17-E
Choctaw Meridian
Map Group 4

Cities & Towns
Lauderdale
Lockhart

Cemeteries
Barrett Cemetery
Clayton Cemetery
Gordon Cemetery
Lockhart Cemetery
Miller Cemetery
Segars Cemetery

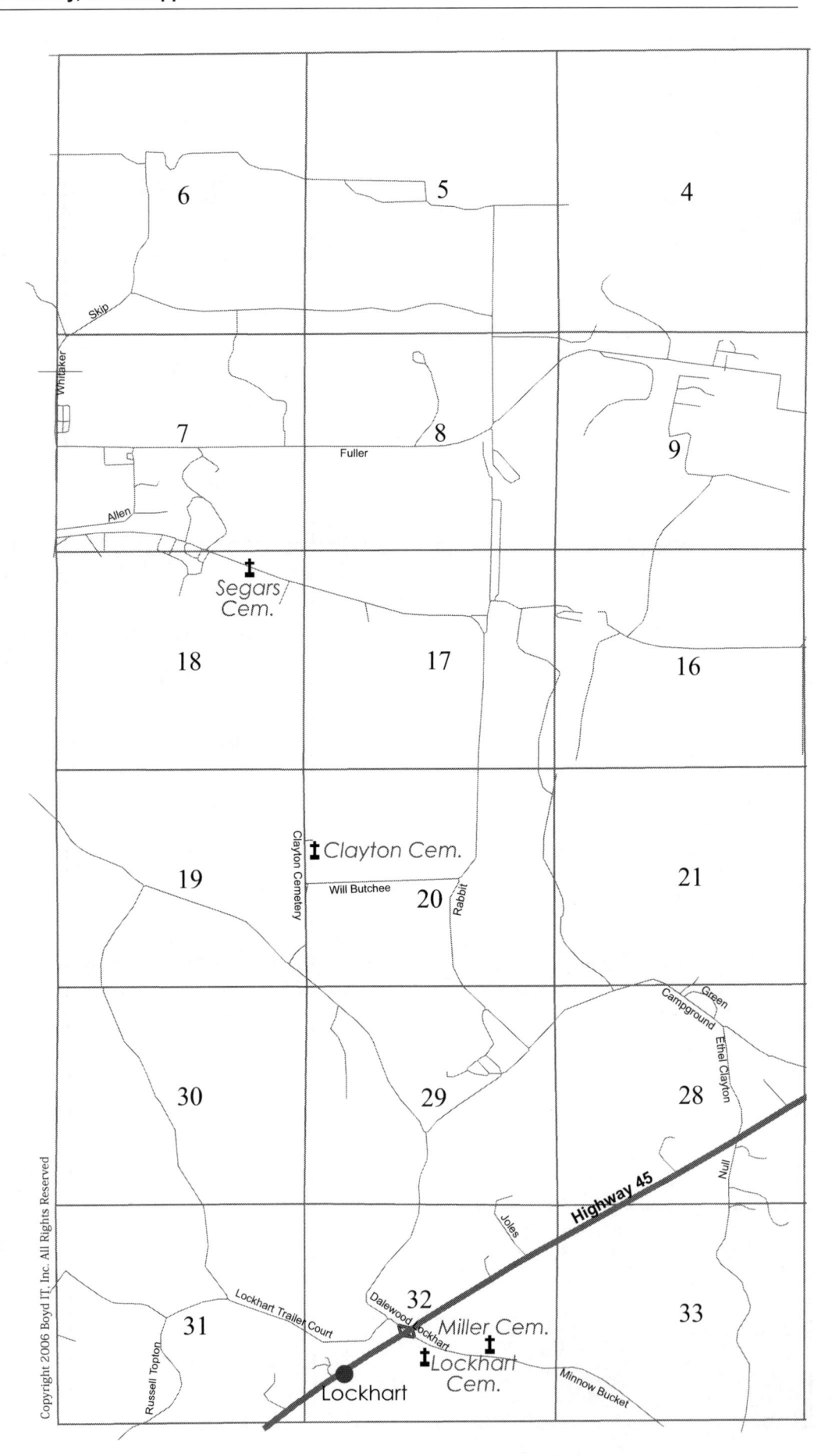

Township 8-N Range 17-E (Choctaw) - Map Group 4

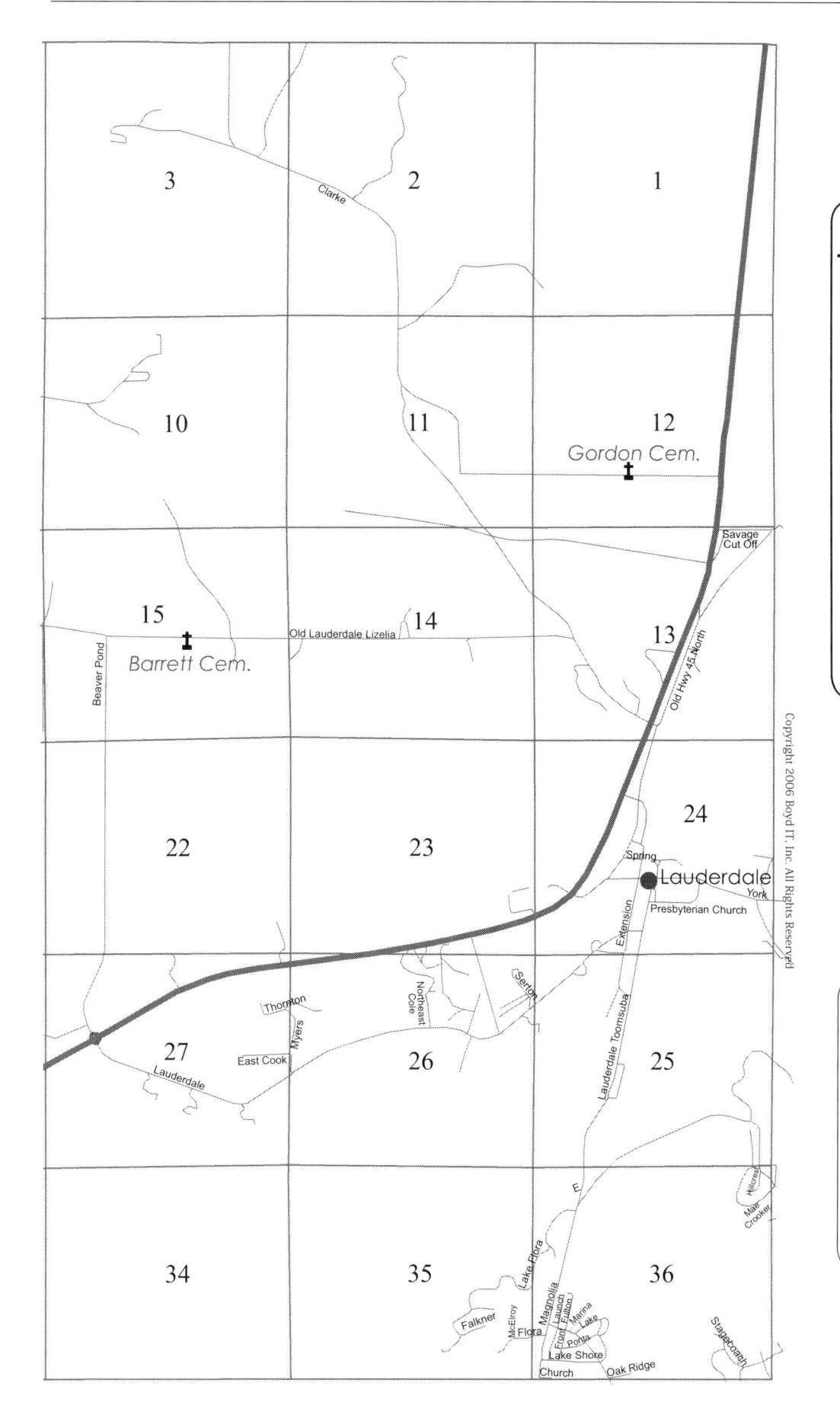

Helpful Hints

1. This road map has a number of uses, but primarily it is to help you: a) find the present location of land owned by your ancestors (at least the general area), b) find cemeteries and city-centers, and c) estimate the route/roads used by Census-takers & tax-assessors.

2. If you plan to travel to Lauderdale County to locate cemeteries or land parcels, please pick up a modern travel map for the area before you do. Mapping old land parcels on modern maps is not as exact a science as you might think. Just the slightest variations in public land survey coordinates, estimates of parcel boundaries, or road-map deviations can greatly alter a map's representation of how a road either does or doesn't cross a particular parcel of land.

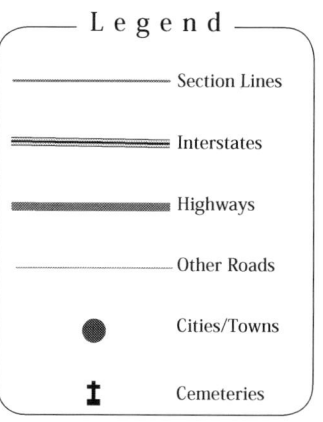

Legend

— Section Lines
≡ Interstates
▧ Highways
— Other Roads
● Cities/Towns
† Cemeteries

Scale: Section = 1 mile X 1 mile
(generally, with some exceptions)

95

Family Maps of Lauderdale County, Mississippi

Historical Map
T8-N R17-E
Choctaw Meridian
Map Group 4

Cities & Towns
Lauderdale
Lockhart

Cemeteries
Barrett Cemetery
Clayton Cemetery
Gordon Cemetery
Lockhart Cemetery
Miller Cemetery
Segars Cemetery

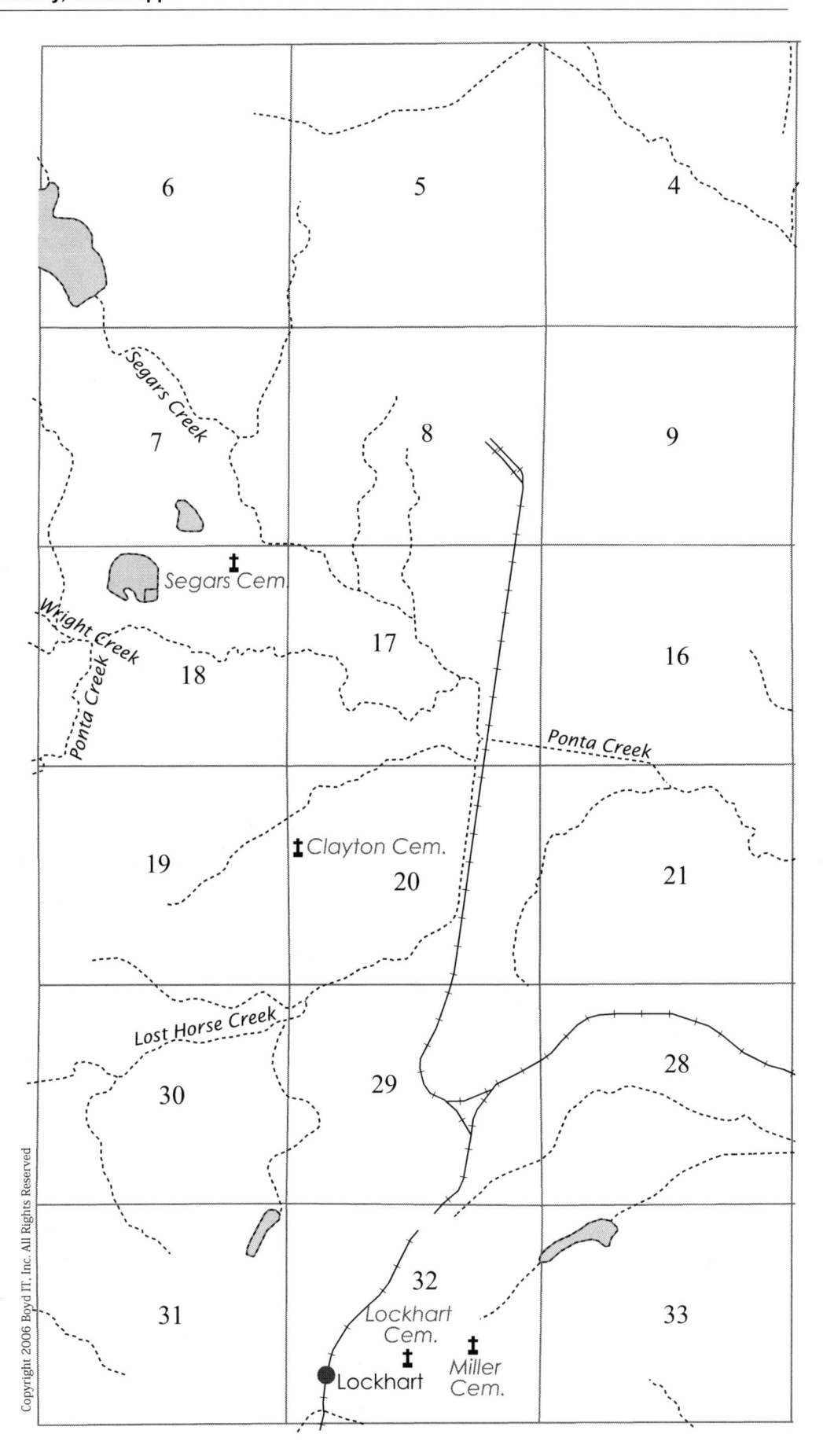

Township 8-N Range 17-E (Choctaw) - Map Group 4

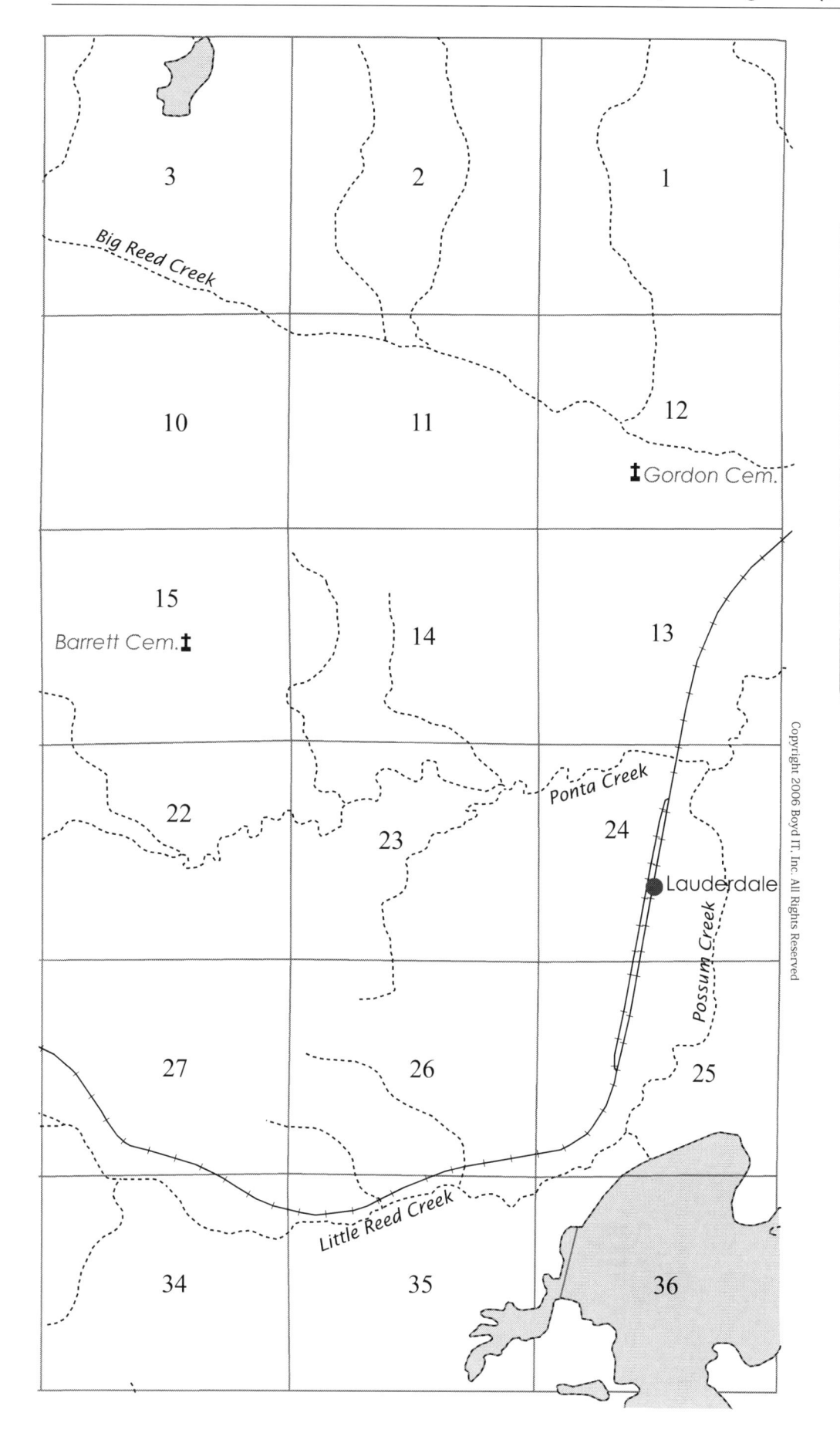

Helpful Hints

1. This Map takes a different look at the same Congressional Township displayed in the preceding two maps. It presents features that can help you better envision the historical development of the area: a) Water-bodies (lakes & ponds), b) Water-courses (rivers, streams, etc.), c) Railroads, d) City/town center-points (where they were oftentimes located when first settled), and e) Cemeteries.

2. Using this "Historical" map in tandem with this Township's Patent Map and Road Map, may lead you to some interesting discoveries. You will often find roads, towns, cemeteries, and waterways are named after nearby landowners: sometimes those names will be the ones you are researching. See how many of these research gems you can find here in Lauderdale County.

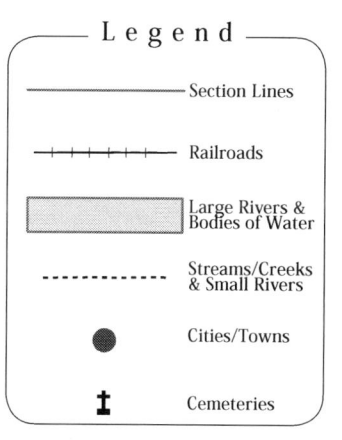

Legend

- Section Lines
- Railroads
- Large Rivers & Bodies of Water
- Streams/Creeks & Small Rivers
- Cities/Towns
- Cemeteries

Scale: Section = 1 mile X 1 mile
(there are some exceptions)

Features shown on map:
- Big Reed Creek
- Gordon Cem.
- Barrett Cem.
- Ponta Creek
- Possum Creek
- Lauderdale
- Little Reed Creek

Copyright 2006 Boyd IT, Inc. All Rights Reserved

97

Family Maps of Lauderdale County, Mississippi

Map Group 5: Index to Land Patents
Township 8-North Range 18-East (Choctaw)

After you locate an individual in this Index, take note of the Section and Section Part then proceed to the Land Patent map on the pages immediately following. You should have no difficulty locating the corresponding parcel of land.

The "For More Info" Column will lead you to more information about the underlying Patents. See the *Legend* at right, and the "How to Use this Book" chapter, for more information.

```
                    LEGEND
           "For More Info . . ." column
   A = Authority (Legislative Act, See Appendix "A")
   B = Block or Lot (location in Section unknown)
   C = Cancelled Patent
   F = Fractional Section
   G = Group (Multi-Patentee Patent, see Appendix "C")
   V = Overlaps another Parcel
   R = Re-Issued (Parcel patented more than once)

   (A & G items require you to look in the Appendixes referred
   to above. All other Letter-designations followed by a number
   require you to locate line-items in this index that possess
   the ID number found after the letter).
```

ID	Individual in Patent	Sec.	Sec. Part	Date Issued	Other Counties	For More Info . . .
1253	AIKINS, Robert	11	NESE	1859-10-01		A1
1254	" "	11	SENE	1859-10-01		A1
1193	ALLEN, John	1	20	1841-02-27		A1 G9
1194	" "	12	E½NE	1841-02-27		A1 G9
1191	" "	9	E½SW	1841-02-27		A1
1192	" "	9	W½SE	1841-02-27		A1
1195	" "	1	3	1844-09-10		A1 G7
1196	" "	1	6	1844-09-10		A1 G7
1197	" "	2	1	1846-12-01		A1 G8
1198	" "	2	8	1846-12-01		A1 G8
1145	ANDERSON, James	25	E½NW	1841-02-27		A1
1146	" "	25	NWSE	1841-02-27		A1
1147	" "	25	W½NE	1841-02-27		A1
1371	ANDERSON, William T	22	SESE	1849-12-01		A1
1256	BAILEY, Robert C	11	NENE	1892-06-30		A4
1209	BALL, Johnson	30	NWNW	1841-02-27		A1
1182	BEESON, Jeremiah S	1	12	1844-09-10		A1 G19
1183	" "	1	13	1844-09-10		A1 G19
1195	" "	1	3	1844-09-10		A1 G7
1184	" "	1	4	1844-09-10		A1 G19
1185	" "	1	5	1844-09-10		A1 G19
1196	" "	1	6	1844-09-10		A1 G7
1324	BEVILL, Thomas L	5	1	1841-02-27		A1
1325	" "	5	2	1841-02-27		A1
1326	" "	5	7	1841-02-27		A1
1327	" "	5	8	1841-02-27		A1
1139	BOND, Hartwell	13	SWSW	1896-12-14		A4
1148	BOND, James	13	SENW	1861-01-01		A1
1103	BOYD, David	21	E½NE	1841-02-27		A1
1104	" "	22	SENW	1841-02-27		A1
1105	" "	22	SW	1841-02-27		A1
1106	" "	22	W½NW	1841-02-27		A1
1125	BOYD, Gordon D	9	E½NW	1842-06-07		A1 G26
1126	" "	9	W½NW	1842-06-07		A1 G26
1124	" "	9	W½SW	1842-06-07		A1 G27
1199	BOYD, John	29	SENE	1841-02-27		A1
1358	BOYD, William L	6	2	1841-02-27		A1
1359	" "	6	3	1841-02-27		A1
1360	" "	6	7	1841-02-27		A1
1255	BROWN, Robert	34	W½NW	1841-02-27		A1
1085	BURTON, Cato	11	SWNW	1912-10-10		A4
1102	BUSBY, David Asbry	11	SWSW	1912-08-26		A4
1197	BUSON, Jeremiah S	2	1	1846-12-01		A1 G8
1198	" "	2	8	1846-12-01		A1 G8
1338	BUSTIN, William	19	W½SW	1848-09-01		A1 G35
1231	CAMPBELL, Neill	11	W½NE	1892-04-29		A4

98

Township 8-N Range 18-E (Choctaw) - Map Group 5

ID	Individual in Patent	Sec.	Sec. Part	Date Issued	Other Counties	For More Info...
1142	CANTERBURY, Isaac	18	SW	1846-09-01		A1
1141	" "	18	NWSE	1851/03/15		A2
1143	" "	20	NWNW	1851/03/15		A2
1140	CHILES, Henry	31	NENW	1911-09-18		A1
1177	CLAY, James T	15	W½SE	1898-02-24		A4
1284	CLAY, Royal G	2	3	1848-09-01		A1
1285	" "	2	4	1848-09-01		A1
1289	CLAY, Samuel	2	13	1848-09-01		A1
1290	" "	2	14	1848-09-01		A1
1291	" "	2	19	1848-09-01		A1
1292	" "	2	20	1848-09-01		A1
1307	CLAY, Stephen W	2	15	1850-12-05		A1
1372	CLAY, William W	3	1	1846-09-01		A1
1373	" "	3	2	1846-09-01		A1
1310	COOPWOOD, Thomas	4	3	1841-02-27		A1 G49
1311	" "	4	4	1841-02-27		A1 G49
1312	" "	4	5	1841-02-27		A1 G49
1313	" "	4	6	1841-02-27		A1 G49
1125	" "	9	E½NW	1842-06-07		A1 G26
1126	" "	9	W½NW	1842-06-07		A1 G26
1086	CRENSHAW, Charles G	27	SENE	1850-12-05		A1
1314	DAVIS, Thomas	7	SW	1841-02-27		A1 G62
1107	DELK, David	4	15	1846-09-01		A1
1108	" "	4	16	1846-09-01		A1
1109	" "	4	17	1846-09-01		A1
1110	" "	4	18	1846-09-01		A1
1114	DELK, Elias A	3	10	1844-09-10		A1
1115	" "	3	7	1844-09-10		A1
1116	" "	3	8	1844-09-10		A1
1117	" "	3	9	1844-09-10		A1
1283	DELK, Robert T	7	SE	1848-12-01		A1
1121	DOVE, Godfrey	36	12	1847-04-01		A1
1122	" "	36	13	1847-04-01		A1
1123	" "	36	5	1847-04-01		A1
1201	DOVE, John	25	SWSW	1860-10-01		A1
1339	EAKINS, William	34	E½SW	1841-02-27		A1
1340	" "	34	W½SE	1841-02-27		A1
1250	EDWARDS, Rigdon	4	12	1841-02-27		A1 G69
1251	" "	4	13	1841-02-27		A1 G69
1252	" "	4	20	1841-02-27		A1 G69
1310	ESTELL, John	4	3	1841-02-27		A1 G49
1311	" "	4	4	1841-02-27		A1 G49
1312	" "	4	5	1841-02-27		A1 G49
1313	" "	4	6	1841-02-27		A1 G49
1072	EZELL, Avris	25	SESW	1841-02-27		A1
1073	" "	25	SWSE	1841-02-27		A1
1074	EZELL, Benjamin	11	E½SW	1841-02-27		A1
1075	" "	11	W½SE	1841-02-27		A1
1076	" "	26	E½NE	1841-02-27		A1
1315	FLEWELLEN, Thomas	10	S½	1840-02-20		A1 G77
1316	" "	15	N½	1840-02-20		A1 G77
1317	" "	15	SW	1840-02-20		A1 G77
1319	" "	30	E½	1840-02-20		A1 G77
1320	" "	31	NE	1840-02-20		A1 G77
1321	" "	32	NW	1840-02-20		A1 G77
1318	" "	29	SW	1841-02-27		A1 G77
1362	GAINES, William M	31	NWSW	1848-09-01		A1 G79
1363	" "	31	W½NW	1848-09-01		A1 G80
1222	GAYLIE, Matt	18	E½NW	1841-02-27		A1 V1113
1223	" "	5	10	1841-02-27		A1
1224	" "	5	15	1841-02-27		A1
1225	" "	5	18	1841-02-27		A1
1308	GEE, Theoderick J	25	W½NW	1854-03-15		A1
1309	" "	27	NENE	1854-03-15		A1
1315	GRANT, David B	10	S½	1840-02-20		A1 G77
1316	" "	15	N½	1840-02-20		A1 G77
1317	" "	15	SW	1840-02-20		A1 G77
1319	" "	30	E½	1840-02-20		A1 G77
1320	" "	31	NE	1840-02-20		A1 G77
1321	" "	32	NW	1840-02-20		A1 G77
1318	" "	29	SW	1841-02-27		A1 G77
1129	GRANT, Green W	36	1	1841-02-27		A1
1130	" "	36	16	1841-02-27		A1

Family Maps of Lauderdale County, Mississippi

ID	Individual in Patent	Sec.	Sec. Part	Date Issued	Other Counties	For More Info . . .
1131	GRANT, Green W (Cont'd)	36	2	1841-02-27		A1
1132	" "	36	7	1841-02-27		A1
1133	" "	36	8	1841-02-27		A1
1134	" "	36	9	1841-02-27		A1
1135	GRANT, Greene W	36	6	1841-02-27		A1 G87
1136	" "	36	SWSE	1841-02-27		A1 G87
1095	GREENE, Daniel	3	16	1841-02-27		A1
1096	" "	3	17	1841-02-27		A1
1097	" "	1	11	1841-12-15		A1 G89
1098	" "	1	14	1841-12-15		A1 G89
1099	" "	1	19	1841-12-15		A1 G89
1035	GRIFFIN, Archibald M	1	10	1841-02-27		A1
1036	" "	1	15	1841-02-27		A1
1037	" "	1	16	1841-02-27		A1
1038	" "	1	17	1841-02-27		A1
1039	" "	1	18	1841-02-27		A1
1193	" "	1	20	1841-02-27		A1 G9
1040	" "	1	9	1841-02-27		A1
1194	" "	12	E½NE	1841-02-27		A1 G9
1041	" "	12	W½NW	1841-02-27		A1
1042	" "	3	15	1841-02-27		A1
1043	" "	3	18	1841-02-27		A1
1044	" "	4	11	1841-02-27		A1
1250	" "	4	12	1841-02-27		A1 G69
1251	" "	4	13	1841-02-27		A1 G69
1045	" "	4	14	1841-02-27		A1
1046	" "	4	19	1841-02-27		A1
1252	" "	4	20	1841-02-27		A1 G69
1047	" "	5	11	1841-02-27		A1
1048	" "	5	12	1841-02-27		A1
1049	" "	5	13	1841-02-27		A1
1050	" "	5	14	1841-02-27		A1
1051	" "	5	16	1841-02-27		A1
1052	" "	5	17	1841-02-27		A1
1053	" "	5	19	1841-02-27		A1
1054	" "	5	20	1841-02-27		A1
1055	" "	5	3	1841-02-27		A1
1056	" "	5	4	1841-02-27		A1
1057	" "	5	5	1841-02-27		A1
1058	" "	5	6	1841-02-27		A1
1059	" "	5	9	1841-02-27		A1
1060	" "	6	1	1841-02-27		A1
1061	" "	6	8	1841-02-27		A1
1062	" "	7	E½NE	1841-02-27		A1
1314	" "	7	SW	1841-02-27		A1 G62
1063	" "	7	W½NE	1841-02-27		A1
1064	" "	8	E½SW	1841-02-27		A1
1065	" "	8	W½SW	1841-02-27		A1 G90
1241	HAND, Ransom	13	NWNE	1859-10-01		A1
1242	" "	13	SENE	1859-10-01		A1
1243	" "	21	NENW	1859-10-01		A1
1244	" "	21	NWNE	1859-10-01		A1
1029	HANNA, Andrew W	6	4	1841-02-27		A1
1030	" "	6	5	1841-02-27		A1
1031	" "	6	6	1841-02-27		A1
1150	HANNA, James	9	E½NE	1841-02-27		A1
1151	" "	9	W½NE	1841-02-27		A1 G100
1259	HANNA, Robert	21	NESW	1841-02-27		A1
1260	" "	21	SESE	1841-02-27		A1
1261	" "	21	W½SE	1841-02-27		A1
1202	HANNER, John	29	NESE	1841-02-27		A1
1315	HARDY, Charles	10	S½	1840-02-20		A1 G77
1316	" "	15	N½	1840-02-20		A1 G77
1317	" "	15	SW	1840-02-20		A1 G77
1319	" "	30	E½	1840-02-20		A1 G77
1320	" "	31	NE	1840-02-20		A1 G77
1321	" "	32	NW	1840-02-20		A1 G77
1318	" "	29	SW	1841-02-27		A1 G77
1081	HARLAN, Benjamin	36	3	1844-09-10		A1
1080	" "	35	E½NW	1896-11-10		A1
1245	HARRIS, Richard	36	10	1841-02-27		A1
1362	HART, James N	31	NWSW	1848-09-01		A1 G79
1125	HENDERSON, John	9	E½NW	1842-06-07		A1 G26

100

Township 8-N Range 18-E (Choctaw) - Map Group 5

ID	Individual in Patent	Sec.	Sec. Part	Date Issued	Other Counties	For More Info...
1126	HENDERSON, John (Cont'd)	9	W½NW	1842-06-07		A1 G26
1124	" "	9	W½SW	1842-06-07		A1 G27
1338	HENDRICK, Bernard G	19	W½SW	1848-09-01		A1 G35
1084	" "	31	SWSW	1851/03/15		A2
1347	HIGGINBOTHAM, William	2	16	1841-02-27		A1
1348	" "	2	17	1841-02-27		A1
1349	" "	2	9	1841-02-27		A1
1023	HINES, Alexander	30	E½NW	1846-09-01		A1
1024	" "	30	SWNW	1846-09-01		A1
1217	HINTON, Joshua H	18	NENE	1849-12-01		A1
1218	" "	18	W½NE	1849-12-01		A1
1205	HODGES, John P	25	SENE	1841-02-27		A1
1262	HODGES, Robert	35	E½SW	1841-02-27		A1
1263	" "	35	SE	1841-02-27		A1
1257	HOUSTON, Robert F	4	1	1841-02-27		A1 G121
1258	" "	4	2	1841-02-27		A1 G121
1097	HUBBARD, Samuel	1	11	1841-12-15		A1 G89
1098	" "	1	14	1841-12-15		A1 G89
1099	" "	1	19	1841-12-15		A1 G89
1294	" "	13	SE	1841-12-15		A1 G123
1295	" "	26	SE	1841-12-15		A1 G123
1204	HUDSON, John M	8	W½NW	1841-02-27		A1
1264	HUDSON, Robert	6	11	1841-02-27		A1
1267	" "	6	12	1841-02-27		A1 G124
1268	" "	6	13	1841-02-27		A1 G124
1265	" "	6	14	1841-02-27		A1
1266	" "	6	19	1841-02-27		A1
1269	" "	6	20	1841-02-27		A1 G124
1113	HUGGINS, Eli	18	NW	1848-09-01		A1 V1222
1232	HULET, Obadiah	8	E½NE	1841-02-27		A1
1233	" "	8	E½SE	1841-02-27		A1
1234	" "	8	W½NE	1841-02-27		A1
1235	" "	8	W½SE	1841-02-27		A1
1065	" "	8	W½SW	1841-02-27		A1 G90
1239	HULET, Obediah	17	NE	1841-02-27		A1
1152	HUMPHREY, James	19	SESE	1841-02-27		A1
1342	HUNT, William H	13	NWSW	1859-10-01		A1
1343	" "	13	SESW	1859-10-01		A1
1341	" "	13	NENW	1860-05-01		A1
1093	IVY, Collin	24	E½SW	1841-02-27		A1
1094	IVY, Cullin	24	NWNW	1845-04-21		A1
1236	JARMAN, Obed F	33	NESE	1859-10-01		A1
1238	" "	33	SWSE	1859-10-01		A1
1237	" "	33	SWNE	1860-05-01		A1
1025	JONES, Alfred	3	3	1846-09-01		A1
1026	" "	3	4	1846-09-01		A1
1153	JONES, James J	26	W½NW	1841-02-27		A1
1154	" "	36	11	1841-02-27		A1
1155	" "	36	14	1841-02-27		A1
1156	" "	36	NWNW	1841-02-27		A1
1344	KEELAND, William H	30	N½SW	1846-09-01		A1
1182	KILLINGSWORTH, Jesse	1	12	1844-09-10		A1 G19
1183	" "	1	13	1844-09-10		A1 G19
1184	" "	1	4	1844-09-10		A1 G19
1185	" "	1	5	1844-09-10		A1 G19
1120	KNOX, George W	12	SWSW	1850-12-05		A1
1158	KNOX, James	13	W½NW	1841-02-27		A1
1159	" "	14	E½NE	1841-02-27		A1
1160	" "	26	E½NW	1841-02-27		A1
1161	" "	26	E½SW	1841-02-27		A1
1162	" "	26	W½NE	1841-02-27		A1
1163	" "	28	NWSW	1841-02-27		A1
1164	" "	28	SWNW	1841-02-27		A1
1354	KNOX, William	26	W½SW	1841-02-27		A1
1355	" "	27	E½SE	1841-02-27		A1
1356	" "	34	NENE	1841-02-27		A1
1357	" "	35	NWNW	1841-02-27		A1
1203	KUNERY, John L	11	NWNW	1854-03-15		A1
1028	LANCASTER, Andrew D	23	NENE	1908-10-29		A1
1077	LANCASTER, Benjamin F	23	E½NW	1899-02-25		A4
1078	" "	23	W½NE	1899-02-25		A4
1157	LANCASTER, James J	25	NWSW	1886-03-20		A4
1257	LANE, Isaac	4	1	1841-02-27		A1 G121

101

Family Maps of Lauderdale County, Mississippi

ID	Individual in Patent	Sec.	Sec. Part	Date Issued	Other Counties	For More Info . . .
1258	LANE, Isaac (Cont'd)	4	2	1841-02-27		A1 G121
1032	LEWIS, Anthony	24	SE	1841-02-27		A1
1033	" "	28	NESW	1841-02-27		A1
1034	" "	28	SWSW	1841-02-27		A1
1097	LEWIS, Rufus G	1	11	1841-12-15		A1 G89
1098	" "	1	14	1841-12-15		A1 G89
1099	" "	1	19	1841-12-15		A1 G89
1286	" "	10	E½NW	1841-12-15		A1 G134
1287	" "	10	NE	1841-12-15		A1 G134
1294	" "	13	SE	1841-12-15		A1 G123
1295	" "	26	SE	1841-12-15		A1 G123
1328	LEWIS, Thomas	28	SESW	1841-02-27		A1
1329	" "	33	NENW	1841-02-27		A1
1333	LEWIS, Warner H	21	NWSW	1860-05-01		A1
1332	" "	21	NESE	1860-10-01		A1
1334	" "	21	SESW	1860-10-01		A1
1179	MAURY, James W	21	SWNW	1859-10-01		A1
1178	" "	11	NWSW	1860-05-01		A1
1168	MCCOWN, James	10	W½NW	1841-02-27		A1
1169	" "	3	13	1841-02-27		A1
1170	" "	3	14	1841-02-27		A1
1171	" "	3	19	1841-02-27		A1
1172	" "	3	20	1841-02-27		A1
1173	" "	34	W½NE	1841-02-27		A1
1267	" "	6	12	1841-02-27		A1 G124
1268	" "	6	13	1841-02-27		A1 G124
1269	" "	6	20	1841-02-27		A1 G124
1174	" "	9	E½SE	1841-02-27		A1
1151	" "	9	W½NE	1841-02-27		A1 G100
1363	MCGRAW, John C	31	W½NW	1848-09-01		A1 G80
1200	MCGREW, John C	19	E½NW	1854-03-15		A1
1220	MCGREW, Mandy	29	NENE	1907-05-13		A4
1374	MCGREW, Willis	29	NWNE	1906-09-14		A4
1100	MCNEILL, Daniel L	23	SESE	1848-09-01		A1
1101	" "	23	SWSW	1855-03-15		A1
1300	MCPHERSON, Sarah	2	10	1846-08-10		A1
1301	" "	2	11	1846-08-10		A1
1302	" "	2	12	1846-08-10		A1
1303	" "	2	5	1846-08-10		A1
1304	" "	2	6	1846-08-10		A1
1361	MILLER, William L	35	S½NE	1896-11-13		A1
1066	MOORE, Arthur	13	NESW	1841-02-27		A1
1067	" "	24	E½NE	1841-02-27		A1
1068	" "	24	E½NW	1841-02-27		A1
1069	" "	24	W½NE	1841-02-27		A1
1219	MOORE, Joshua	17	NESW	1860-05-01		A1
1331	MOORE, Tyre G	2	7	1849-12-01		A1
1330	" "	2	2	1850-12-05		A1
1111	MORGAN, Edgar W	25	NENE	1901-03-23		A4
1322	MORGAN, Thomas J	23	NESW	1892-04-29		A4
1323	" "	33	NWNE	1906-06-16		A4
1175	MURPHY, James P	17	NWNW	1906-10-18		A1
1176	" "	17	S½NW	1906-10-18		A1
1216	NASH, Joseph W	19	NWNW	1858-07-15		A1
1082	NEWBERRY, Benjamin	7	E½NW	1841-02-27		A1
1083	" "	7	W½NW	1841-02-27		A1
1221	NUNNARY, Mary	33	E½NE	1841-02-27		A1
1119	PAYNE, George	22	W½SE	1844-09-10		A1
1186	PEAVY, Jerry	21	SWSW	1905-08-26		A4
1288	PEYTON, Sam	27	NWSW	1901-12-17		A4
1144	PIGFORD, Jacob O	23	SENE	1849-12-01		A1
1206	PITTS, John	13	SWNE	1898-01-19		A4
1346	PLUMMER, William H	19	NWSE	1860-10-01		A1
1345	" "	19	NESE	1875-08-20		A4
1135	REYNOLDS, John M	36	6	1841-02-27		A1 G87
1136	" "	36	SWSE	1841-02-27		A1 G87
1227	REYNOLDS, Moses	31	E½SW	1841-02-27		A1
1228	" "	31	W½SE	1841-02-27		A1
1226	ROGERS, Mitchell	19	NE	1899-09-30		A4
1350	SEALE, William J	32	E½NE	1841-02-27		A1
1351	" "	32	SWNE	1841-02-27		A1
1352	" "	33	W½NW	1841-02-27		A1
1299	SHELBY, Samuel	27	SWSW	1860-10-01		A1

Township 8-N Range 18-E (Choctaw) - Map Group 5

ID	Individual in Patent	Sec.	Sec. Part	Date Issued	Other Counties	For More Info . . .
1240	SILLIMAN, Peter	17	NESE	1905-02-13		A4
1070	SIMMONS, Asa	29	NWSE	1859-10-01		A1
1071	" "	29	SWNE	1859-10-01		A1
1127	SIMMONS, Green B	33	E½SW	1859-10-01		A1
1128	" "	33	W½SW	1859-10-01		A1
1166	SIMMONS, James L	31	NESE	1859-10-01		A1
1167	" "	33	NWSE	1859-10-01		A1
1165	" "	29	SENW	1860-05-01		A1
1270	SIMMONS, Robert	19	SWNW	1848-09-01		A1
1296	SMITH, Samuel S	17	NWSE	1860-05-01		A1
1112	SMITHWICK, Edward W	17	SWSW	1859-10-01		A1
1297	SMITHWICK, Samuel S	17	SESE	1859-10-01		A1
1298	" "	17	SWSE	1859-10-01		A1
1246	STANTON, Richard	1	1	1841-02-27		A1
1247	" "	1	2	1841-02-27		A1
1248	" "	1	7	1841-02-27		A1
1249	" "	1	8	1841-02-27		A1
1271	STRAIT, Robert	14	W½NW	1841-02-27		A1
1272	" "	15	E½SE	1841-02-27		A1
1273	" "	22	E½NE	1841-02-27		A1
1274	" "	22	W½NE	1841-02-27		A1
1275	" "	23	E½SE	1841-02-27		A1
1276	" "	25	NESE	1841-02-27		A1
1277	" "	27	E½NW	1841-02-27		A1
1278	" "	27	E½SW	1841-02-27		A1
1279	" "	27	W½NE	1841-02-27		A1
1280	" "	27	W½SE	1841-02-27		A1
1281	" "	34	E½NW	1841-02-27		A1
1282	" "	35	N½NE	1841-02-27		A1
1367	STRAIT, William	27	W½NW	1841-02-27		A1
1368	" "	28	E½NE	1841-02-27		A1
1369	" "	28	E½SE	1841-02-27		A1
1370	" "	28	W½NE	1841-02-27		A1
1365	SWAIN, William S	23	NWSW	1888-02-25		A4
1366	" "	23	SWNW	1888-02-25		A4
1364	" "	23	NWNW	1899-01-23		A4
1207	SWIFT, John	12	E½NW	1841-02-27		A1
1208	" "	12	W½NE	1841-02-27		A1
1097	TAPPAN, John	1	11	1841-12-15		A1 G89
1098	" "	1	14	1841-12-15		A1 G89
1099	" "	1	19	1841-12-15		A1 G89
1294	" "	13	SE	1841-12-15		A1 G123
1295	" "	26	SE	1841-12-15		A1 G123
1149	TARTT, James Elnathan	21	NWNW	1919-05-22		A1
1079	THOMPSON, Benjamin F	13	NENE	1896-07-11		A4
1087	TUTON, Charles	6	10	1841-02-27		A1
1088	" "	6	15	1841-02-27		A1
1089	" "	6	16	1841-02-27		A1
1090	" "	6	17	1841-02-27		A1
1091	" "	6	18	1841-02-27		A1
1092	" "	6	9	1841-02-27		A1
1137	TUTT, Harry	29	N½NW	1897-11-22		A4
1138	" "	29	SWNW	1897-11-22		A4
1181	WALKER, James	34	W½SW	1841-02-27		A1
1189	WALKER, Joel P	31	SESE	1849-12-01		A1
1190	" "	33	SENW	1854-03-15		A1
1188	" "	19	SESW	1859-10-01		A1
1187	" "	19	NESW	1860-05-01		A1
1214	WARREN, Joseph M	3	11	1841-02-27		A1 G179
1215	" "	3	6	1841-02-27		A1 G179
1210	" "	4	10	1841-02-27		A1
1211	" "	4	7	1841-02-27		A1
1212	" "	4	8	1841-02-27		A1
1213	" "	4	9	1841-02-27		A1
1118	WATSON, Ephram	29	S½SE	1916-05-31		A4
1214	WATTS, Haden	3	11	1841-02-27		A1 G179
1215	" "	3	6	1841-02-27		A1 G179
1305	WATTS, Sealy	3	12	1841-02-27		A1
1306	" "	3	5	1841-02-27		A1
1293	WELLS, Samuel G	11	E½NW	1841-02-27		A1
1027	WESTBROOK, Alonzo L	25	NESW	1906-05-01		A4
1315	WHITE, James F	10	S½	1840-02-20		A1 G77
1316	" "	15	N½	1840-02-20		A1 G77

Family Maps of Lauderdale County, Mississippi

ID	Individual in Patent	Sec.	Sec. Part	Date Issued	Other Counties	For More Info . . .
1317	WHITE, James F (Cont'd)	15	SW	1840-02-20		A1 G77
1319	" "	30	E½	1840-02-20		A1 G77
1320	" "	31	NE	1840-02-20		A1 G77
1321	" "	32	NW	1840-02-20		A1 G77
1318	" "	29	SW	1841-02-27		A1 G77
1286	WHITSETT, John C	10	E½NW	1841-12-15		A1 G134
1287	" "	10	NE	1841-12-15		A1 G134
1335	WILKINSON, William B	17	NENW	1860-05-01		A1
1336	" "	17	NWSW	1860-05-01		A1
1337	" "	17	SESW	1860-05-01		A1
1180	WRIGHT, James W	11	SESE	1884-12-30		A4
1353	WRIGHT, William J	35	W½SW	1850-12-05		A1
1229	YARBOROUGH, Nathan	21	SENW	1906-10-24		A4
1230	" "	21	SWNE	1906-10-24		A4

Family Maps of Lauderdale County, Mississippi

Patent Map

T8-N R18-E
Choctaw Meridian

Map Group 5

Township Statistics

Parcels Mapped	:	352
Number of Patents	:	250
Number of Individuals	:	154
Patentees Identified	:	150
Number of Surnames	:	111
Multi-Patentee Parcels	:	49
Oldest Patent Date	:	2/20/1840
Most Recent Patent	:	5/22/1919
Block/Lot Parcels	:	133
Parcels Re - Issued	:	0
Parcels that Overlap	:	2
Cities and Towns	:	0
Cemeteries	:	4

Lots-Sec. 6
1. GRIFFIN, Archibald M 1841
2. BOYD, William L 1841
3. BOYD, William L 1841
4. HANNA, Andrew W 1841
5. HANNA, Andrew W 1841
6. HANNA, Andrew W 1841
7. BOYD, William L 1841
8. GRIFFIN, Archibald M 1841
9. TUTON, Charles 1841
10. TUTON, Charles 1841
11. HUDSON, Robert 1841
12. HUDSON, Robert [124] 1841
13. HUDSON, Robert [124] 1841
14. HUDSON, Robert 1841
15. TUTON, Charles 1841
16. TUTON, Charles 1841
17. TUTON, Charles 1841
18. TUTON, Charles 1841
19. HUDSON, Robert 1841
20. HUDSON, Robert [124] 1841

Lots-Sec. 5
1. BEVILL, Thomas L 1841
2. BEVILL, Thomas L 1841
3. GRIFFIN, Archibald M 1841
4. GRIFFIN, Archibald M 1841
5. GRIFFIN, Archibald M 1841
6. GRIFFIN, Archibald M 1841
7. BEVILL, Thomas L 1841
8. BEVILL, Thomas L 1841
9. GRIFFIN, Archibald M 1841
10. GAYLIE, Matt 1841
11. GRIFFIN, Archibald M 1841
12. GRIFFIN, Archibald M 1841
13. GRIFFIN, Archibald M 1841
14. GRIFFIN, Archibald M 1841
15. GAYLIE, Matt 1841
16. GRIFFIN, Archibald M 1841
17. GRIFFIN, Archibald M 1841
18. GAYLIE, Matt 1841
19. GRIFFIN, Archibald M 1841
20. GRIFFIN, Archibald M 1841

Lots-Sec. 4
1. HOUSTON, Robert [121] 1841
2. HOUSTON, Robert [121] 1841
3. COOPWOOD, Thomas [49] 1841
4. COOPWOOD, Thomas [49] 1841
5. COOPWOOD, Thomas [49] 1841
6. COOPWOOD, Thomas [49] 1841
7. WARREN, Joseph M 1841
8. WARREN, Joseph M 1841
9. WARREN, Joseph M 1841
10. WARREN, Joseph M 1841
11. GRIFFIN, Archibald M 1841
12. EDWARDS, Rigdon [69] 1841
13. EDWARDS, Rigdon [69] 1841
14. GRIFFIN, Archibald M 1841
15. DELK, David 1846
16. DELK, David 1846
17. DELK, David 1846
18. DELK, David 1846
19. GRIFFIN, Archibald M 1841
20. EDWARDS, Rigdon [69] 1841

106

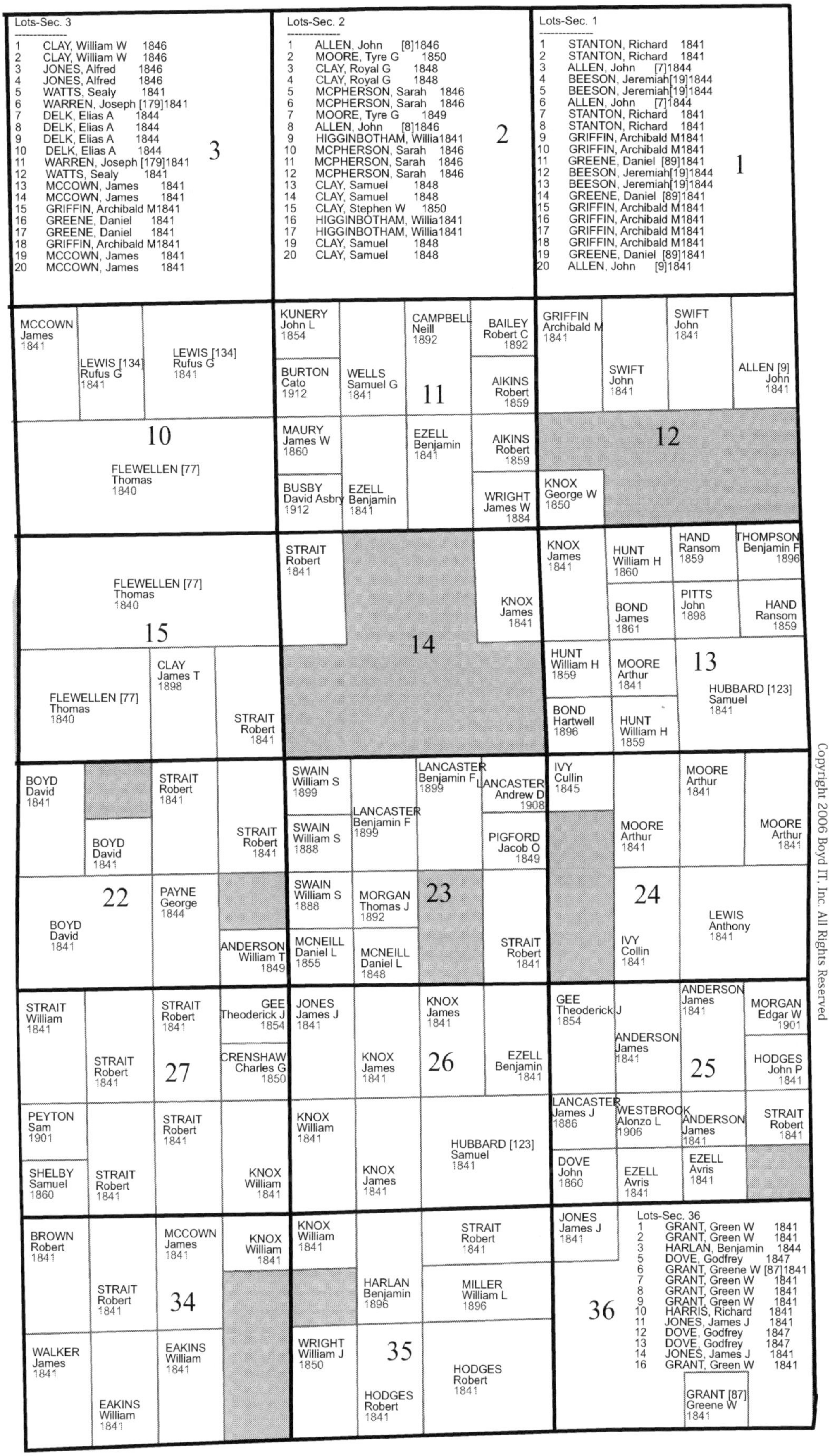

Township 8-N Range 18-E (Choctaw) - Map Group 5

Lots-Sec. 3
1. CLAY, William W 1846
2. CLAY, William W 1846
3. JONES, Alfred 1846
4. JONES, Alfred 1846
5. WATTS, Sealy 1841
6. WARREN, Joseph [179]1841
7. DELK, Elias A 1844
8. DELK, Elias A 1844
9. DELK, Elias A 1844
10. DELK, Elias A 1844
11. WARREN, Joseph [179]1841
12. WATTS, Sealy 1841
13. MCCOWN, James 1841
14. MCCOWN, James 1841
15. GRIFFIN, Archibald M 1841
16. GREENE, Daniel 1841
17. GREENE, Daniel 1841
18. GRIFFIN, Archibald M 1841
19. MCCOWN, James 1841
20. MCCOWN, James 1841

Lots-Sec. 2
1. ALLEN, John [8]1846
2. MOORE, Tyre G 1850
3. CLAY, Royal G 1848
4. CLAY, Royal G 1848
5. MCPHERSON, Sarah 1846
6. MCPHERSON, Sarah 1846
7. MOORE, Tyre G 1849
8. ALLEN, John [8]1846
9. HIGGINBOTHAM, Willia 1841
10. MCPHERSON, Sarah 1846
11. MCPHERSON, Sarah 1846
12. MCPHERSON, Sarah 1846
13. CLAY, Samuel 1848
14. CLAY, Samuel 1848
15. CLAY, Stephen W 1850
16. HIGGINBOTHAM, Willia 1841
17. HIGGINBOTHAM, Willia 1841
18. CLAY, Samuel 1848
19. CLAY, Samuel 1848
20. CLAY, Samuel 1848

Lots-Sec. 1
1. STANTON, Richard 1841
2. STANTON, Richard 1841
3. ALLEN, John [7]1844
4. BEESON, Jeremiah [19]1844
5. BEESON, Jeremiah [19]1844
6. ALLEN, John [7]1844
7. STANTON, Richard 1841
8. STANTON, Richard 1841
9. GRIFFIN, Archibald M 1841
10. GRIFFIN, Archibald M 1841
11. GREENE, Daniel [89]1841
12. BEESON, Jeremiah [19]1844
13. BEESON, Jeremiah [19]1844
14. GREENE, Daniel [89]1841
15. GRIFFIN, Archibald M 1841
16. GRIFFIN, Archibald M 1841
17. GRIFFIN, Archibald M 1841
18. GRIFFIN, Archibald M 1841
19. GREENE, Daniel [89]1841
20. ALLEN, John [9]1841

Lots-Sec. 36
1. GRANT, Green W 1841
2. GRANT, Green W 1841
3. HARLAN, Benjamin 1844
4. DOVE, Godfrey 1847
5. DOVE, Godfrey 1847
6. GRANT, Greene W [87]1841
7. GRANT, Green W 1841
8. GRANT, Green W 1841
9. GRANT, Green W 1841
10. HARRIS, Richard 1841
11. JONES, James J 1841
12. DOVE, Godfrey 1847
13. DOVE, Godfrey 1847
14. JONES, James J 1841
15. GRANT, Green W 1841
16. GRANT, Green W 1841

Helpful Hints

1. This Map's INDEX can be found on the preceding pages.

2. Refer to Map "C" to see where this Township lies within Lauderdale County, Mississippi.

3. Numbers within square brackets [ ] denote a multi-patentee land parcel (multi-owner). Refer to Appendix "C" for a full list of members in this group.

4. Areas that look to be crowded with Patentees usually indicate multiple sales of the same parcel (Re-issues) or Overlapping parcels. See this Township's Index for an explanation of these and other circumstances that might explain "odd" groupings of Patentees on this map.

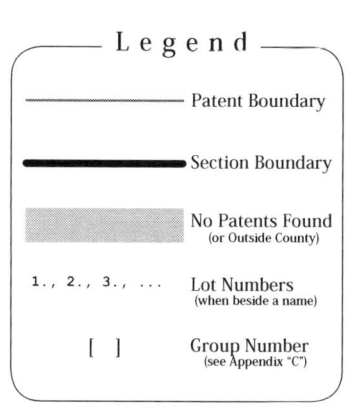

Copyright 2006 Boyd IT, Inc. All Rights Reserved

Legend

— Patent Boundary
— Section Boundary
▓ No Patents Found (or Outside County)
1., 2., 3., ... Lot Numbers (when beside a name)
[ ] Group Number (see Appendix "C")

Scale: Section = 1 mile X 1 mile (generally, with some exceptions)

107

Family Maps of Lauderdale County, Mississippi

Road Map
T8-N R18-E
Choctaw Meridian
Map Group 5

Cities & Towns
None

Cemeteries
Anna York Cemetery
Center Ridge Cemetery
Lauderdale Cemetery
Lauderdale Springs Cemetery

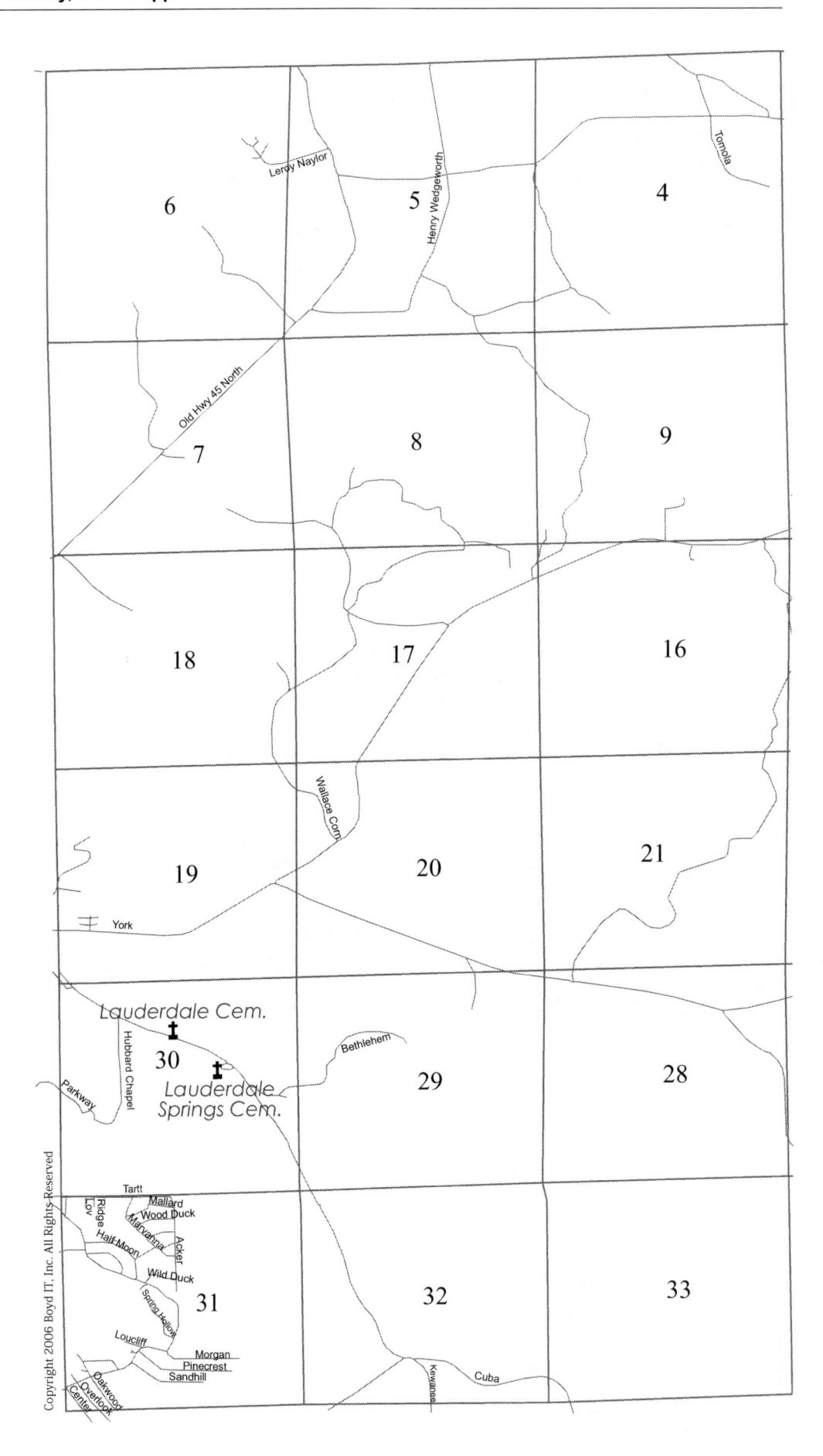

Township 8-N Range 18-E (Choctaw) - Map Group 5

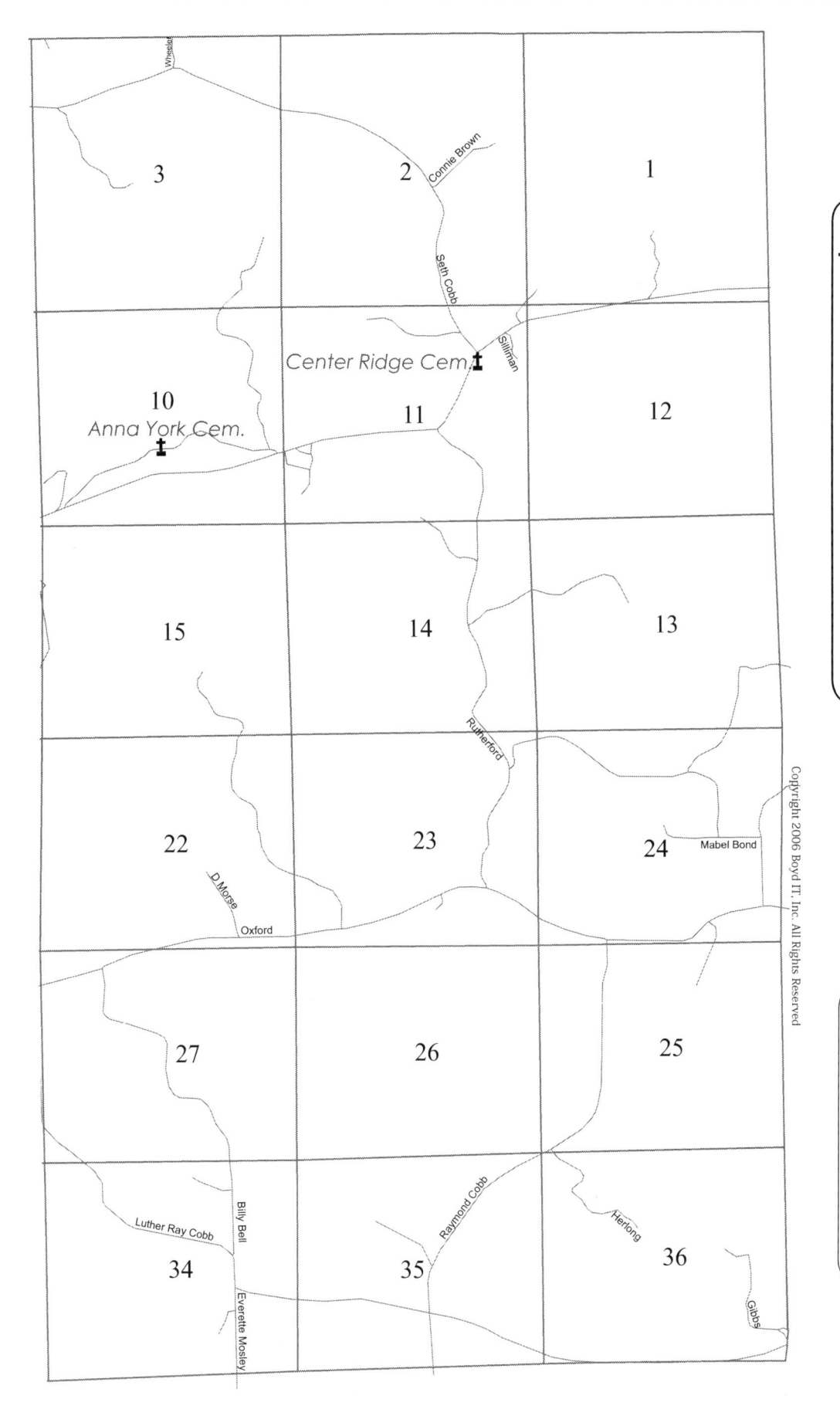

Helpful Hints

1. This road map has a number of uses, but primarily it is to help you: a) find the present location of land owned by your ancestors (at least the general area), b) find cemeteries and city-centers, and c) estimate the route/roads used by Census-takers & tax-assessors.

2. If you plan to travel to Lauderdale County to locate cemeteries or land parcels, please pick up a modern travel map for the area before you do. Mapping old land parcels on modern maps is not as exact a science as you might think. Just the slightest variations in public land survey coordinates, estimates of parcel boundaries, or road-map deviations can greatly alter a map's representation of how a road either does or doesn't cross a particular parcel of land.

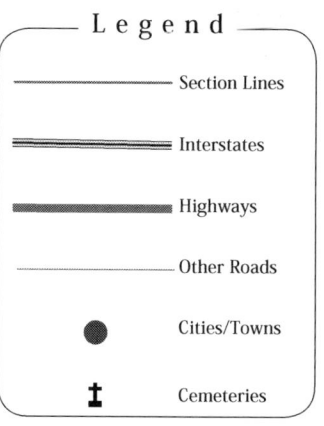

Legend
— Section Lines
═ Interstates
▬ Highways
— Other Roads
● Cities/Towns
✝ Cemeteries

Scale: Section = 1 mile X 1 mile
(generally, with some exceptions)

109

Family Maps of Lauderdale County, Mississippi

Historical Map
T8-N R18-E
Choctaw Meridian
Map Group 5

Cities & Towns
None

Cemeteries
Anna York Cemetery
Center Ridge Cemetery
Lauderdale Cemetery
Lauderdale Springs Cemetery

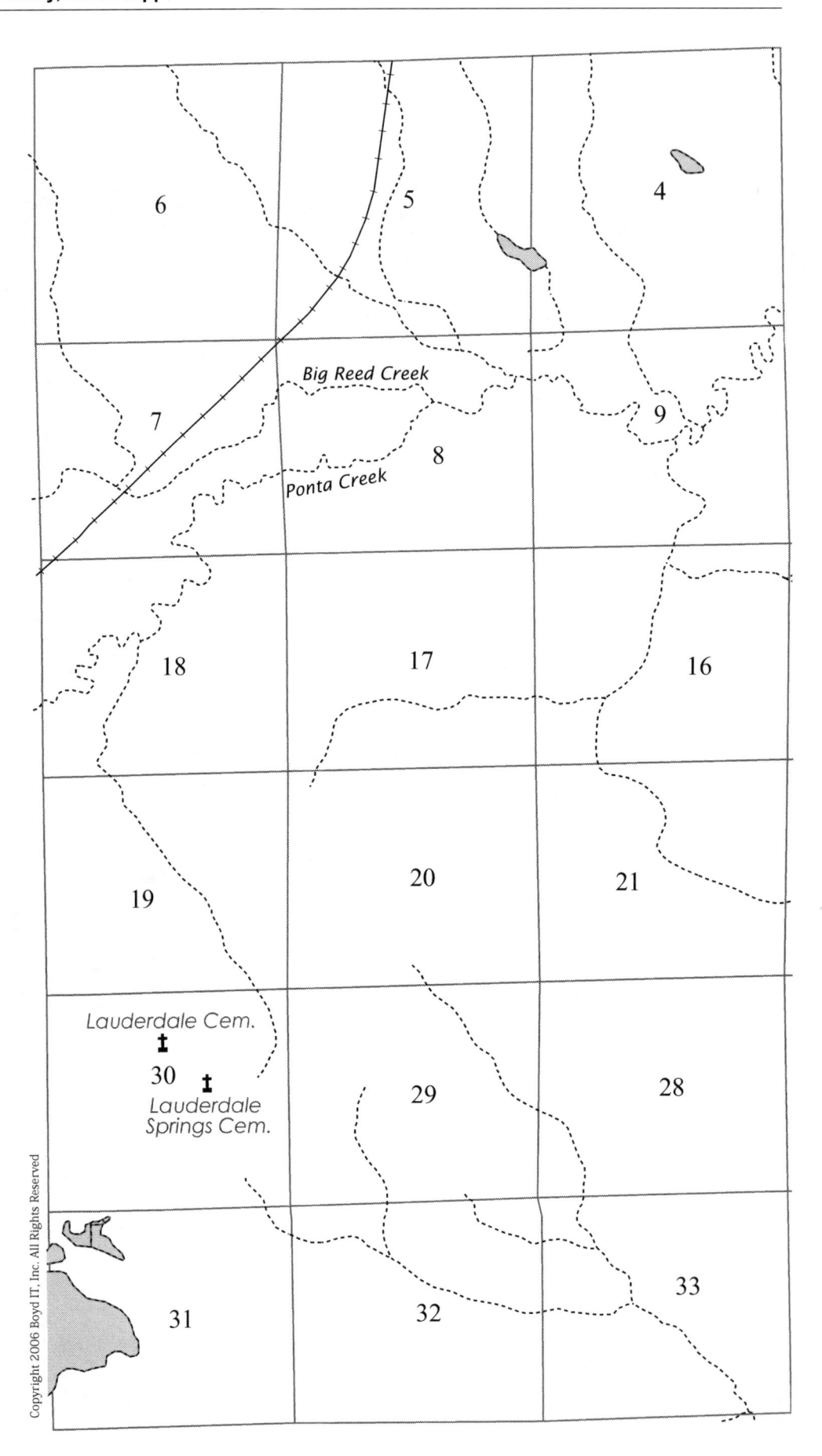

Big Reed Creek
Ponta Creek
Lauderdale Cem.
Lauderdale Springs Cem.

Copyright 2006 Boyd IT, Inc. All Rights Reserved

Township 8-N Range 18-E (Choctaw) - Map Group 5

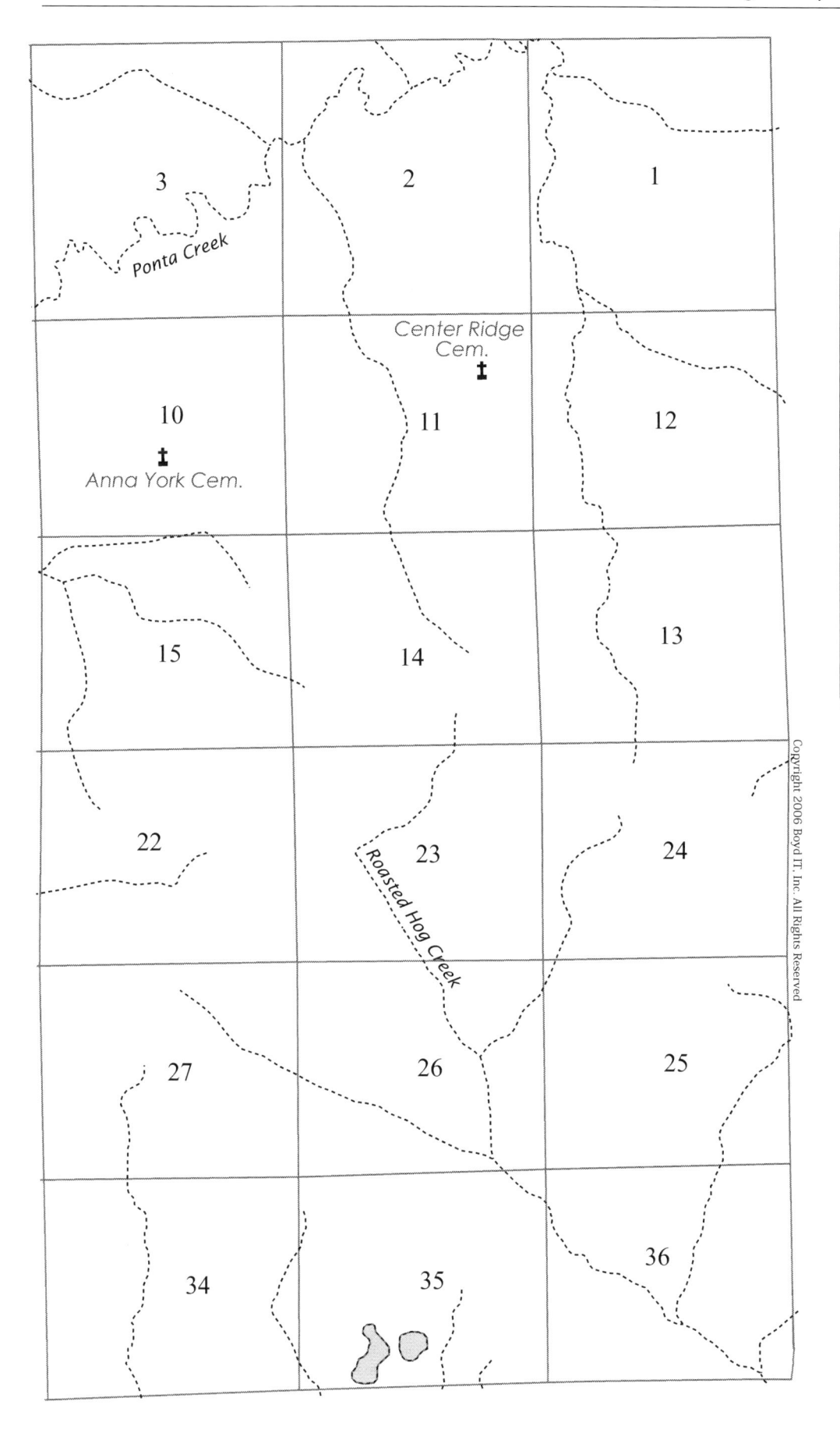

Helpful Hints

1. This Map takes a different look at the same Congressional Township displayed in the preceding two maps. It presents features that can help you better envision the historical development of the area: a) Water-bodies (lakes & ponds), b) Water-courses (rivers, streams, etc.), c) Railroads, d) City/town center-points (where they were oftentimes located when first settled), and e) Cemeteries.

2. Using this "Historical" map in tandem with this Township's Patent Map and Road Map, may lead you to some interesting discoveries. You will often find roads, towns, cemeteries, and waterways are named after nearby landowners: sometimes those names will be the ones you are researching. See how many of these research gems you can find here in Lauderdale County.

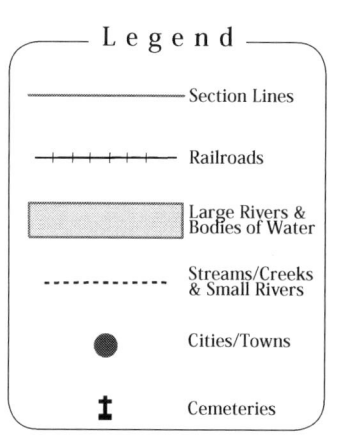

Features shown on map:
- Ponta Creek (sections 3, 10)
- Center Ridge Cem. (section 11)
- Anna York Cem. (section 10)
- Roasted Hog Creek (section 23)

Legend

- Section Lines
- Railroads
- Large Rivers & Bodies of Water
- Streams/Creeks & Small Rivers
- Cities/Towns (●)
- Cemeteries (†)

Scale: Section = 1 mile X 1 mile
(there are some exceptions)

Copyright 2006 Boyd IT, Inc. All Rights Reserved

Family Maps of Lauderdale County, Mississippi

Map Group 6: Index to Land Patents
Township 8-North Range 19-East (Choctaw)

After you locate an individual in this Index, take note of the Section and Section Part then proceed to the Land Patent map on the pages immediately following. You should have no difficulty locating the corresponding parcel of land.

The "For More Info" Column will lead you to more information about the underlying Patents. See the *Legend* at right, and the "How to Use this Book" chapter, for more information.

```
          LEGEND
    "For More Info . . ." column
A = Authority (Legislative Act, See Appendix "A")
B = Block or Lot (location in Section unknown)
C = Cancelled Patent
F = Fractional Section
G = Group  (Multi-Patentee Patent, see Appendix "C")
V = Overlaps another Parcel
R = Re-Issued (Parcel patented more than once)

(A & G items require you to look in the Appendixes referred
to above. All other Letter-designations followed by a number
require you to locate line-items in this index that possess
the ID number found after the letter).
```

ID	Individual in Patent	Sec.	Sec. Part	Date Issued	Other Counties	For More Info . . .
1390	BURTON, Linu H	7	1	1912-10-10		A4
1388	COBB, Joshua J	7	2	1894-12-17		A4
1389	" "	7	3	1894-12-17		A4
1391	LARD, Will	7	4	1920-02-04		A4
1375	LEWIS, Anthony	19	3	1841-02-27		A1
1376	" "	19	4	1841-02-27		A1
1379	MCCOWN, James	30		1841-02-27		A1 F
1380	" "	31		1841-02-27		A1 F
1381	" "	6	2	1841-02-27		A1
1382	" "	6	3	1841-02-27		A1
1383	" "	6	4	1841-02-27		A1
1384	" "	6	5	1841-02-27		A1
1385	" "	6	6	1841-02-27		A1
1386	" "	7	5	1841-02-27		A1
1387	" "	7	6	1841-02-27		A1
1377	MOORE, Arthur	19	1	1841-02-27		A1
1378	" "	19	2	1841-02-27		A1
1392	STRAIT, William	18	3	1844-09-10		A1
1393	" "	18	4	1844-09-10		A1

Township 8-N Range 19-E (Choctaw) - Map Group 6

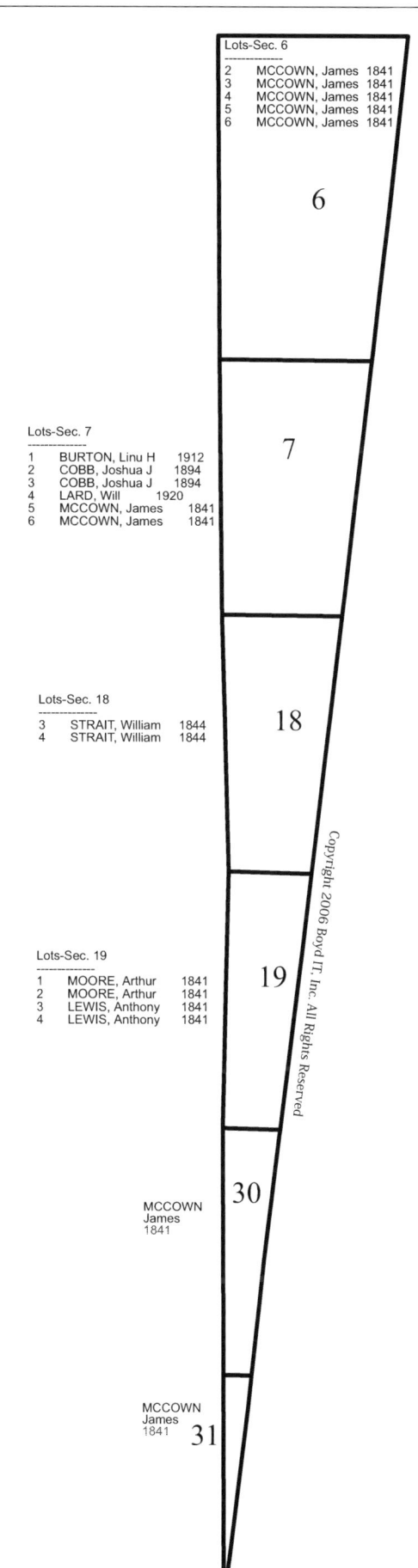

Patent Map

T8-N R19-E
Choctaw Meridian

Map Group 6

Township Statistics

Parcels Mapped	:	19
Number of Patents	:	12
Number of Individuals	:	7
Patentees Identified	:	7
Number of Surnames	:	7
Multi-Patentee Parcels	:	0
Oldest Patent Date	:	2/27/1841
Most Recent Patent	:	2/4/1920
Block/Lot Parcels	:	17
Parcels Re-Issued	:	0
Parcels that Overlap	:	0
Cities and Towns	:	0
Cemeteries	:	0

Note: the area contained in this map amounts to far less than a full Township. Therefore, its contents are completely on this single page (instead of a "normal" 2-page spread).

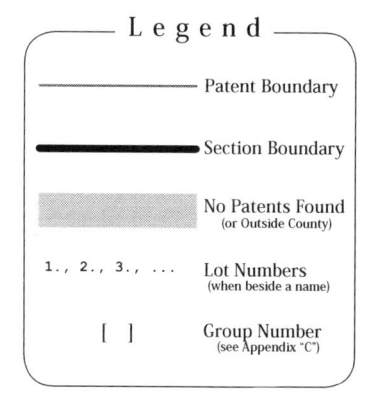

Lots-Sec. 6
2	MCCOWN, James	1841
3	MCCOWN, James	1841
4	MCCOWN, James	1841
5	MCCOWN, James	1841
6	MCCOWN, James	1841

Section 6

Lots-Sec. 7
1	BURTON, Linu H	1912
2	COBB, Joshua J	1894
3	COBB, Joshua J	1894
4	LARD, Will	1920
5	MCCOWN, James	1841
6	MCCOWN, James	1841

Section 7

Lots-Sec. 18
3	STRAIT, William	1844
4	STRAIT, William	1844

Section 18

Lots-Sec. 19
1	MOORE, Arthur	1841
2	MOORE, Arthur	1841
3	LEWIS, Anthony	1841
4	LEWIS, Anthony	1841

Section 19

Section 30 — MCCOWN, James 1841

Section 31 — MCCOWN, James 1841

Copyright 2006 Boyd IT, Inc. All Rights Reserved

Legend

— Patent Boundary
— Section Boundary
(shaded) No Patents Found (or Outside County)
1., 2., 3., ... Lot Numbers (when beside a name)
[ ] Group Number (see Appendix "C")

Scale: Section = 1 mile X 1 mile (generally, with some exceptions)

113

Family Maps of Lauderdale County, Mississippi

Road Map

T8-N R19-E
Choctaw Meridian

Map Group 6

Note: the area contained in this map amounts to far less than a full Township. Therefore, its contents are completely on this single page (instead of a "normal" 2-page spread).

Cities & Towns
None

Cemeteries
None

Legend
- Section Lines
- Interstates
- Highways
- Other Roads
- ● Cities/Towns
- † Cemeteries

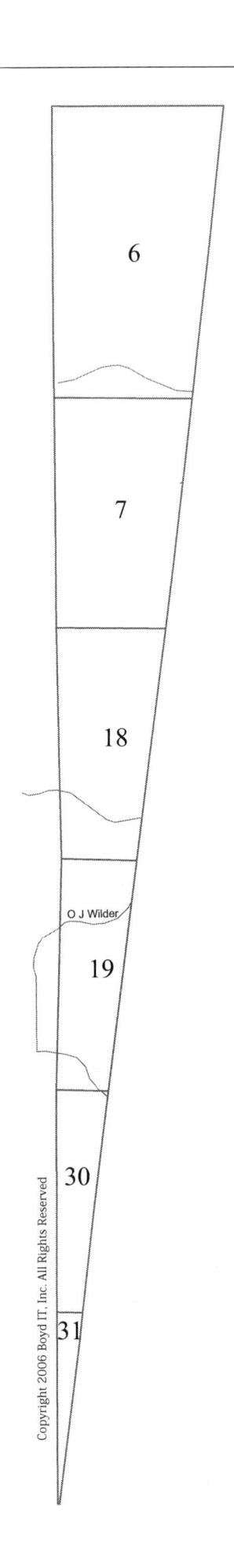

Scale: Section = 1 mile X 1 mile
(generally, with some exceptions)

6

7

18

O J Wilder

19

30

31

Copyright 2006 Boyd IT, Inc. All Rights Reserved

114

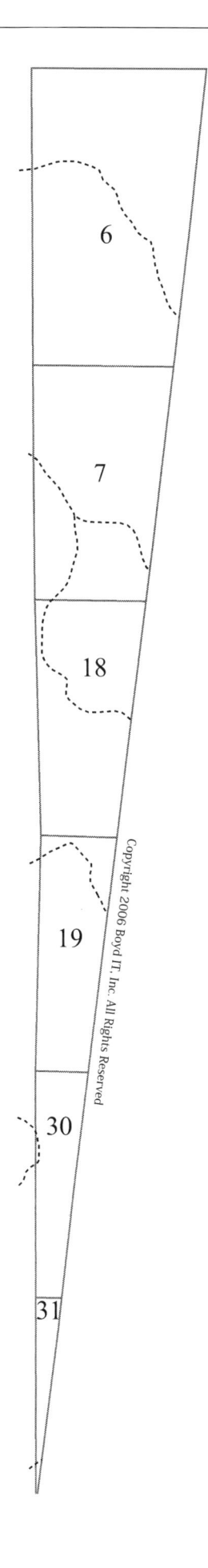

Township 8-N Range 19-E (Choctaw) - Map Group 6

Historical Map

T8-N R19-E
Choctaw Meridian

Map Group 6

Note: the area contained in this map amounts to far less than a full Township. Therefore, its contents are completely on this single page (instead of a "normal" 2-page spread).

Cities & Towns
None

Cemeteries
None

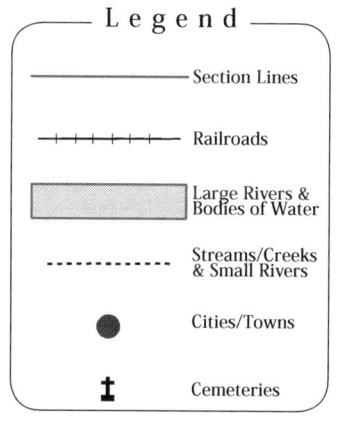

Legend

- Section Lines
- Railroads
- Large Rivers & Bodies of Water
- Streams/Creeks & Small Rivers
- Cities/Towns
- Cemeteries

Scale: Section = 1 mile X 1 mile
(there are some exceptions)

Copyright 2006 Boyd IT, Inc. All Rights Reserved

Family Maps of Lauderdale County, Mississippi

Map Group 7: Index to Land Patents
Township 7-North Range 14-East (Choctaw)

After you locate an individual in this Index, take note of the Section and Section Part then proceed to the Land Patent map on the pages immediately following. You should have no difficulty locating the corresponding parcel of land.

The "For More Info" Column will lead you to more information about the underlying Patents. See the *Legend* at right, and the "How to Use this Book" chapter, for more information.

LEGEND
"For More Info . . ." column

- **A** = Authority (Legislative Act, See Appendix "A")
- **B** = Block or Lot (location in Section unknown)
- **C** = Cancelled Patent
- **F** = Fractional Section
- **G** = Group (Multi-Patentee Patent, see Appendix "C")
- **V** = Overlaps another Parcel
- **R** = Re-Issued (Parcel patented more than once)

(A & G items require you to look in the Appendixes referred to above. All other Letter-designations followed by a number require you to locate line-items in this index that possess the ID number found after the letter).

ID	Individual in Patent	Sec.	Sec. Part	Date Issued	Other Counties	For More Info . . .
1420	BALL, Jeremiah	3	SW	1841-01-05		A1
1436	BARNES, Richmond	2	SWSE	1860-05-02		A1
1422	BLACKMAN, John A	4	NWSW	1859-05-02		A1
1424	" "	4	SESW	1859-05-02		A1
1421	" "	4	NESE	1860-07-02		A1
1423	" "	4	SENE	1860-07-02		A1
1425	CALHOUN, John	13	W½SE	1841-01-05		A1
1426	" "	24	W½NW	1841-01-05		A1
1397	CARPENTER, Benjamin	14	NESW	1859-05-02		A1
1398	" "	14	NW	1859-05-02		A1
1432	CHA, Koo	31		1845/12/06		A2
1439	COATS, William	12	E½SE	1841-01-05		A1
1396	COCHRUM, Ardis	4	S½SE	1859-05-02		A1
1433	COOPER, Moses	23	W½SE	1841-01-05		A1
1407	CRORY, James C	1	E½SE	1841-01-05		A1
1399	EDWARDS, Edmund W	2	E½SW	1841-01-05		A1
1408	FORD, James C	10	E½NE	1841-01-05		A1
1409	" "	11	E½SW	1841-01-05		A1
1410	" "	11	N½	1841-01-05		A1
1411	" "	11	SE	1841-01-05		A1
1412	" "	13	NENE	1841-01-05		A1
1413	" "	13	NW	1841-01-05		A1
1414	" "	13	W½NE	1841-01-05		A1
1415	" "	2	W½SW	1841-01-05		A1
1416	" "	3	N½	1846-09-01		A1
1417	" "	3	SE	1846-09-01		A1
1402	HOLIMAN, Enoch	14	NENE	1841-01-05		A1
1395	HONAH, Anthle	32	NW	1846/09/04		A2
1431	HUGGINS, Kader	4	SENW	1859-06-01		A1
1434	JONES, Richard	30	E½NE	1841-01-05		A1
1435	" "	30	E½SE	1841-01-05		A1
1403	MAHAN, Hiram D	14	SE	1859-11-10		A1
1404	" "	14	SESW	1859-11-10		A1
1405	" "	24	E½SW	1859-11-10		A1
1406	" "	24	NESE	1859-11-10		A1
1394	MOTT, Absalom	6	N½NW	1859-05-02		A1
1400	OSBORN, Edward	23	NESE	1841-01-05		A1
1401	" "	23	W½NE	1841-01-05		A1
1427	ROBERTSON, John	13	SENE	1841-01-05		A1
1418	RUSSELL, James	24	E½NW	1841-01-05		A1
1419	" "	24	NE	1841-01-05		A1
1428	SAVELL, John	23	E½NE	1841-01-05		A1
1429	TRUSSELL, John	4	NESW	1859-05-02		A1
1430	" "	4	NWSE	1859-05-02		A1
1438	TRUSSELL, William C	4	N½NE	1859-11-10		A1
1437	WELLS, Thomas	12	E½NE	1859-05-02		A1

Family Maps of Lauderdale County, Mississippi

Patent Map

T7-N R14-E
Choctaw Meridian

Map Group 7

Township Statistics

Parcels Mapped	:	46
Number of Patents	:	32
Number of Individuals	:	25
Patentees Identified	:	25
Number of Surnames	:	24
Multi-Patentee Parcels	:	0
Oldest Patent Date	:	1/5/1841
Most Recent Patent	:	7/2/1860
Block/Lot Parcels	:	0
Parcels Re - Issued	:	0
Parcels that Overlap	:	0
Cities and Towns	:	3
Cemeteries	:	2

Section 6: MOTT Absalom 1859

Section 4: TRUSSELL William C 1859; HUGGINS Kader 1859; BLACKMAN John A 1860; BLACKMAN John A 1859; TRUSSELL John 1859; TRUSSELL John 1859; BLACKMAN John A 1860; BLACKMAN John A 1859; COCHRUM Ardis 1859

Section 29/30: JONES Richard 1841; JONES Richard 1841

Section 31: CHA Koo 1845

Section 32: HONAH Anthle 1846

Copyright 2006 Boyd IT, Inc. All Rights Reserved

118

Township 7-N Range 14-E (Choctaw) - Map Group 7

Section 3
- FORD, James C — 1846
- BALL, Jeremiah — 1841
- FORD, James C — 1846

Section 2
- FORD, James C — 1841
- EDWARDS, Edmund W — 1841
- BARNES, Richmond — 1860

Section 1
- CRORY, James C — 1841

Section 10

Section 11
- FORD, James C — 1841
- FORD, James C — 1841
- FORD, James C — 1841
- FORD, James C — 1841

Section 12
- WELLS, Thomas — 1859
- COATS, William — 1841

Section 15

Section 14
- CARPENTER, Benjamin — 1859
- HOLIMAN, Enoch — 1841
- CARPENTER, Benjamin — 1859
- MAHAN, Hiram D — 1859
- MAHAN, Hiram D — 1859

Section 13
- FORD, James C — 1841
- FORD, James C — 1841
- FORD, James C — 1841
- ROBERTSON, John — 1841
- CALHOUN, John — 1841

Section 22

Section 23
- OSBORN, Edward — 1841
- SAVELL, John — 1841
- COOPER, Moses — 1841
- OSBORN, Edward — 1841

Section 24
- CALHOUN, John — 1841
- RUSSELL, James — 1841
- RUSSELL, James — 1841
- MAHAN, Hiram D — 1859
- MAHAN, Hiram D — 1859

Section 27

Section 26

Section 25

Section 34

Section 35

Section 36

Helpful Hints

1. This Map's INDEX can be found on the preceding pages.
2. Refer to Map "C" to see where this Township lies within Lauderdale County, Mississippi.
3. Numbers within square brackets [ ] denote a multi-patentee land parcel (multi-owner). Refer to Appendix "C" for a full list of members in this group.
4. Areas that look to be crowded with Patentees usually indicate multiple sales of the same parcel (Re-issues) or Overlapping parcels. See this Township's Index for an explanation of these and other circumstances that might explain "odd" groupings of Patentees on this map.

Copyright 2006 Boyd IT, Inc. All Rights Reserved

Legend
- Patent Boundary
- Section Boundary
- No Patents Found (or Outside County)
- 1., 2., 3., ... Lot Numbers (when beside a name)
- [ ] Group Number (see Appendix "C")

Scale: Section = 1 mile X 1 mile (generally, with some exceptions)

Family Maps of Lauderdale County, Mississippi

Road Map
T7-N R14-E
Choctaw Meridian
Map Group 7

Cities & Towns
Moseley
Schamberville
Suqualena

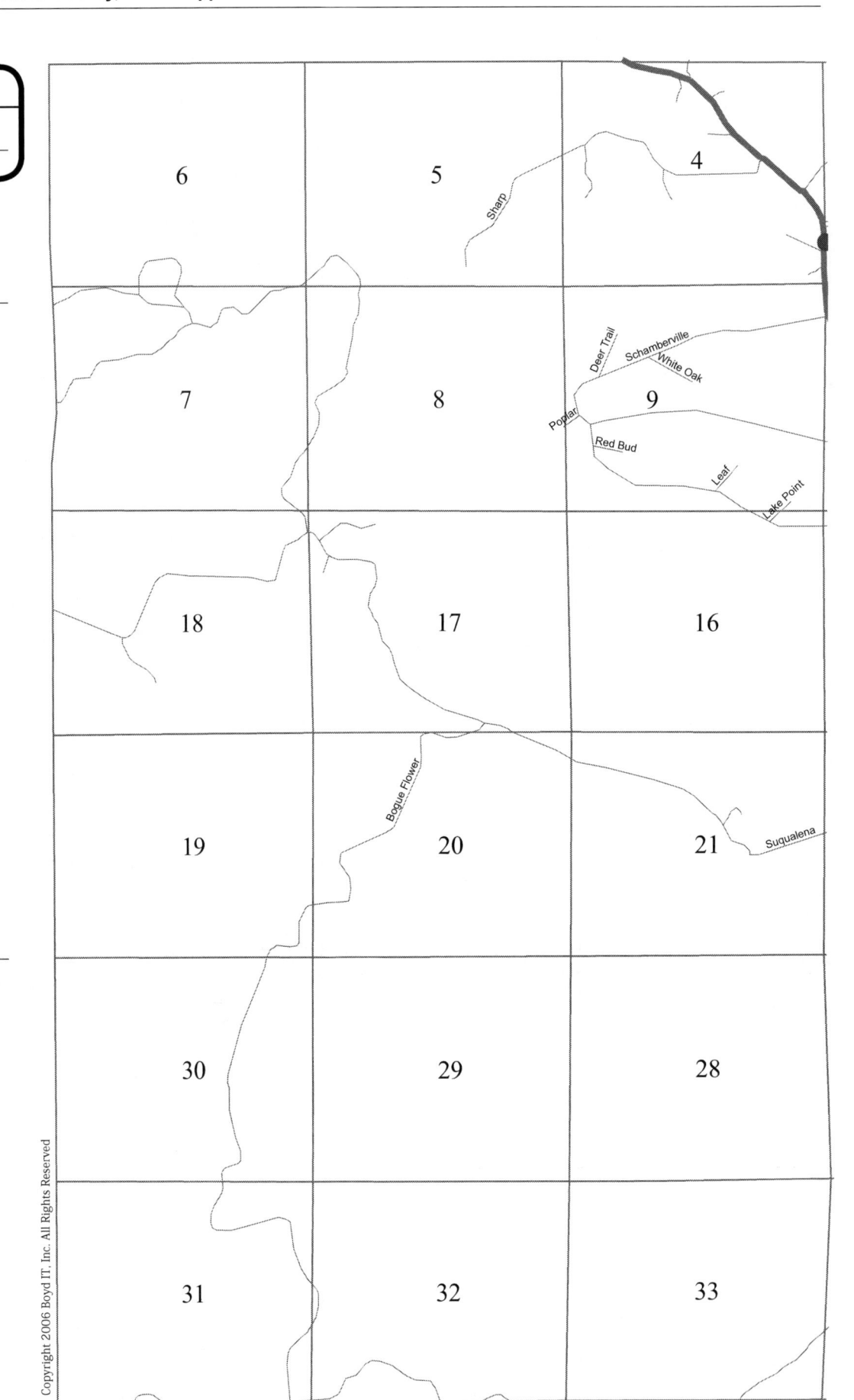

Cemeteries
Cedar Grove Cemetery
Suqualena Cemetery

Township 7-N Range 14-E (Choctaw) - Map Group 7

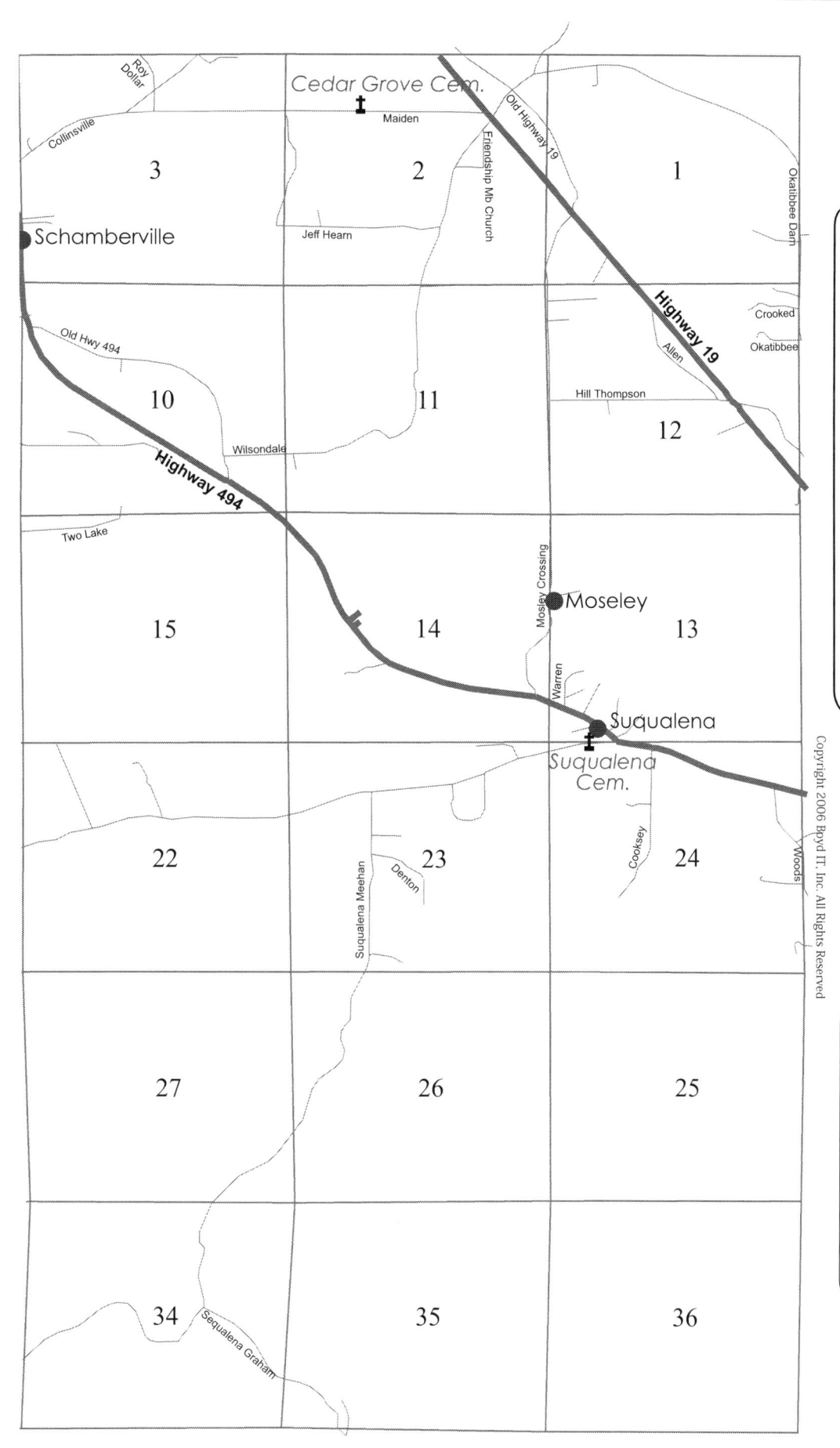

Helpful Hints

1. This road map has a number of uses, but primarily it is to help you: a) find the present location of land owned by your ancestors (at least the general area), b) find cemeteries and city-centers, and c) estimate the route/roads used by Census-takers & tax-assessors.

2. If you plan to travel to Lauderdale County to locate cemeteries or land parcels, please pick up a modern travel map for the area before you do. Mapping old land parcels on modern maps is not as exact a science as you might think. Just the slightest variations in public land survey coordinates, estimates of parcel boundaries, or road-map deviations can greatly alter a map's representation of how a road either does or doesn't cross a particular parcel of land.

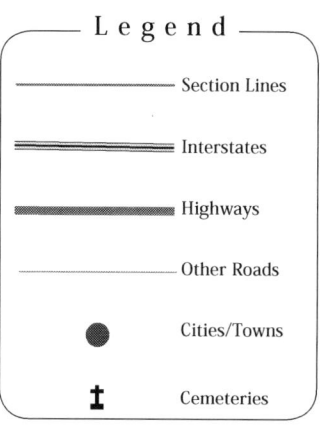

Legend

- Section Lines
- Interstates
- Highways
- Other Roads
- ● Cities/Towns
- ✝ Cemeteries

Scale: Section = 1 mile X 1 mile
(generally, with some exceptions)

121

Family Maps of Lauderdale County, Mississippi

Historical Map
T7-N R14-E
Choctaw Meridian
Map Group 7

Cities & Towns
Moseley
Schamberville
Suqualena

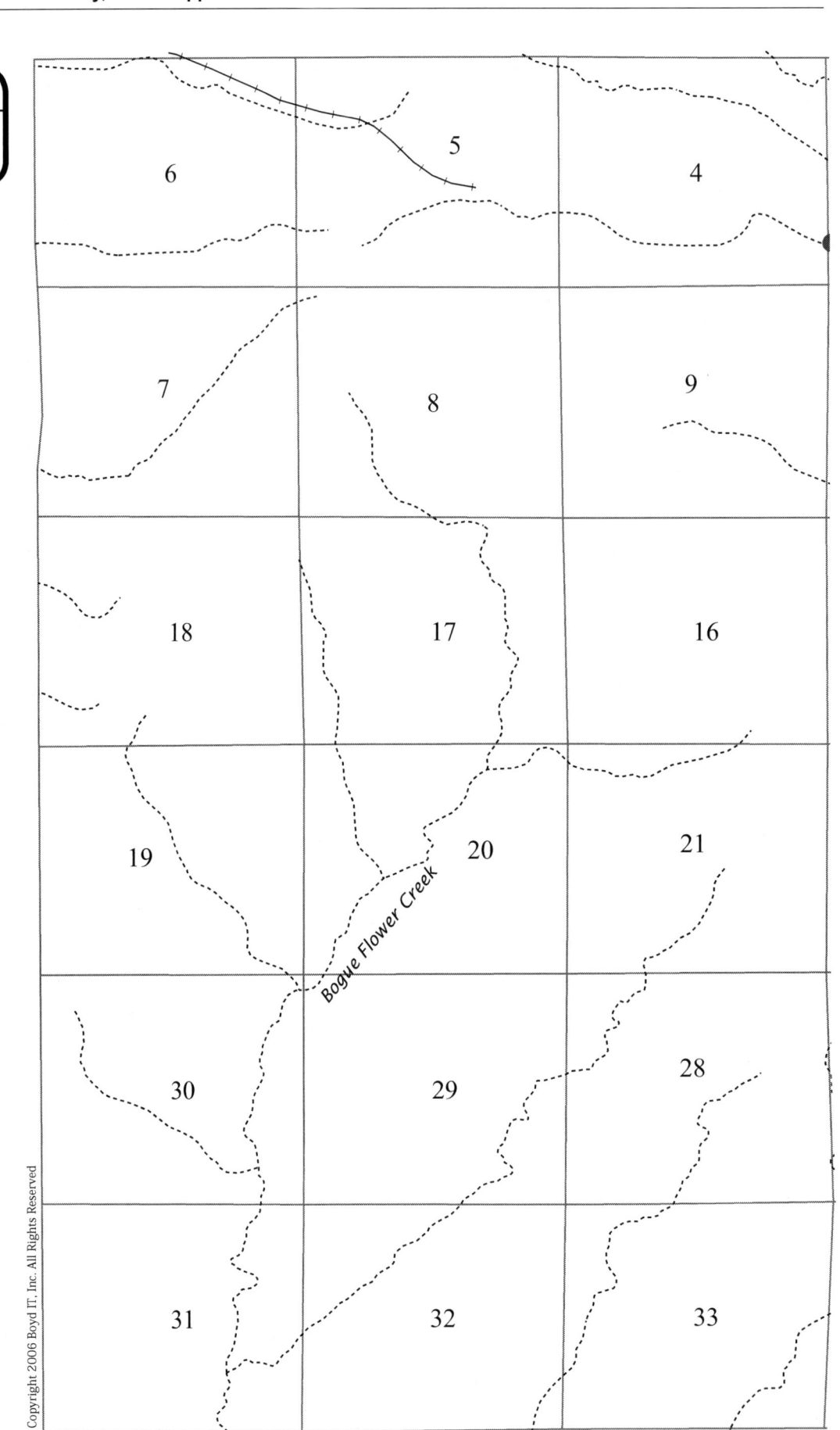

Cemeteries
Cedar Grove Cemetery
Suqualena Cemetery

Township 7-N Range 14-E (Choctaw) - Map Group 7

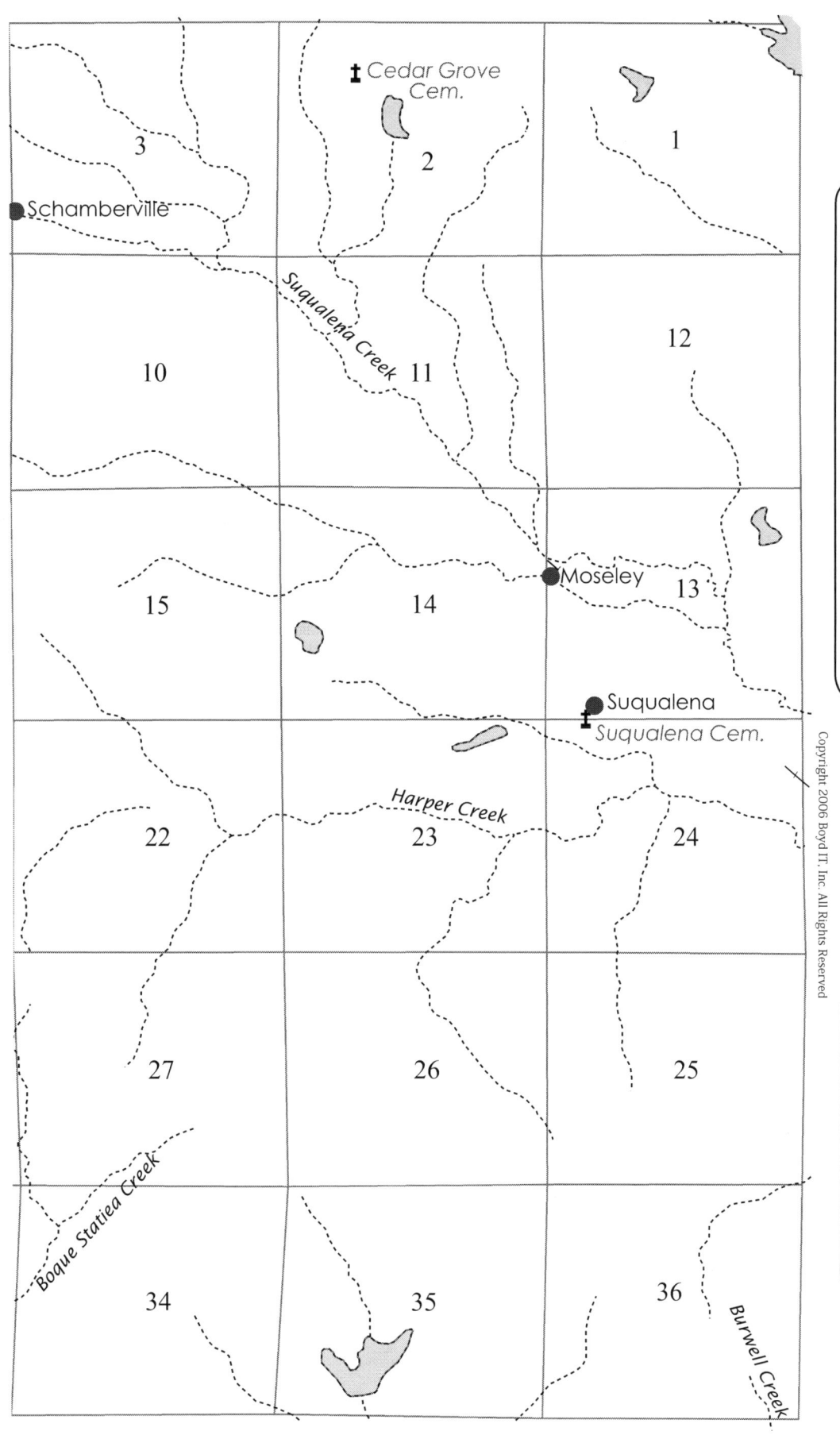

Helpful Hints

1. This Map takes a different look at the same Congressional Township displayed in the preceding two maps. It presents features that can help you better envision the historical development of the area: a) Water-bodies (lakes & ponds), b) Water-courses (rivers, streams, etc.), c) Railroads, d) City/town center-points (where they were oftentimes located when first settled), and e) Cemeteries.

2. Using this "Historical" map in tandem with this Township's Patent Map and Road Map, may lead you to some interesting discoveries. You will often find roads, towns, cemeteries, and waterways are named after nearby landowners: sometimes those names will be the ones you are researching. See how many of these research gems you can find here in Lauderdale County.

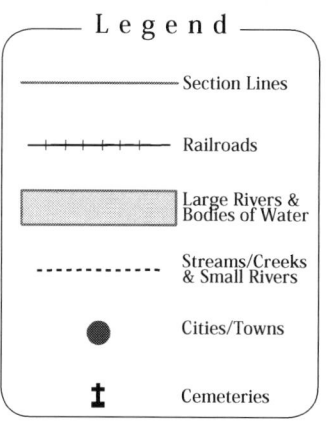

Copyright 2006 Boyd IT, Inc. All Rights Reserved

Legend

— Section Lines
+++++ Railroads
▒ Large Rivers & Bodies of Water
- - - Streams/Creeks & Small Rivers
● Cities/Towns
✝ Cemeteries

Scale: Section = 1 mile X 1 mile
(there are some exceptions)

123

Family Maps of Lauderdale County, Mississippi

Map Group 8: Index to Land Patents
Township 7-North Range 15-East (Choctaw)

After you locate an individual in this Index, take note of the Section and Section Part then proceed to the Land Patent map on the pages immediately following. You should have no difficulty locating the corresponding parcel of land.

The "For More Info" Column will lead you to more information about the underlying Patents. See the *Legend* at right, and the "How to Use this Book" chapter, for more information.

LEGEND
"For More Info . . ." column

- **A** = Authority (Legislative Act, See Appendix "A")
- **B** = Block or Lot (location in Section unknown)
- **C** = Cancelled Patent
- **F** = Fractional Section
- **G** = Group (Multi-Patentee Patent, see Appendix "C")
- **V** = Overlaps another Parcel
- **R** = Re-Issued (Parcel patented more than once)

(A & G items require you to look in the Appendixes referred to above. All other Letter-designations followed by a number require you to locate line-items in this index that possess the ID number found after the letter).

ID	Individual in Patent	Sec.	Sec. Part	Date Issued	Other Counties	For More Info . . .
1513	ALFORD, Jacob	8	SENE	1841-01-05		A1
1617	ALFORD, Sherod H	8	NENE	1841-01-05		A1
1440	ALLISON, Abner	25	E½SW	1892-03-23		A4
1454	BABERS, Andrew J	11	SESE	1859-11-10		A1
1581	BAILY, Moses	1	SENW	1900-07-21		A1
1582	" "	1	SWNE	1900-07-21		A1
1540	BARDEN, Jerry	11	SWNE	1902-02-12		A4
1629	BARNET, William	23	W½SE	1841-01-05		A1
1543	BARNETT, John	23	W½SW	1841-01-05		A1
1560	BARNETT, Joseph H	26	NWNE	1846-09-01		A1
1488	BATTLE, Elizabeth W	15	NENW	1848-09-01		A1
1489	" "	15	NWSW	1848-09-01		A1
1490	" "	15	W½NW	1848-09-01		A1
1628	BEASON, William B	25	SENE	1859-11-10		A1
1650	BEASON, William R	13	NENW	1875-07-01		A4
1622	BLANKS, Thomas Y	1	NWNE	1859-06-01		A1
1448	BOON, Alen	30	E½NE	1841-01-05		A1
1462	BROWN, David M	2	W½NW	1841-01-05		A1
1463	" "	3	NESW	1841-01-05		A1
1464	" "	3	W½SE	1841-01-05		A1
1493	BROWN, George	21	NWSW	1848-09-01		A1
1516	BROWN, James	5	NWSW	1848-09-01		A1
1517	" "	5	SWNW	1848-09-01		A1
1514	" "	36	NESW	1849-12-01		A1
1515	" "	36	SWSW	1849-12-01		A1
1544	BROWN, John	23	NENE	1841-01-05		A1
1548	" "	27	SWNW	1848-09-01		A1
1547	" "	27	NWNW	1859-06-01		A1 R1533
1546	" "	27	NENW	1859-11-10		A1
1545	" "	25	NWNW	1884-12-30		A4
1630	BROWN, William	12	E½NE	1841-01-05		A1
1631	" "	12	E½SE	1841-01-05		A1
1632	" "	23	E½SE	1841-01-05		A1
1633	" "	23	SENE	1841-01-05		A1
1634	" "	24	NWNW	1841-01-05		A1
1635	" "	26	E½NW	1841-01-05		A1
1636	" "	33	E½NE	1841-01-05		A1
1455	CALHOUN, Archibald G	34	NWNE	1841-01-05		A1
1467	CALHOUN, Duncan	34	E½NE	1841-01-05		A1
1468	" "	34	NWSE	1841-01-05		A1
1469	" "	34	SWNE	1841-01-05		A1
1470	" "	34	SWSE	1841-01-05		A1
1502	CALHOUN, Hugh	27	SESE	1841-01-05		A1
1503	" "	27	W½SE	1841-01-05		A1
1549	CALHOUN, John	19	W½NW	1841-01-05		A1
1550	" "	3	NE	1841-01-05		A1

124

Township 7-N Range 15-E (Choctaw) - Map Group 8

ID	Individual in Patent	Sec.	Sec. Part	Date Issued	Other Counties	For More Info...
1651	CAMPBELL, William R	3	NW	1841-01-05		A1 G37
1652	" "	4	E½NE	1841-01-05		A1 G37
1551	CANADY, John	13	SESW	1841-01-05		A1
1523	CASTLES, James	12	NWNW	1841-01-05		A1 G42 R1604
1521	" "	2	W½SW	1841-01-05		A1
1522	" "	3	E½SE	1841-01-05		A1
1619	CLAY, Stephen G	4	N½SW	1859-05-02		A1
1620	" "	4	NW	1860-04-02		A1
1637	COATS, William	17	NWSW	1841-01-05		A1
1638	" "	18	NESE	1841-01-05		A1
1639	" "	7	W½SW	1841-01-05		A1
1662	COLE, Wyatt	31	S½SE	1898-12-01		A4
1663	" "	31	S½SW	1898-12-01		A4
1564	CRAIN, Lewis	13	NWNW	1841-01-05		A1
1537	CRANE, Jeremiah	12	W½SE	1841-01-05		A1
1565	CRANE, Lewis	12	SWSW	1841-01-05		A1
1573	CRANE, Martin	12	NENW	1841-01-05		A1
1574	" "	12	W½NE	1841-01-05		A1
1535	DANIEL, James W	33	E½SE	1841-01-05		A1
1552	DANIEL, John	1	SWSE	1859-06-01		A1
1500	DEEN, Hillard	35	NWNE	1859-11-10		A1
1501	" "	35	NWSE	1859-11-10		A1
1584	DEEN, Nathan P	35	NENW	1848-09-01		A1
1585	" "	35	W½NW	1848-09-01		A1
1586	DEEN, Nathaniel P	35	NWSW	1859-06-01		A1
1506	EUBANKS, Isaac	26	W½SW	1841-01-05		A1
1507	" "	27	NESE	1841-01-05		A1
1508	" "	34	NESE	1841-01-05		A1
1509	" "	34	SENW	1841-01-05		A1
1504	" "	20	E½SE	1848-09-01		A1
1505	" "	26	SWSE	1848-09-01		A1
1510	" "	35	SWNE	1848-09-01		A1
1624	EVANS, Vincent A	8	NWNE	1841-01-05		A1
1518	FORD, James C	18	NW	1841-01-05		A1
1519	" "	19	E½NW	1841-01-05		A1
1520	" "	19	SW	1841-01-05		A1
1605	FOUNTAIN, Robert A	28	E½NW	1841-01-05		A1
1606	" "	28	W½NE	1841-01-05		A1
1626	GAINS, Wiley	15	S½SE	1882-03-30		A4
1466	GILBERT, Dock	25	NWNE	1895-05-11		A4 G82
1466	GILBERT, Francis	25	NWNE	1895-05-11		A4 G82
1441	GILLASPIE, Absolem W	13	SESE	1848-09-01		A1
1621	HAMBRICK, Thomas	14	E½SW	1841-01-05		A1
1527	HAMRICK, James	27	SESW	1841-01-05		A1
1528	" "	34	NENW	1841-01-05		A1
1529	" "	36	E½SE	1841-01-05		A1
1473	HARPER, Edward	19	SE	1841-01-05		A1
1474	" "	20	SESW	1841-01-05		A1
1475	" "	20	W½SW	1841-01-05		A1
1476	" "	29	E½NE	1841-01-05		A1
1477	" "	29	E½NW	1841-01-05		A1
1478	" "	29	E½SE	1841-01-05		A1
1479	" "	29	W½NE	1841-01-05		A1
1480	" "	29	W½NW	1841-01-05		A1
1481	" "	29	W½SE	1841-01-05		A1
1482	" "	32	E½NE	1841-01-05		A1
1483	" "	32	NWNE	1841-01-05		A1
1487	HARPER, Elizabeth E	19	SWNE	1841-01-05		A1
1591	HARPER, Richard B	33	E½SW	1841-01-05		A1
1592	" "	33	SENW	1841-01-05		A1
1593	" "	33	W½NE	1841-01-05		A1
1594	" "	33	W½SW	1841-01-05		A1
1607	HATCHER, Samuel C	8	E½SE	1841-01-05		A1
1608	" "	8	SWNE	1841-01-05		A1
1609	" "	9	NWNE	1841-01-05		A1
1610	" "	9	NWSE	1841-01-05		A1
1611	" "	9	W½SW	1841-01-05		A1
1616	HATCHER, Samuel T	4	SWSW	1859-05-02		A1
1453	HOLLAND, Amanda	25	SWNW	1906-06-21		A4
1443	HORSEY, Adaline	25	SENW	1898-09-28		A4
1561	HOUSTON, Joseph	1	SESW	1859-11-10		A1
1538	HOWELL, Jeremiah	11	W½NW	1841-01-05		A1
1539	" "	4	SESE	1841-01-05		A1

125

Family Maps of Lauderdale County, Mississippi

ID	Individual in Patent	Sec.	Sec. Part	Date Issued	Other Counties	For More Info . . .
1612	HUBBARD, Samuel	10	E½NW	1841-01-05		A1 G123
1613	" "	10	SWNW	1841-01-05		A1 G123
1640	HUNTER, William H	21	E½SW	1841-01-05		A1
1641	" "	28	NWSE	1841-01-05		A1
1642	" "	28	SWSW	1841-01-05		A1
1643	" "	28	W½NW	1841-01-05		A1
1451	HUTTON, Alice	25	SWSW	1895-05-11		A4
1491	JACKSON, Emily	1	E½NE	1906-06-30		A4 G125
1530	JACKSON, James	25	NWSW	1884-12-30		A4
1491	JACKSON, William	1	E½NE	1906-06-30		A4 G125
1583	JOLLY, Nathan C	18	SWNE	1848-09-01		A1
1625	KALSAW, Wat	15	NESE	1896-04-28		A4
1536	KENNEDY, Jane E	35	SWSE	1891-05-20		A4
1623	KENNEDY, Uriah	13	NESW	1841-01-05		A1
1553	KERBY, John J	9	SWSE	1841-01-05		A1
1456	KILLENS, Benjamin	31	SENE	1874-04-10		A4
1580	KILLENS, Milly	31	NENE	1874-04-10		A4
1444	KILLIN, Adam	31	NESW	1895-02-21		A4
1445	" "	31	NWSE	1895-02-21		A4
1446	" "	31	SENW	1895-02-21		A4
1447	" "	31	SWNE	1895-02-21		A4
1449	KYLES, Alexander	9	SWNE	1841-01-05		A1
1562	LEE, Levi	21	NWSE	1848-09-01		A1
1563	" "	21	W½NE	1848-09-01		A1
1572	LEE, Martha	33	SWNW	1860-04-02		A1
1576	LEE, Mathew	27	NESW	1841-01-05		A1
1578	" "	28	E½SE	1841-01-05		A1
1577	" "	27	NWSW	1859-11-10		A1
1579	" "	33	N½NW	1859-11-10		A1
1614	LEE, Samuel	10	NWNE	1841-01-05		A1
1615	" "	4	W½NE	1841-01-05		A1
1612	LEWIS, Rufus G	10	E½NW	1841-01-05		A1 G123
1613	" "	10	SWNW	1841-01-05		A1 G123
1644	LEWIS, William M	6	W½SE	1841-01-05		A1
1645	" "	9	E½NE	1841-01-05		A1
1651	MARSHALL, John R	3	NW	1841-01-05		A1 G37
1652	" "	4	E½NE	1841-01-05		A1 G37
1661	MATHEWS, Winston	1	NWSW	1861-05-01		A1
1484	MCDONALD, Eli	23	NENW	1860-10-01		A1
1531	MCDONALD, James	13	SENW	1859-11-10		A1
1532	" "	13	W½SW	1859-11-10		A1
1533	MCGEE, James	27	NWNW	1889-12-19		A4 R1547
1450	MCMELLON, Alexander	3	SESW	1841-01-05		A1
1554	MCWOOTON, John	21	N½NW	1841-01-05		A1
1590	MITCHELL, Rebecca	27	SENE	1841-01-05		A1
1618	MOBLEY, Solomon	13	SWNW	1905-05-02		A4
1523	MOTH, Lovelace	12	NWNW	1841-01-05		A1 G42 R1604
1442	MOTT, Absolom	13	NESE	1841-01-05		A1
1567	MOTT, Lovelace	12	SWNW	1841-01-05		A1
1568	" "	13	NENE	1841-01-05		A1
1569	" "	13	SENE	1841-01-05		A1
1570	" "	13	W½NE	1841-01-05		A1
1566	MOTT, Lovelace H	11	SESW	1841-01-05		A1
1465	PACE, Dempsey	11	E½NE	1841-01-05		A1
1511	PACE, Isham	23	E½SW	1841-01-05		A1
1555	POOL, John	30	E½NW	1859-05-02		A1
1556	" "	30	SWNW	1859-05-02		A1
1648	PRIER, William	5	NENW	1841-01-05		A1
1649	" "	5	NWNE	1841-01-05		A1
1534	RAY, James	35	E½SW	1846-09-01		A1
1471	REASON, Edward F	25	NENE	1860-07-02		A1
1472	" "	25	SWNE	1860-07-02		A1
1497	RODGERS, Hays	2	E½SW	1859-05-02		A1
1498	" "	2	SENW	1859-05-02		A1
1499	ROGERS, Hays	10	E½NE	1841-01-05		A1
1458	RUSHING, Charles E	1	NESW	1859-11-10		A1
1459	" "	1	NWSE	1859-11-10		A1
1457	" "	1	NENW	1860-04-02		A1
1460	" "	35	SESE	1860-04-02		A1
1452	RUSSELL, Allen	33	W½SE	1841-01-05		A1
1494	SANDYFORD, Gray	4	W½SE	1841-01-05		A1
1495	" "	8	W½SE	1841-01-05		A1
1496	" "	9	E½NW	1841-01-05		A1

Township 7-N Range 15-E (Choctaw) - Map Group 8

ID	Individual in Patent	Sec.	Sec. Part	Date Issued	Other Counties	For More Info . . .
1602	SCRUGS, Richard	11	E½NW	1841-01-05		A1
1557	SMITH, John	26	NWNW	1841-01-05		A1
1575	SMITH, Mary	26	SWNW	1841-01-05		A1
1558	STEELE, John	5	E½SW	1848-09-01		A1
1512	STEPHENSON, Jackson	25	NENW	1884-12-30		A4
1492	STOKES, George A	4	NESE	1841-01-05		A1
1653	STOKES, William	10	NWNW	1841-01-05		A1
1654	" "	3	W½SW	1841-01-05		A1
1655	STRINGER, William	5	NWNW	1848-09-01		A1
1656	" "	5	SWSE	1848-09-01		A1
1612	TAPPAN, John	10	E½NW	1841-01-05		A1 G123
1613	" "	10	SWNW	1841-01-05		A1 G123
1485	THORNTON, Eli S	11	NESW	1841-01-05		A1
1486	" "	11	W½SE	1841-01-05		A1
1526	TOLSON, James D	23	SWNE	1859-05-02		A1
1525	" "	23	S½NW	1859-06-01		A1
1524	" "	23	NWNW	1860-10-01		A1
1646	TOWNSEND, William M	6	NWNW	1859-11-10		A1
1647	" "	6	SWNW	1859-11-10		A1
1595	TUTT, Richard B	6	E½NW	1841-01-05		A1
1596	" "	6	E½SW	1841-01-05		A1
1597	" "	6	W½SW	1841-01-05		A1
1598	" "	9	E½SE	1841-01-05		A1
1599	" "	9	E½SW	1841-01-05		A1
1587	VAUGHN, Pinkney	20	NESW	1859-11-10		A1
1588	" "	29	E½SW	1861-05-01		A1
1589	" "	29	SWSW	1861-05-01		A1
1600	WALKER, Richard S	23	NWNE	1841-01-05		A1
1601	" "	24	NWSW	1841-01-05		A1
1559	WARD, John	36	SWNE	1841-01-05		A1
1660	WARD, William	36	NWNE	1839-09-02		A1
1657	" "	25	E½SE	1841-01-05		A1
1658	" "	25	W½SE	1841-01-05		A1
1659	" "	36	E½NE	1841-01-05		A1
1603	WATSON, Richard	1	E½SE	1906-06-21		A4
1461	WILLIAMS, David H	35	SENW	1860-10-01		A1
1571	WILSON, Marcus	31	NESE	1879-05-06		A4
1541	WOMACK, Jesse	1	SWSW	1841-01-05		A1
1542	" "	2	SESE	1841-01-05		A1
1604	WORMACK, Richard	12	NWNW	1841-01-05		A1 R1523
1627	YOUNG, Wiley	11	NESE	1841-01-05		A1

Family Maps of Lauderdale County, Mississippi

Patent Map

T7-N R15-E
Choctaw Meridian

Map Group 8

Township Statistics

Parcels Mapped	:	224
Number of Patents	:	187
Number of Individuals	:	125
Patentees Identified	:	120
Number of Surnames	:	91
Multi-Patentee Parcels	:	7
Oldest Patent Date	:	9/2/1839
Most Recent Patent	:	6/30/1906
Block/Lot Parcels	:	0
Parcels Re-Issued	:	2
Parcels that Overlap	:	0
Cities and Towns	:	4
Cemeteries	:	6

128

Township 7-N Range 15-E (Choctaw) - Map Group 8

Section 3
- CAMPBELL [37] William R 1841
- CALHOUN John 1841
- STOKES William 1841
- BROWN David M 1841
- BROWN David M 1841
- MCMELLON Alexander 1841
- CASTLES James 1841

Section 2
- BROWN David M 1841
- RODGERS Hays 1859
- CASTLES James 1841
- RODGERS Hays 1859
- WOMACK Jesse 1841

Section 1
- RUSHING Charles E 1860
- BLANKS Thomas Y 1859
- JACKSON [125] Emily 1906
- BAILY Moses 1900
- BAILY Moses 1900
- MATHEWS Winston 1861
- RUSHING Charles E 1859
- RUSHING Charles E 1859
- WOMACK Jesse 1841
- HOUSTON Joseph 1859
- DANIEL John 1859
- WATSON Richard 1906

Section 10
- STOKES William 1841
- HUBBARD [23] Samuel 1841
- HUBBARD [23] Samuel 1841
- LEE Samuel 1841
- ROGERS Hays 1841

Section 11
- HOWELL Jeremiah 1841
- SCRUGS Richard 1841
- BARDEN Jerry 1902
- PACE Dempsey 1841
- THORNTON Eli S 1841
- THORNTON Eli S 1841
- YOUNG Wiley 1841
- MOTT Lovelace H 1841
- BABERS Andrew J 1859

Section 12
- CASTLES [42] James 1841
- WORMACK Richard 1841
- CRANE Martin 1841
- CRANE Martin 1841
- MOTT Lovelace 1841
- CRANE Lewis 1841
- CRANE Jeremiah 1841
- BROWN William 1841
- BROWN William 1841

Section 15
- BATTLE Elizabeth W 1848
- BATTLE Elizabeth W 1848
- BATTLE Elizabeth W 1848
- KALSAW Wat 1896
- GAINS Wiley 1882

Section 14
- HAMBRICK Thomas 1841

Section 13
- CRAIN Lewis 1841
- BEASON William R 1875
- MOTT Lovelace 1841
- MOTT Lovelace 1841
- MOBLEY Solomon 1905
- MCDONALD James 1859
- MOTT Lovelace 1841
- KENNEDY Uriah 1841
- MCDONALD James 1859
- CANADY John 1841
- MOTT Absolom 1841
- GILLASPIE Absolem W 1848

Section 22

Section 23
- TOLSON James D 1860
- MCDONALD Eli 1860
- WALKER Richard S 1841
- BROWN John 1841
- TOLSON James D 1859
- TOLSON James D 1859
- BROWN William 1841
- BARNETT John 1841
- BARNET William 1841
- PACE Isham 1841
- BROWN William 1841

Section 24
- BROWN William 1841
- WALKER Richard S 1841

Section 27
- BROWN John 1859
- MCGEE James 1889
- BROWN John 1859
- BROWN John 1848
- MITCHELL Rebecca 1841
- LEE Mathew 1859
- LEE Mathew 1841
- CALHOUN Hugh 1841
- EUBANKS Isaac 1841
- HAMRICK James 1841
- CALHOUN Hugh 1841

Section 26
- SMITH John 1841
- BARNETT Joseph H 1846
- SMITH Mary 1841
- BROWN William 1841
- EUBANKS Isaac 1841
- EUBANKS Isaac 1848

Section 25
- BROWN John 1884
- STEPHENSON Jackson 1884
- GILBERT [82] Dock 1895
- REASON Edward F 1860
- HOLLAND Amanda 1906
- HORSEY Adaline 1898
- REASON Edward F 1860
- BEASON William B 1859
- JACKSON James 1884
- WARD William 1841
- HUTTON Alice 1895
- ALLISON Abner 1892
- WARD William 1841

Section 34
- HAMRICK James 1841
- CALHOUN Archibald G 1841
- EUBANKS Isaac 1841
- CALHOUN Duncan 1841
- CALHOUN Duncan 1841
- CALHOUN Duncan 1841
- EUBANKS Isaac 1841

Section 35
- DEEN Nathan P 1848
- DEEN Nathan P 1848
- DEEN Hillard 1859
- WILLIAMS David H 1860
- EUBANKS Isaac 1848
- DEEN Nathaniel P 1859
- DEEN Hillard 1859
- RAY James 1846
- KENNEDY Jane E 1891
- RUSHING Charles E 1860

Section 36
- WARD William 1839
- WARD John 1841
- WARD William 1841
- BROWN James 1849
- BROWN James 1849
- HAMRICK James 1841

Helpful Hints

1. This Map's INDEX can be found on the preceding pages.
2. Refer to Map "C" to see where this Township lies within Lauderdale County, Mississippi.
3. Numbers within square brackets [ ] denote a multi-patentee land parcel (multi-owner). Refer to Appendix "C" for a full list of members in this group.
4. Areas that look to be crowded with Patentees usually indicate multiple sales of the same parcel (Re-issues) or Overlapping parcels. See this Township's Index for an explanation of these and other circumstances that might explain "odd" groupings of Patentees on this map.

Copyright 2006 Boyd IT, Inc. All Rights Reserved

Legend

- Patent Boundary
- Section Boundary
- No Patents Found (or Outside County)
- 1., 2., 3., ... Lot Numbers (when beside a name)
- [ ] Group Number (see Appendix "C")

Scale: Section = 1 mile X 1 mile (generally, with some exceptions)

129

Family Maps of Lauderdale County, Mississippi

Road Map

T7-N R15-E
Choctaw Meridian

Map Group 8

Cities & Towns
Bailey
Hookston
Nellieburg
Pine Springs

Cemeteries
Bethel Cemetery
Lad Cemetery
Oak Grove Cemetery
Pine Springs Cemetery
Stenis Cemetery
Wilson Cemetery

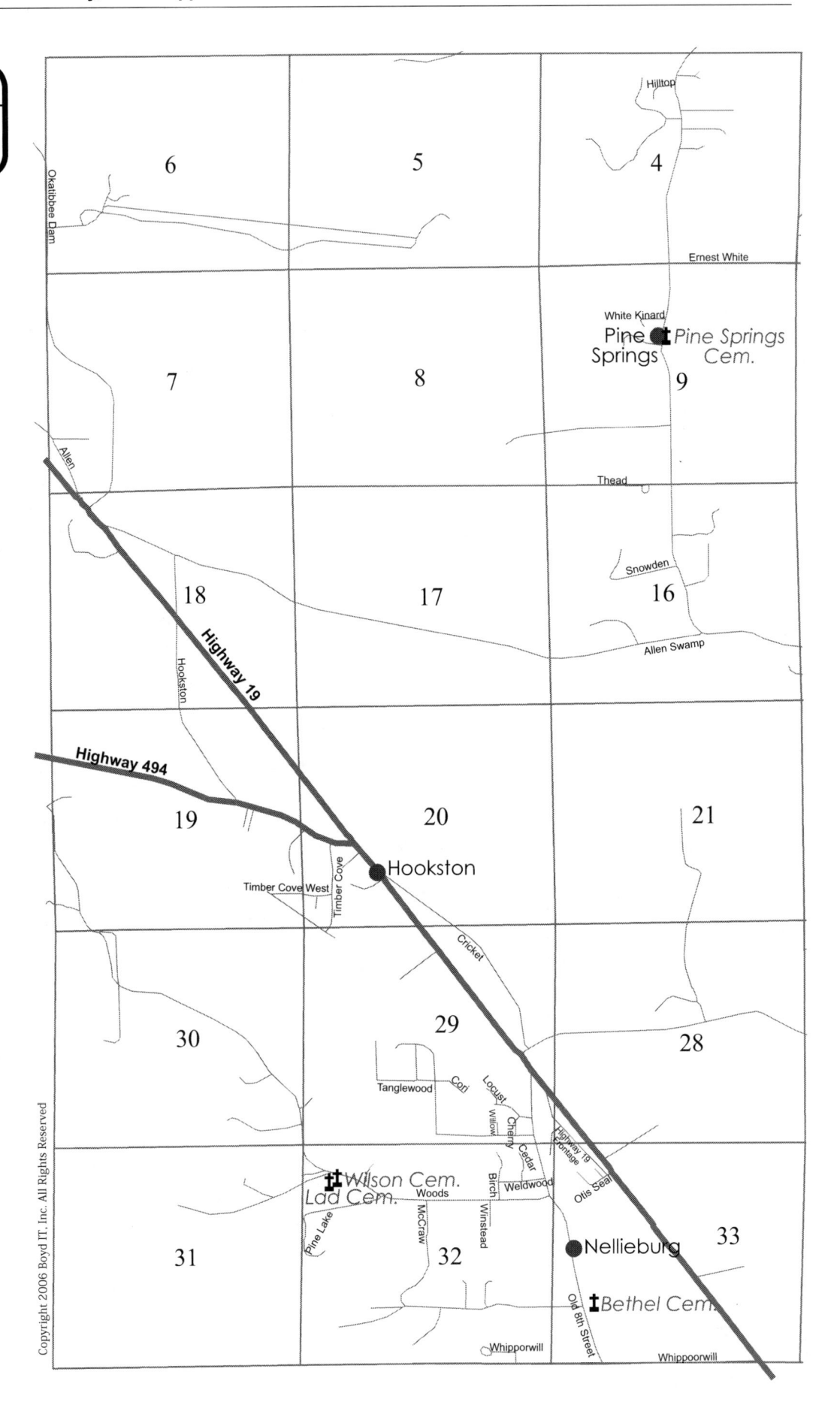

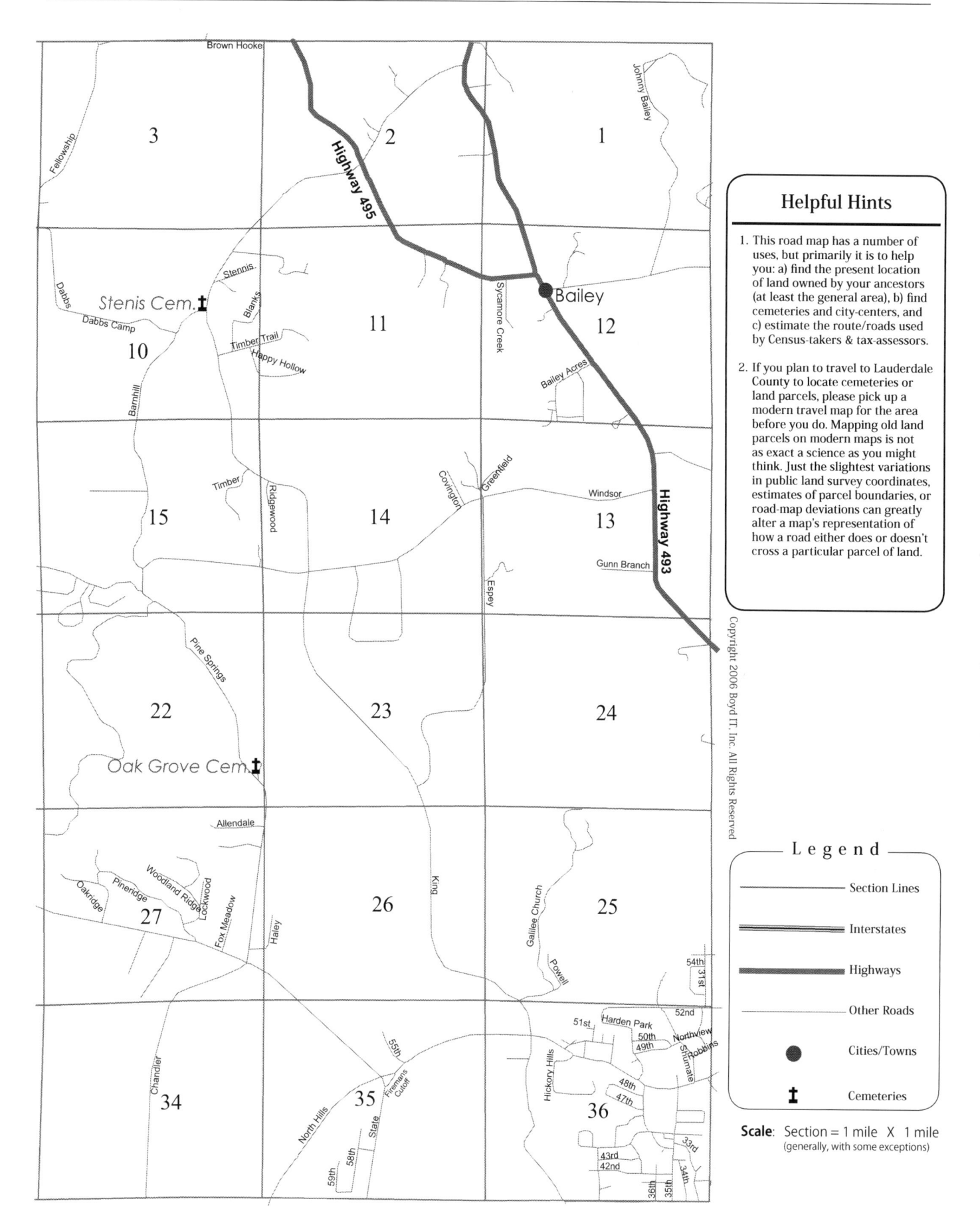

Township 7-N Range 15-E (Choctaw) - Map Group 8

Helpful Hints

1. This road map has a number of uses, but primarily it is to help you: a) find the present location of land owned by your ancestors (at least the general area), b) find cemeteries and city-centers, and c) estimate the route/roads used by Census-takers & tax-assessors.

2. If you plan to travel to Lauderdale County to locate cemeteries or land parcels, please pick up a modern travel map for the area before you do. Mapping old land parcels on modern maps is not as exact a science as you might think. Just the slightest variations in public land survey coordinates, estimates of parcel boundaries, or road-map deviations can greatly alter a map's representation of how a road either does or doesn't cross a particular parcel of land.

Legend

- Section Lines
- Interstates
- Highways
- Other Roads
- ● Cities/Towns
- ✝ Cemeteries

Scale: Section = 1 mile X 1 mile
(generally, with some exceptions)

Copyright 2006 Boyd IT, Inc. All Rights Reserved

131

Family Maps of Lauderdale County, Mississippi

Historical Map
T7-N R15-E
Choctaw Meridian
Map Group 8

Cities & Towns
Bailey
Hookston
Nellieburg
Pine Springs

Cemeteries
Bethel Cemetery
Lad Cemetery
Oak Grove Cemetery
Pine Springs Cemetery
Stenis Cemetery
Wilson Cemetery

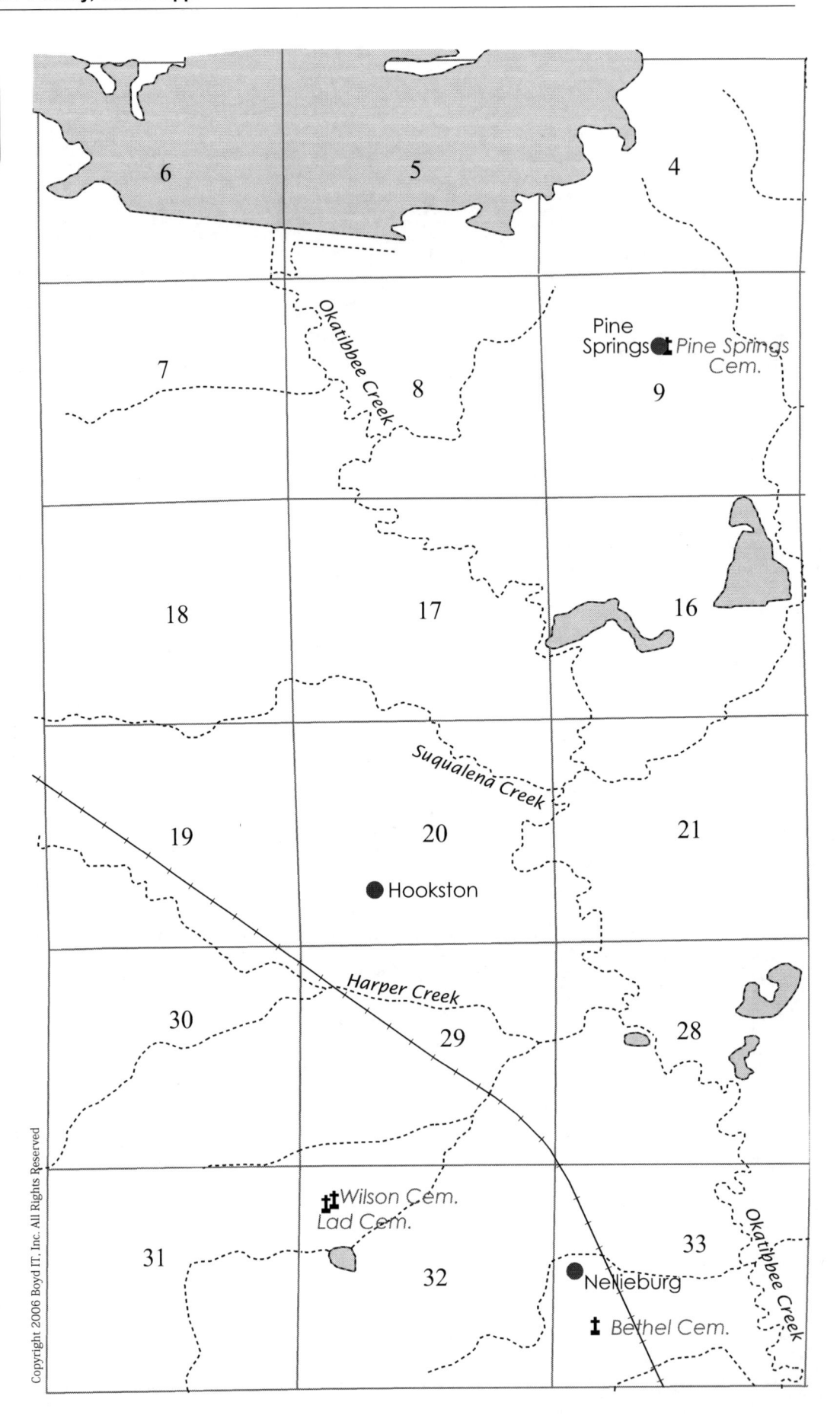

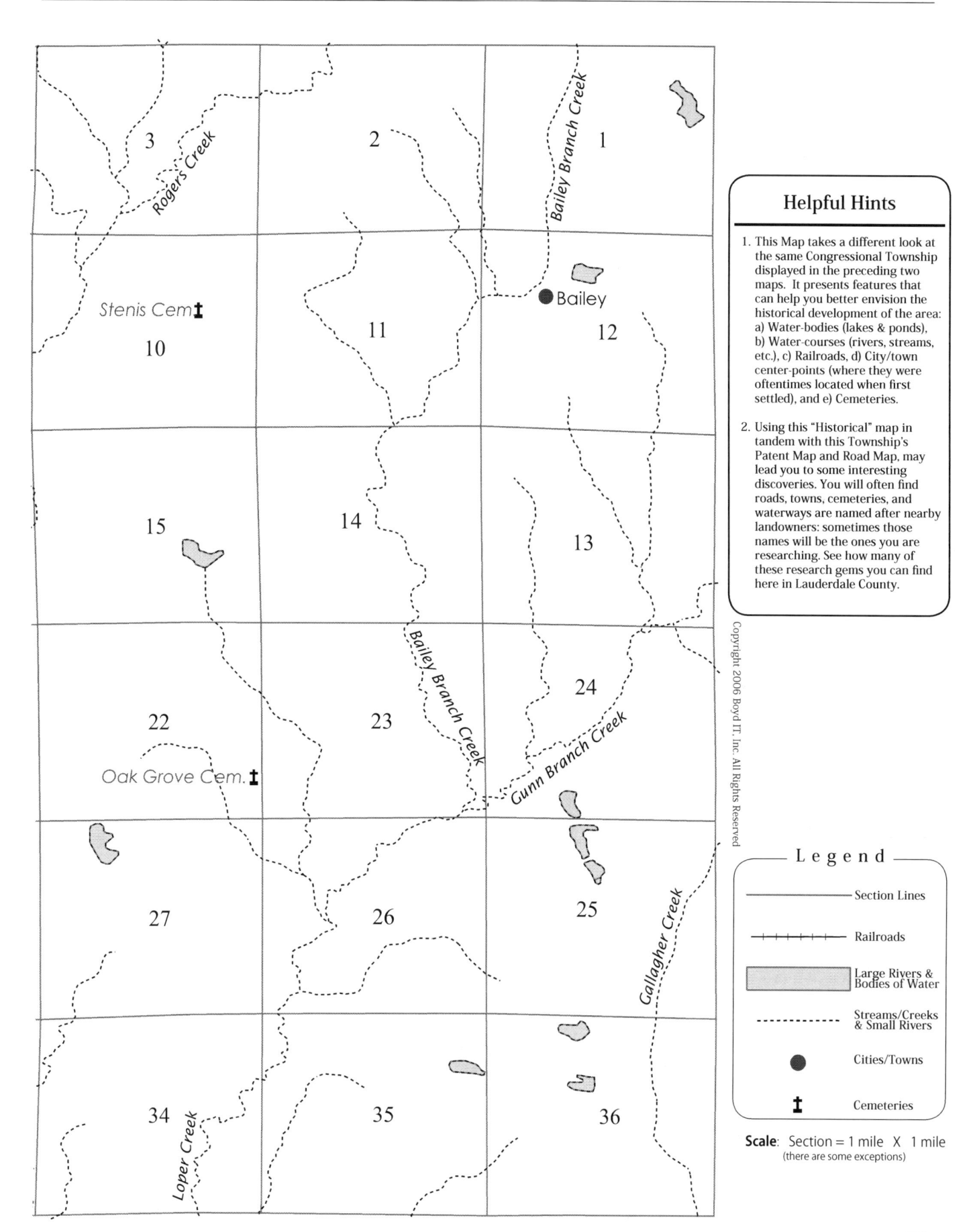

Township 7-N Range 15-E (Choctaw)   -   Map Group 8

Helpful Hints

1. This Map takes a different look at the same Congressional Township displayed in the preceding two maps. It presents features that can help you better envision the historical development of the area: a) Water-bodies (lakes & ponds), b) Water-courses (rivers, streams, etc.), c) Railroads, d) City/town center-points (where they were oftentimes located when first settled), and e) Cemeteries.

2. Using this "Historical" map in tandem with this Township's Patent Map and Road Map, may lead you to some interesting discoveries. You will often find roads, towns, cemeteries, and waterways are named after nearby landowners: sometimes those names will be the ones you are researching. See how many of these research gems you can find here in Lauderdale County.

Copyright 2006 Boyd IT, Inc. All Rights Reserved

Legend

— Section Lines
+++++ Railroads
▬ Large Rivers & Bodies of Water
- - - Streams/Creeks & Small Rivers
● Cities/Towns
✝ Cemeteries

Scale: Section = 1 mile X 1 mile
(there are some exceptions)

Features shown on map:
- Sections 1–3, 10–15, 22–27, 34–36
- Rogers Creek
- Bailey Branch Creek
- Stenis Cem (Section 10)
- Bailey (town, Section 12)
- Oak Grove Cem. (Section 22)
- Gunn Branch Creek
- Gallagher Creek
- Loper Creek

Family Maps of Lauderdale County, Mississippi

Map Group 9: Index to Land Patents
Township 7-North Range 16-East (Choctaw)

After you locate an individual in this Index, take note of the Section and Section Part then proceed to the Land Patent map on the pages immediately following. You should have no difficulty locating the corresponding parcel of land.

The "For More Info" Column will lead you to more information about the underlying Patents. See the *Legend* at right, and the "How to Use this Book" chapter, for more information.

```
                         LEGEND
              "For More Info . . . " column
    A = Authority (Legislative Act, See Appendix "A")
    B = Block or Lot (location in Section unknown)
    C = Cancelled Patent
    F = Fractional Section
    G = Group (Multi-Patentee Patent, see Appendix "C")
    V = Overlaps another Parcel
    R = Re-Issued (Parcel patented more than once)

   (A & G items require you to look in the Appendixes referred
   to above. All other Letter-designations followed by a number
   require you to locate line-items in this index that possess
   the ID number found after the letter).
```

ID	Individual in Patent	Sec.	Sec. Part	Date Issued	Other Counties	For More Info . . .
1754	ALEXANDER, Isaac	23	NESW	1841-01-05		A1
1788	ALEXANDER, Joel	21	SWSW	1841-01-05		A1
1789	" "	23	SENW	1841-01-05		A1
1790	" "	24	W½SW	1841-01-05		A1
1791	ALEXANDER, John	21	NENW	1841-01-05		A1
1792	" "	21	SWNW	1841-01-05		A1
1819	ALEXANDER, Joseph	15	NESW	1841-01-05		A1
1820	" "	15	NWSE	1841-01-05		A1
1821	" "	15	SESW	1841-01-05		A1
1822	" "	15	W½NE	1841-01-05		A1
1823	" "	21	E½SW	1841-01-05		A1
1824	" "	21	W½SE	1841-01-05		A1
1825	" "	8	E½SE	1841-01-05		A1
1826	" "	9	E½SW	1841-01-05		A1
1827	" "	9	W½NE	1841-01-05		A1
1832	ALEXANDER, Josiah H	27	NWSW	1860-04-02		A1
1848	ALEXANDER, Mathew	22	SESE	1841-01-05		A1
1849	" "	23	W½SW	1841-01-05		A1
1850	" "	26	NWNW	1841-01-05		A1
1887	ALEXANDER, Thomas	27	SENE	1841-01-05		A1
1898	ALEXANDER, Wesley	22	NESE	1841-01-05		A1
1899	" "	23	SESW	1841-01-05		A1
1901	ALEXANDER, William	10	SWSE	1841-01-05		A1
1902	" "	9	NESE	1841-01-05		A1
1793	ALFORD, John C	22	NWNE	1841-01-05		A1
1794	" "	23	SESE	1841-01-05		A1
1795	" "	26	NENW	1841-01-05		A1
1687	ARRINGTON, Arthur	26	E½NE	1841-01-05		A1
1669	BALLARD, Alexander	10	SWNW	1841-01-05		A1
1670	" "	9	E½NE	1841-01-05		A1
1903	BEASON, William B	31	NWSW	1859-11-10		A1
1772	BERRIEN, James W	25	E½SE	1840-06-10		A1 G21
1773	" "	25	NESW	1840-06-10		A1 G21
1774	" "	25	NWSE	1840-06-10		A1 G21
1776	" "	33	E½SE	1840-06-10		A1 G21
1777	" "	34	NENE	1840-06-10		A1 G21
1778	" "	34	NW	1840-06-10		A1 G21
1779	" "	34	S½NE	1840-06-10		A1 G21
1780	" "	34	W½SE	1840-06-10		A1 G21
1781	" "	35	W½NW	1840-06-10		A1 G21
1782	" "	35	W½SE	1840-06-10		A1 G21
1783	" "	36	W½SW	1840-06-10		A1 G21
1775	" "	27	E½SE	1841-01-05		A1 G21
1888	BISHOP, Thomas	14	W½SW	1841-01-05		A1
1889	" "	15	E½NE	1841-01-05		A1
1890	" "	15	E½SE	1841-01-05		A1

134

Township 7-N Range 16-E (Choctaw) - Map Group 9

ID	Individual in Patent	Sec.	Sec. Part	Date Issued	Other Counties	For More Info . . .
1772	BOTHWELL, David E	25	E½SE	1840-06-10		A1 G21
1773	" "	25	NESW	1840-06-10		A1 G21
1774	" "	25	NWSE	1840-06-10		A1 G21
1776	" "	33	E½SE	1840-06-10		A1 G21
1777	" "	34	NENE	1840-06-10		A1 G21
1778	" "	34	NW	1840-06-10		A1 G21
1779	" "	34	S½NE	1840-06-10		A1 G21
1780	" "	34	W½SE	1840-06-10		A1 G21
1781	" "	35	W½NW	1840-06-10		A1 G21
1782	" "	35	W½SE	1840-06-10		A1 G21
1783	" "	36	W½SW	1840-06-10		A1 G21
1775	" "	27	E½SE	1841-01-05		A1 G21
1847	BROWN, Martha Louisa	1	E½NE	1916-11-01		A4
1861	BROWN, Randolph W	7	SENW	1859-11-10		A1
1904	BROWN, William	7	W½SW	1841-01-05		A1
1671	BULLARD, Alexander	10	E½NW	1841-01-05		A1
1799	BUSBY, John F	28	SW	1841-01-05		A1
1885	BUSBY, Shepherd	27	W½SE	1841-01-05		A1
1855	CAINS, Musem D	23	SENE	1841-01-05		A1
1753	CALHOUN, Hugh	28	E½SE	1841-01-05		A1
1689	CARROL, Benjamin	26	SESE	1841-01-05		A1
1851	CHANDLER, Meredith	26	NWSW	1841-01-05		A1
1852	" "	26	SESW	1841-01-05		A1
1905	CHANDLER, William	26	SWSW	1841-01-05		A1
1906	" "	27	E½SW	1841-01-05		A1
1907	" "	35	NENW	1841-01-05		A1
1908	" "	35	SENW	1841-01-05		A1
1854	CHANEY, Mollie	19	NWNE	1908-10-26		A4
1800	CHESTER, John F	5	SWSW	1841-01-05		A1 R1685
1801	" "	6	NWSE	1841-01-05		A1
1802	" "	7	NWNE	1841-01-05		A1
1829	CLINTON, Joseph	27	SENW	1848-09-01		A1
1828	" "	27	NENW	1859-05-02		A1
1909	COLVIN, William	19	NESW	1890-08-16		A4
1841	COVINGTON, Lewis H	5	SWNE	1885-06-12		A4
1797	CRANE, John	9	SESE	1841-01-05		A1
1798	" "	9	W½SE	1841-01-05		A1
1836	CRANE, Lewis	10	NWNW	1841-01-05		A1
1837	" "	3	W½SW	1841-01-05		A1
1838	" "	4	E½SE	1841-01-05		A1
1839	" "	4	NWNE	1841-01-05		A1
1840	" "	4	NWNW	1841-01-05		A1
1816	CUNNINGHAM, John S	7	NENW	1912-10-10		A4
1856	DEEN, Nathan P	27	SWSW	1846-09-01		A1
1668	DUNN, Albert	3	E½SW	1841-01-05		A1
1755	EUBANKS, Isaac	23	NWSE	1841-01-05		A1
1672	FINNEN, Alexander	4	W½SW	1841-01-05		A1
1673	" "	5	E½NE	1841-01-05		A1
1674	" "	5	SE	1841-01-05		A1
1762	FINNEN, James	4	SESW	1841-01-05		A1
1763	" "	4	SWSE	1841-01-05		A1
1730	FLOYD, George	29	NWNE	1882-03-04		A4
1758	FORD, James C	22	E½SW	1841-01-05		A1
1759	" "	22	W½SE	1841-01-05		A1
1760	" "	27	NENE	1841-01-05		A1 C
1761	" "	27	W½NE	1841-01-05		A1 C
1733	FREELAND, George W	36	E½SW	1841-01-05		A1
1718	GARRETT, Edward T	35	NESE	1841-01-05		A1
1719	" "	35	SESE	1841-01-05		A1
1771	GARRETT, James T	34	E½SE	1841-01-05		A1 G81
1857	GILBERT, Nelson	19	SWSW	1891-08-19		A4
1664	GILLASPIE, Absalom	18	SENW	1896-04-30		A1
1665	GILLASPIE, Absolam	18	NWNW	1841-01-05		A1
1666	" "	18	SWNW	1841-01-05		A1
1667	GILLASPIE, Absolem W	18	NWSW	1848-09-01		A1
1884	GILLASPIE, Sandy	19	NWSW	1894-03-17		A4
1833	GLENN, Julius J	29	SESE	1859-06-01		A1
1834	" "	33	W½NE	1859-06-01		A1
1737	GRANT, Greene W	25	E½NW	1841-01-05		A1 G85
1738	" "	25	NE	1841-01-05		A1 G85
1739	" "	25	SWNW	1841-01-05		A1 G85
1740	" "	26	W½NE	1841-01-05		A1 G85
1736	" "	3	W½NW	1841-01-05		A1

Family Maps of Lauderdale County, Mississippi

ID	Individual in Patent	Sec.	Sec. Part	Date Issued	Other Counties	For More Info . . .
1741	GRANT, Greene W (Cont'd)	4	SWNW	1841-01-05		A1 G86
1830	GRAY, Joseph P	26	NWSE	1846-09-01		A1
1831	"	26	SWSE	1846-09-01		A1
1743	HAILS, Henry	15	W½NW	1841-01-05		A1
1744	"	2	W½SW	1841-01-05		A1
1745	"	3	E½SE	1841-01-05		A1
1910	HALL, William G	17	NWNW	1841-01-05		A1
1911	"	33	E½SW	1841-01-05		A1
1912	"	33	W½SE	1841-01-05		A1
1864	HARPER, Richard B	17	NESW	1859-11-10		A1
1865	"	17	NWSE	1859-11-10		A1
1896	HAYES, Ulysses A	7	NESW	1841-01-05		A1
1897	"	7	NWSE	1841-01-05		A1
1707	HAYS, Daniel C	6	E½SE	1841-01-05		A1
1705	"	19	SENW	1860-10-01		A1
1706	"	5	NWNE	1860-10-01		A1
1734	HAYS, Gill B	5	E½SW	1841-01-05		A1
1735	"	7	E½NE	1841-01-05		A1
1805	HENDERSON, John	36	NENW	1841-01-05		A1
1806	"	36	NWNE	1841-01-05		A1
1807	"	36	SENW	1841-01-05		A1
1808	"	36	W½NW	1841-01-05		A1
1913	HENDERSON, William	35	NE	1841-01-05		A1
1914	"	36	W½SE	1841-01-05		A1
1929	HENDERSON, Wilson	36	E½SE	1841-01-05		A1
1737	HERNDON, Thomas H	25	E½NW	1841-01-05		A1 G85
1738	"	25	NE	1841-01-05		A1 G85
1739	"	25	SWNW	1841-01-05		A1 G85
1740	"	26	W½NE	1841-01-05		A1 G85
1853	HOLMES, Miles B	17	W½SW	1892-03-17		A4
1726	HOOKS, Elizabeth P	5	NWNW	1884-12-30		A4
1727	"	5	SWNW	1909-08-16		A4
1729	HOOKS, George C	1	S½NW	1878-06-24		A4
1741	HOUZE, William J	4	SWNW	1841-01-05		A1 G86
1787	HOWELL, Jeremiah	23	NESE	1841-01-05		A1
1876	HUBBARD, Samuel	12	E½SE	1841-01-05		A1 G123
1877	"	13	S½NE	1841-01-05		A1 G123
1878	"	13	SW	1841-01-05		A1 G123
1879	"	14	SE	1841-01-05		A1 G123
1880	"	24	NW	1841-01-05		A1 G123
1711	HUSSEY, Edward G	1	SE	1841-01-05		A1
1712	"	12	E½NE	1841-01-05		A1
1713	"	12	W½SE	1841-01-05		A1
1714	"	14	E½NW	1841-01-05		A1
1715	"	14	W½NW	1841-01-05		A1
1716	"	2	W½NE	1841-01-05		A1
1717	"	4	E½NE	1841-01-05		A1
1732	JOHNSON, George	33	SWSW	1882-03-30		A4
1725	JORDAN, Elias	4	NWSE	1841-01-05		A1
1809	JORDAN, John	28	E½NW	1841-01-05		A1
1810	"	4	E½NW	1841-01-05		A1
1811	"	4	NESW	1841-01-05		A1
1812	"	4	SWNE	1841-01-05		A1
1817	JORDAN, John S	15	NWSW	1841-01-05		A1
1930	JORDAN, Zachariah	10	SW	1841-01-05		A1
1931	"	15	E½NW	1841-01-05		A1
1932	"	20	NESE	1841-01-05		A1
1933	"	20	SENE	1841-01-05		A1
1934	"	20	W½NE	1841-01-05		A1
1935	"	27	W½NW	1841-01-05		A1
1813	JORDON, John	21	W½NE	1841-01-05		A1
1688	KEETON, Austin	22	SENE	1841-01-05		A1
1764	KEETON, James	33	NENW	1841-01-05		A1
1858	KEETON, Obadiah	33	W½NW	1841-01-05		A1
1709	KINERD, David C	15	SWSW	1860-04-02		A1
1728	KITTRELL, Franklin T	1	SWNE	1911-01-12		A1
1881	LEE, Samuel	24	E½SW	1841-01-05		A1
1882	"	25	NWNW	1841-01-05		A1
1876	LEWIS, Rufus G	12	E½SE	1841-01-05		A1 G123
1877	"	13	S½NE	1841-01-05		A1 G123
1878	"	13	SW	1841-01-05		A1 G123
1879	"	14	SE	1841-01-05		A1 G123
1880	"	24	NW	1841-01-05		A1 G123

Township 7-N Range 16-E (Choctaw) - Map Group 9

ID	Individual in Patent	Sec.	Sec. Part	Date Issued	Other Counties	For More Info . . .
1922	LEWIS, William M	1	E½SW	1841-01-05		A1
1923	" "	14	E½SW	1841-01-05		A1
1924	" "	17	E½NE	1841-01-05		A1
1757	MALONE, Isham C	21	SENW	1859-05-02		A1
1756	" "	21	NWNW	1860-10-01		A1
1925	MANN, William	23	SWSE	1841-01-05		A1
1765	MASON, James	21	E½NE	1859-05-02		A1
1926	MASON, William	29	SENE	1860-10-01		A1 R1927
1927	MASON, William S	29	SENE	1897-08-09		A1 R1926
1862	MCCANN, Randsom D	15	SWSE	1841-01-05		A1
1863	" "	23	W½NW	1841-01-05		A1
1786	MCDONALD, Jefferson	19	NWSE	1888-02-25		A4
1916	MCDOW, William L	1	W½SW	1841-01-05		A1
1917	" "	11	E½NE	1841-01-05		A1
1918	" "	11	E½SE	1841-01-05		A1
1919	" "	12	E½SW	1841-01-05		A1
1920	" "	12	W½NE	1841-01-05		A1
1921	" "	2	E½SE	1841-01-05		A1
1867	MCLAMORE, Richard	34	NWNE	1841-01-05		A1
1868	" "	34	SW	1841-01-05		A1
1697	MCMILLON, Columbus James	1	NWNW	1919-05-26		A4
1675	MCMULLUN, Alexander	23	SWNE	1841-01-05		A1
1721	MILLER, Eli	12	W½SW	1841-01-05		A1
1722	" "	22	NENE	1841-01-05		A1
1723	" "	23	N½NE	1841-01-05		A1
1844	MOTT, Lovelace	19	W½NW	1841-01-05		A1
1676	MURPHY, Alexander	19	S½NE	1896-10-31		A4
1710	NAYLOR, Dock	19	E½SE	1897-11-01		A4
1690	NICHOLSON, Berry	19	SESW	1881-05-10		A4
1691	" "	19	SWSE	1881-05-10		A4
1766	ODOM, James	17	SENW	1841-01-05		A1
1767	" "	7	E½SE	1841-01-05		A1
1768	" "	7	SWNE	1841-01-05		A1
1769	" "	8	NWSW	1841-01-05		A1
1770	" "	8	SWSW	1841-01-05		A1
1845	ODOM, Malachi	7	SESW	1841-01-05		A1
1846	ODOM, Malechi	7	SWSE	1841-01-05		A1
1860	ODOM, Randol	33	E½NE	1841-01-05		A1
1731	PACE, George J	31	W½SE	1860-04-02		A1
1684	PALMER, Alfred	29	SWSW	1894-03-12		A4
1869	PHILIPS, Robert	3	E½NW	1841-01-05		A1
1891	PHILIPS, Thomas	3	W½SE	1841-01-05		A1
1892	" "	5	E½NW	1841-01-05		A1
1893	" "	9	NW	1841-01-05		A1
1677	RAMSAY, Alexander	12	NW	1841-01-05		A1
1678	" "	13	NW	1841-01-05		A1
1679	" "	14	NE	1841-01-05		A1
1742	REID, Hannah	29	NENW	1873-06-10		A4
1746	REID, Henry	29	NENE	1873-06-10		A4
1698	RICHARDSON, Daniel B	11	W½	1841-01-05		A1
1699	" "	11	W½NE	1841-01-05		A1
1700	" "	11	W½SE	1841-01-05		A1
1701	" "	2	NESW	1841-01-05		A1
1702	" "	2	NW	1841-01-05		A1
1703	" "	2	W½SE	1841-01-05		A1
1704	" "	3	NE	1841-01-05		A1
1835	ROBERTS, Lawrence	33	SENW	1894-12-17		A4
1692	RUSHING, Charles E	21	E½SE	1859-05-02		A1
1695	" "	7	W½NW	1859-05-02		A1
1693	" "	29	NESE	1860-04-02		A1
1694	" "	29	W½SE	1860-04-02		A1
1685	RUSSELL, Allen	5	SWSW	1841-01-05		A1 R1800
1686	" "	6	SWSE	1841-01-05		A1
1747	SIKES, Henry	18	W½SE	1841-01-05		A1
1748	" "	31	E½SE	1841-01-05		A1
1749	" "	33	NWSW	1841-01-05		A1
1814	SIKES, John R	19	NENW	1841-01-05		A1
1784	SMITH, James W	1	NENW	1895-08-30		A4
1785	" "	1	NWNE	1895-08-30		A4
1843	SMITH, Little J	35	SW	1841-01-05		A1
1842	STANSEL, Lewis	29	SESW	1904-07-02		A4
1696	STANTON, Christian B	6	E½NE	1841-01-05		A1
1894	STOKES, Thomas	17	SESW	1859-05-02		A1

Family Maps of Lauderdale County, Mississippi

ID	Individual in Patent	Sec.	Sec. Part	Date Issued	Other Counties	For More Info . . .
1895	STOKES, Thomas (Cont'd)	17	SWSE	1859-05-02		A1
1871	TABB, Samuel G	17	NENW	1841-01-05		A1
1872	" "	17	SWNW	1841-01-05		A1
1873	" "	17	W½NE	1841-01-05		A1
1874	" "	18	NE	1841-01-05		A1
1875	" "	8	W½SE	1841-01-05		A1
1876	TAPPAN, John	12	E½SE	1841-01-05		A1 G123
1877	" "	13	S½NE	1841-01-05		A1 G123
1878	" "	13	SW	1841-01-05		A1 G123
1879	" "	14	SE	1841-01-05		A1 G123
1880	" "	24	NW	1841-01-05		A1 G123
1724	THORNTON, Eli S	32	W½SW	1841-01-05		A1
1680	TINNIN, Alexander	10	E½SE	1841-01-05		A1
1681	" "	10	NE	1841-01-05		A1
1682	" "	17	E½SE	1841-01-05		A1
1683	" "	9	W½SW	1841-01-05		A1
1818	TINNIN, John	10	NWSE	1841-01-05		A1
1815	TORRANS, John R	29	W½NW	1860-10-01		A1
1866	TUTT, Richard B	25	W½SW	1841-01-05		A1
1900	TUTT, Wilkins N	22	SWNE	1846-09-01		A1
1803	ULRICK, John G	13	N½NE	1841-01-05		A1
1804	" "	13	SE	1841-01-05		A1
1859	ULRICK, Peter	24	N½NE	1841-01-05		A1
1720	VANCE, Edward	19	NENE	1860-10-01		A1
1750	WABINGTON, Horatio B	36	SWNE	1841-01-05		A1
1751	WARBINGTON, Horatio B	25	SESW	1841-01-05		A1
1771	" "	34	E½SE	1841-01-05		A1 G81
1752	" "	36	E½NE	1841-01-05		A1
1870	WELSH, Roena	29	N½SW	1881-09-17		A4
1708	WHITEHEAD, Daniel	2	SESW	1841-01-05		A1
1796	WILLIAMSON, John C	31	NWNW	1875-07-01		A4
1886	WILSON, Sherod	23	NENW	1841-01-05		A1
1915	WILSON, William J	30	NWNW	1841-01-05		A1
1928	WOOD, William	31	E½NE	1841-01-05		A1
1883	WORBINGTON, Samuel	25	SWSE	1841-01-05		A1

Family Maps of Lauderdale County, Mississippi

Patent Map

T7-N R16-E
Choctaw Meridian

Map Group 9

Township Statistics

Parcels Mapped	:	272
Number of Patents	:	237
Number of Individuals	:	142
Patentees Identified	:	139
Number of Surnames	:	101
Multi-Patentee Parcels	:	23
Oldest Patent Date	:	6/10/1840
Most Recent Patent	:	5/26/1919
Block/Lot Parcels	:	0
Parcels Re - Issued	:	2
Parcels that Overlap	:	0
Cities and Towns	:	3
Cemeteries	:	7

Section 6
STANTON Christian B
HOOKS Elizabeth P 1884
HOOKS Elizabeth P 1909
CHESTER John F 1841
RUSSELL Allen 1841
HAYS Daniel C 1841

Section 5
PHILIPS Thomas 1841
COVINGTON Lewis H 1885
HAYS Daniel C 1860
FINNEN Alexander
RUSSELL Allen 1841
CHESTER John F 1841
HAYS Gill B 1841
FINNEN Alexander 1841

Section 4
CRANE Lewis 1841
JORDAN John 1841
CRANE Lewis 1841
GRANT [86] Greene W 1841
FINNEN Alexander 1841
JORDAN John 1841
HUSSEY Edward G 1841
JORDAN Elias 1841
FINNEN James 1841
CRANE Lewis 1841

Section 7
RUSHING Charles E 1859
CUNNINGHAM John S 1912
CHESTER John F 1841
BROWN Randolph W 1859
ODOM James 1841
HAYS Gill B 1841
BROWN William 1841
HAYES Ulysses A 1841
HAYES Ulysses A 1841
ODOM Malachi 1841
ODOM Malechi 1841
ODOM James 1841

Section 8
ODOM James 1841
ODOM James 1841
TABB Samuel G 1841
ALEXANDER Joseph 1841

Section 9
ALEXANDER Joseph 1841
PHILIPS Thomas 1841
BALLARD Alexander 1841
TINNIN Alexander 1841
CRANE John 1841
ALEXANDER William 1841
ALEXANDER Joseph 1841
CRANE John 1841

Section 18
GILLASPIE Absolam 1841
GILLASPIE Absolam 1841
GILLASPIE Absalom 1896
GILLASPIE Absolem W 1848
TABB Samuel G 1841
SIKES Henry 1841

Section 17
HALL William G 1841
TABB Samuel G 1841
TABB Samuel G 1841
TABB Samuel G 1841
ODOM James 1841
LEWIS William M 1841
HOLMES Miles B 1892
HARPER Richard B 1859
HARPER Richard B 1859
STOKES Thomas 1859
STOKES Thomas 1859
TINNIN Alexander 1841

Section 16

Section 19
MOTT Lovelace 1841
SIKES John R 1841
CHANEY Mollie 1908
VANCE Edward 1860
HAYS Daniel C 1860
MURPHY Alexander 1896
GILLASPIE Sandy 1894
COLVIN William 1890
MCDONALD Jefferson 1888
GILBERT Nelson 1891
NICHOLSON Berry 1881
NICHOLSON Berry 1881
NAYLOR Dock 1897

Section 20
JORDAN Zachariah 1841
JORDAN Zachariah 1841
JORDAN Zachariah 1841

Section 21
MALONE Isham C 1860
ALEXANDER John 1841
JORDON John 1841
ALEXANDER John 1841
MALONE Isham C 1859
MASON James 1859
ALEXANDER Joseph 1841
ALEXANDER Joel 1841
ALEXANDER Joseph 1841
RUSHING Charles E 1859

Section 30
WILSON William J 1841

Section 29
TORRANS John R 1860
REID Hannah 1873
FLOYD George 1882
REID Henry 1873
MASON William 1860
MASON William S 1897
WELSH Roena 1881
RUSHING Charles E 1860
RUSHING Charles E 1860
PALMER Alfred 1894
STANSEL Lewis 1904
GLENN Julius J 1859

Section 28
JORDAN John 1841
BUSBY John F 1841
CALHOUN Hugh 1841

Section 31
WILLIAMSON John C 1875
WOOD William 1841
BEASON William B 1859
PACE George J 1860

Section 32
THORNTON Eli S 1841
SIKES Henry 1841

Section 33
KEETON Obadiah 1841
KEETON James 1841
GLENN Julius J 1859
ROBERTS Lawrence 1894
ODOM Randol 1841
SIKES Henry 1841
HALL William G 1841
JOHNSON George 1882
HALL William G 1841
BERRIEN [21] James W 1840

Copyright 2006 Boyd IT, Inc. All Rights Reserved

140

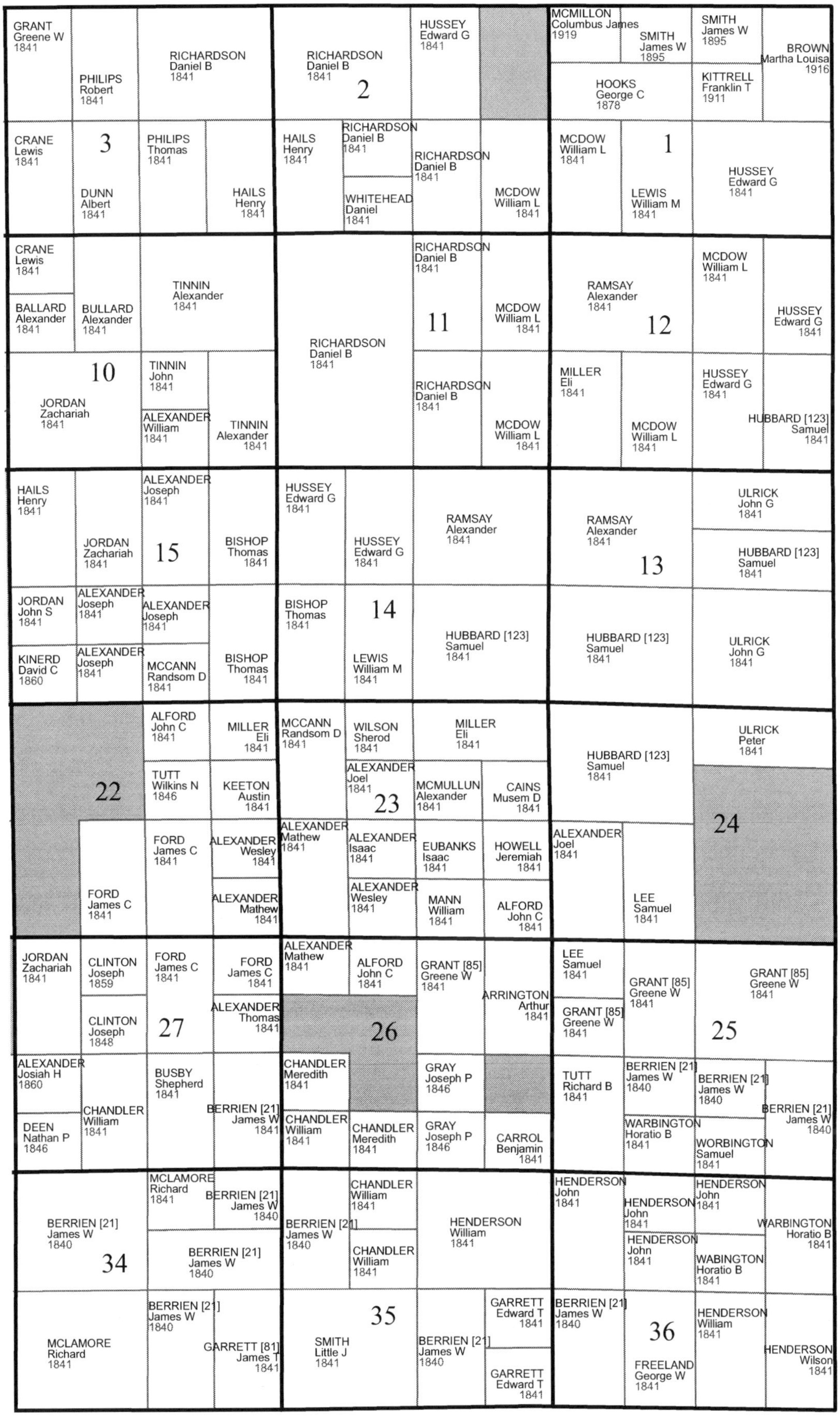

Family Maps of Lauderdale County, Mississippi

Road Map

T7-N R16-E
Choctaw Meridian

Map Group 9

Cities & Towns
Houston
Marion
Poplar Springs

Cemeteries
Barker Cemetery
Barker Cemetery
Confederate Cemetery
Forest Lawn Garden
Kinard Cemetery
Old Marion Cemetery
Tinnin Cemetery

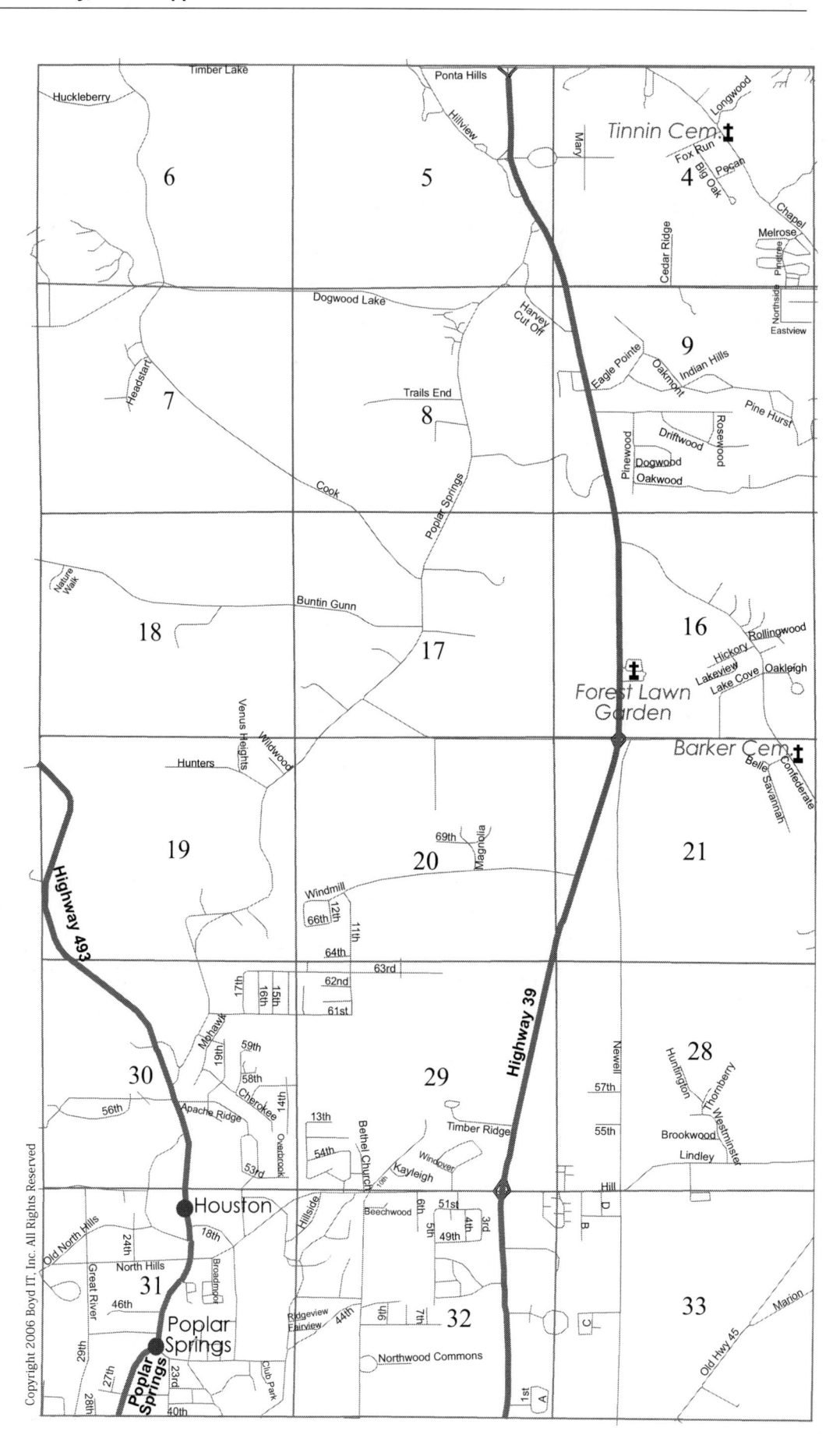

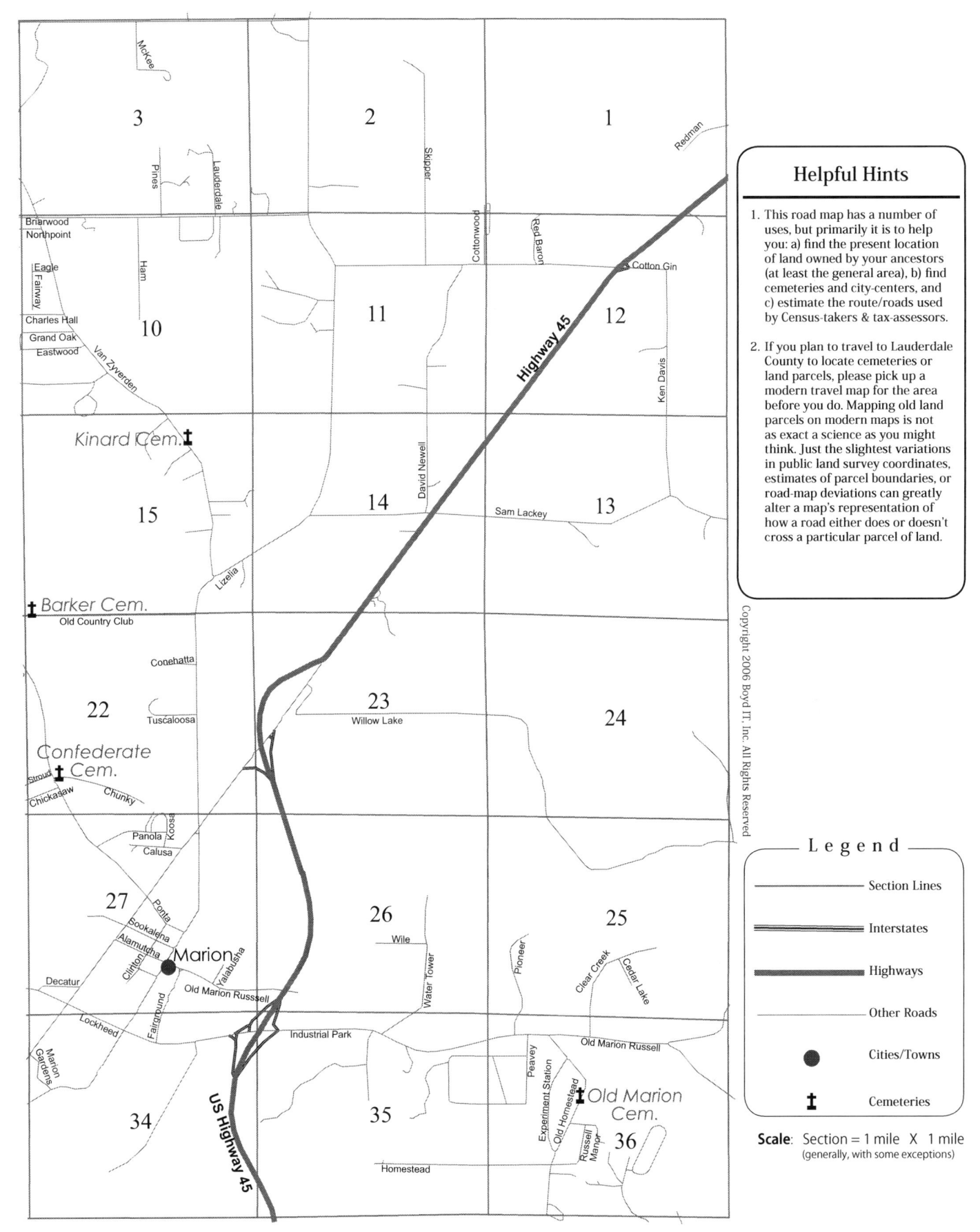

Township 7-N Range 16-E (Choctaw) - Map Group 9

Helpful Hints

1. This road map has a number of uses, but primarily it is to help you: a) find the present location of land owned by your ancestors (at least the general area), b) find cemeteries and city-centers, and c) estimate the route/roads used by Census-takers & tax-assessors.

2. If you plan to travel to Lauderdale County to locate cemeteries or land parcels, please pick up a modern travel map for the area before you do. Mapping old land parcels on modern maps is not as exact a science as you might think. Just the slightest variations in public land survey coordinates, estimates of parcel boundaries, or road-map deviations can greatly alter a map's representation of how a road either does or doesn't cross a particular parcel of land.

Legend

- Section Lines
- Interstates
- Highways
- Other Roads
- ● Cities/Towns
- ✝ Cemeteries

Scale: Section = 1 mile X 1 mile
(generally, with some exceptions)

Copyright 2006 Boyd IT, Inc. All Rights Reserved

143

Family Maps of Lauderdale County, Mississippi

Historical Map
T7-N R16-E
Choctaw Meridian
Map Group 9

Cities & Towns
Houston
Marion
Poplar Springs

Cemeteries
Barker Cemetery
Barker Cemetery
Confederate Cemetery
Forest Lawn Garden
Kinard Cemetery
Old Marion Cemetery
Tinnin Cemetery

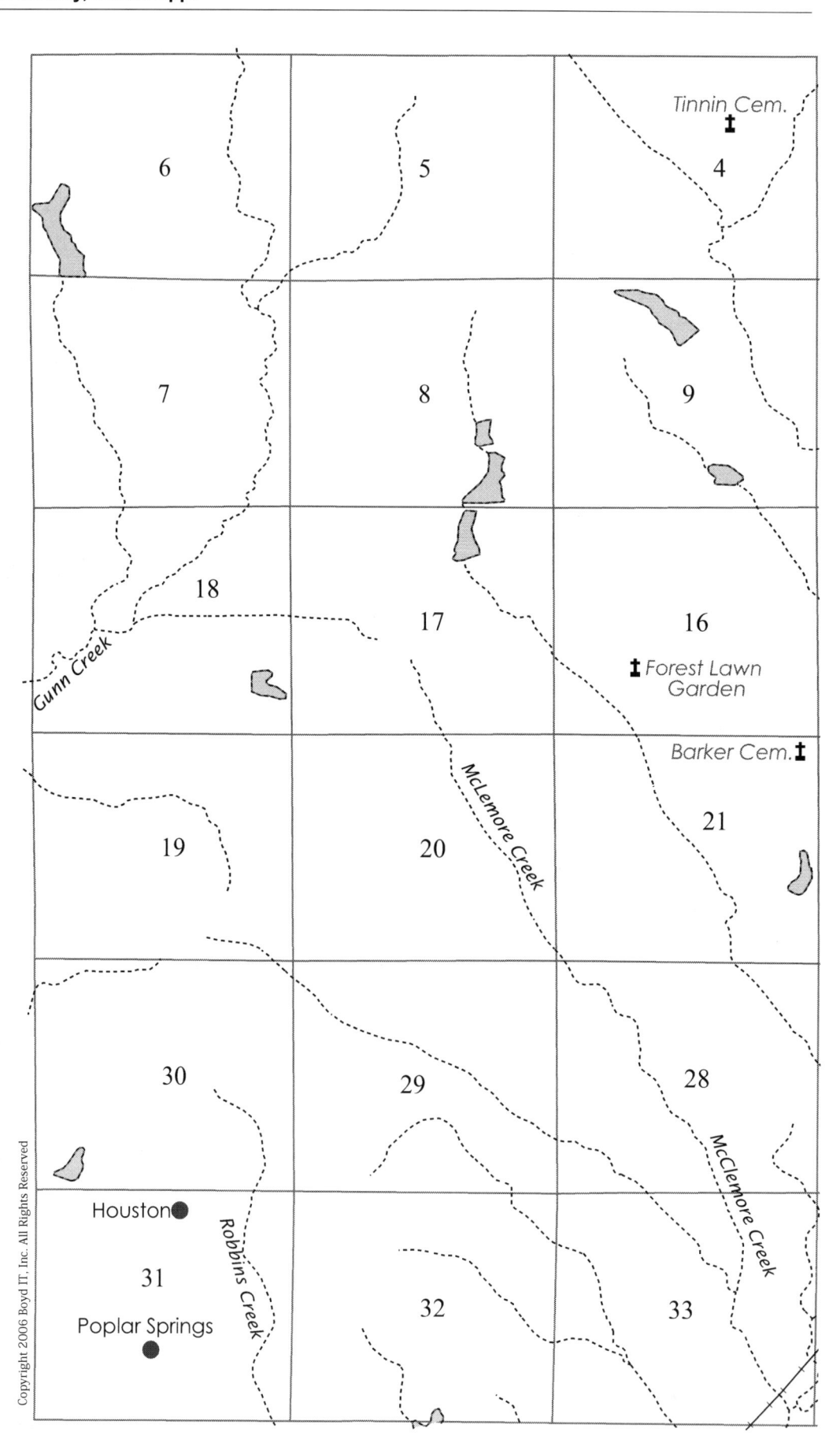

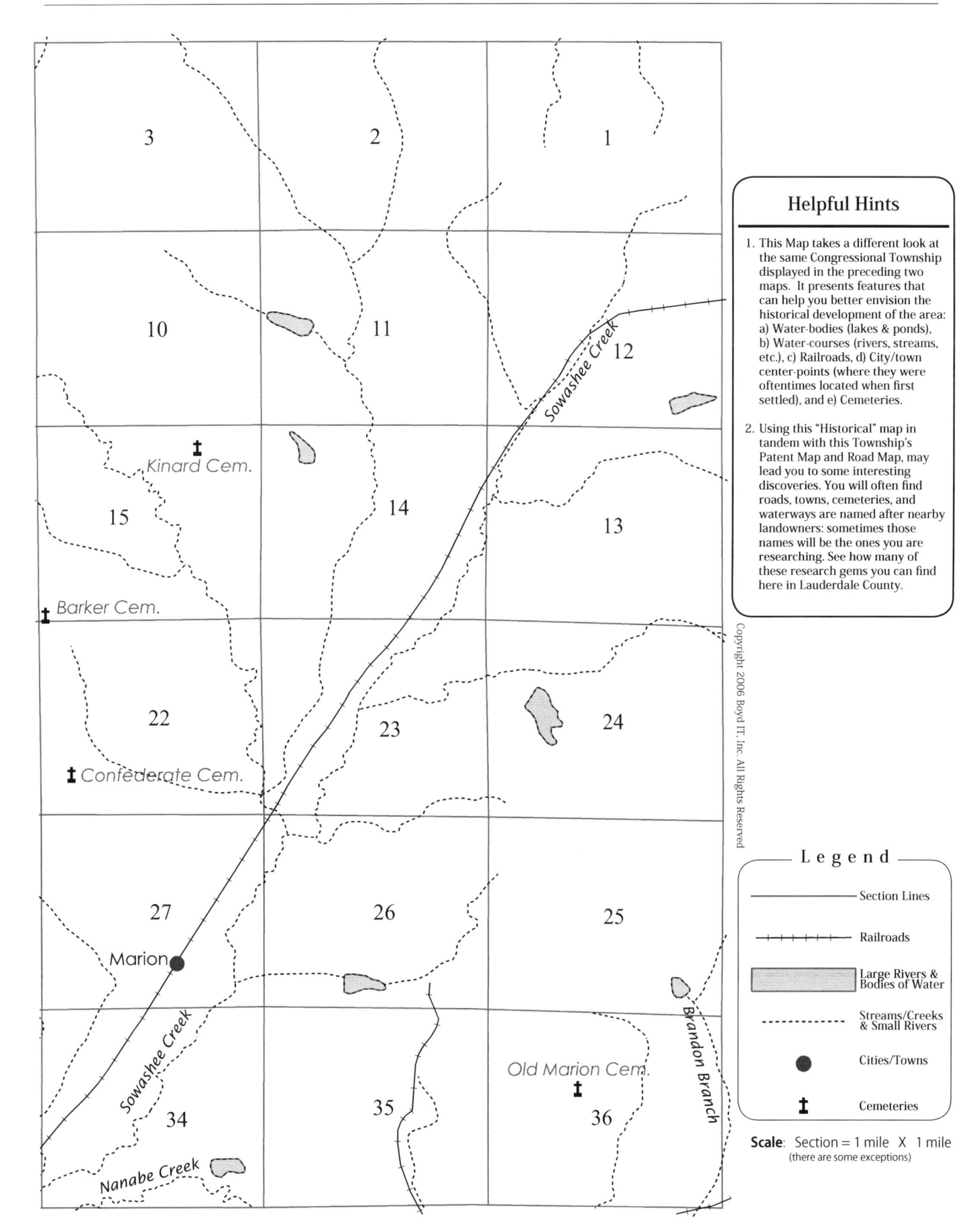

Township 7-N Range 16-E (Choctaw) - Map Group 9

Helpful Hints

1. This Map takes a different look at the same Congressional Township displayed in the preceding two maps. It presents features that can help you better envision the historical development of the area: a) Water-bodies (lakes & ponds), b) Water-courses (rivers, streams, etc.), c) Railroads, d) City/town center-points (where they were oftentimes located when first settled), and e) Cemeteries.

2. Using this "Historical" map in tandem with this Township's Patent Map and Road Map, may lead you to some interesting discoveries. You will often find roads, towns, cemeteries, and waterways are named after nearby landowners: sometimes those names will be the ones you are researching. See how many of these research gems you can find here in Lauderdale County.

Legend

- Section Lines
- Railroads
- Large Rivers & Bodies of Water
- Streams/Creeks & Small Rivers
- Cities/Towns
- Cemeteries

Scale: Section = 1 mile X 1 mile
(there are some exceptions)

Copyright 2006 Boyd IT, Inc. All Rights Reserved

Map Group 10: Index to Land Patents
Township 7-North Range 17-East (Choctaw)

After you locate an individual in this Index, take note of the Section and Section Part then proceed to the Land Patent map on the pages immediately following. You should have no difficulty locating the corresponding parcel of land.

The "For More Info" Column will lead you to more information about the underlying Patents. See the *Legend* at right, and the "How to Use this Book" chapter, for more information.

LEGEND
"For More Info . . . " column

- **A** = Authority (Legislative Act, See Appendix "A")
- **B** = Block or Lot (location in Section unknown)
- **C** = Cancelled Patent
- **F** = Fractional Section
- **G** = Group (Multi-Patentee Patent, see Appendix "C")
- **V** = Overlaps another Parcel
- **R** = Re-Issued (Parcel patented more than once)

(A & G items require you to look in the Appendixes referred to above. All other Letter-designations followed by a number require you to locate line-items in this index that possess the ID number found after the letter).

ID	Individual in Patent	Sec.	Sec. Part	Date Issued	Other Counties	For More Info . . .
2101	ADAMS, Washington B	26	E½SE	1841-01-05		A1 G2
2015	AGNEW, James	4	E½SW	1841-01-05		A1 G3
2013	" "	4	SE	1841-01-05		A1
2014	" "	9	NE	1846-09-01		A1 G4
1999	ALEXANDER, Henry	25	E½SW	1841-01-05		A1
2000	" "	25	W½SE	1841-01-05		A1
1956	ANDING, David	5	W½NE	1846-09-01		A1
2125	ANDREWS, William P	19	NENW	1860-04-02		A1
2109	ASKEW, William D	13	NENW	1899-04-17		A4
2110	" "	13	NWNE	1899-04-17		A4
2111	" "	13	S½NW	1899-04-17		A4
1945	BALLARD, Charles	31	W½NW	1841-01-05		A1 G16
2099	BARNES, Thomas	27	E½NE	1841-01-05		A1
1936	BATCHEE, Abraham	5	E½NE	1849-12-01		A1
2126	BOSWELL, William S	30	SWSW	1846-09-01		A1
2083	BOYKIN, Rowell	17	W½SW	1846-09-01		A1
2057	BOZEMAN, Lula	15	SESE	1919-05-26		A4 G28
2001	BRANNAN, Henry	34	N½NW	1846-09-01		A1
2097	BRASSIELL, Stephen	15	E½NE	1895-06-28		A4
2031	BROWN, John A	6	W½NE	1841-01-05		A1
2044	BROWN, John R	25	W½NW	1846-09-01		A1 G32
2043	" "	26	E½NE	1846-09-01		A1 G33
2074	BROWN, Primus	13	W½SE	1907-05-13		A4
2077	BROWN, Robert	27	SWSW	1860-04-02		A1
2079	" "	33	SENE	1860-04-02		A1
2078	" "	33	NWNE	1860-10-01		A1
2105	BROWN, William	33	SESW	1859-06-01		A1
2106	BRUNER, William	22	SW	1846-09-01		A1
1994	BUNYARD, George G	21	SESW	1875-11-20		A4
1966	BURKHALTER, Elias	3	SE	1846-09-01		A1
2090	BUSBY, Shepherd	10	E½NE	1846-09-01		A1
2091	" "	10	W½NE	1846-09-01		A1
2092	BUSBY, Sheppard	3	SESW	1859-05-02		A1
2093	" "	3	SWNE	1859-05-02		A1
2096	BUSBY, Spencer	10	W½NW	1846-09-01		A1
2098	BUSBY, Stephen	3	W½NW	1860-07-02		A1
1937	BUTCHEE, Abraham	9	W½NW	1859-05-02		A1
2062	CAINS, Mucindine D	17	E½SW	1846-09-01		A1
2063	" "	17	W½NE	1846-09-01		A1
2064	" "	17	W½SE	1846-09-01		A1
2003	CARMICHAEL, Hugh	11	SW	1846-09-01		A1
2082	CARPENTER, Robert L	23	SENW	1890-06-25		A4
1998	CLAYTON, Henry A	5	W½NW	1841-01-05		A1 G43
2044	COURTNEY, James	25	W½NW	1846-09-01		A1 G32
2033	CULBRAITH, John	21	E½NE	1846-09-01		A1
2034	" "	22	W½NW	1846-09-01		A1

Township 7-N Range 17-E (Choctaw) - Map Group 10

ID	Individual in Patent	Sec.	Sec. Part	Date Issued	Other Counties	For More Info . . .
2035	DOVE, John	25	W½SW	1841-01-05		A1 G65
2101	" "	26	E½SE	1841-01-05		A1 G2
2100	DUBOSE, Wade H	30	NENE	1846-09-01		A1
2127	DUBOSE, William W	29	W½NW	1861-05-01		A1
1974	DUN, Emanuel A	29	NWNE	1859-05-02		A1
2128	DUPREE, William W	17	E½NE	1846-09-01		A1
1975	DURR, Emanuel A	15	SESW	1852/05/01		A2
1976	" "	20	E½NE	1852/05/01		A2
1977	" "	20	E½SW	1852/05/01		A2
1978	" "	20	NESE	1852/05/01		A2
1979	" "	20	NWSE	1852/05/01		A2
1980	" "	21	N½W½SE	1852/05/01		A2
1982	" "	21	NWNE	1852/05/01		A2
1983	" "	21	NWNW	1852/05/01		A2
1984	" "	21	S½W½SW	1852/05/01		A2
1986	" "	23	SWSE	1852/05/01		A2
1987	" "	24	SWSW	1852/05/01		A2
1988	" "	27	NWNE	1852/05/01		A2
1989	" "	35	W½SW	1852/05/01		A2
1981	" "	21	NESW	1859-11-10		A1
1985	" "	21	SWSE	1859-11-10		A1
1993	DURR, Frank	21	S½NW	1885-12-19		A4
2049	DURR, Joseph A	15	E½NW	1905-03-30		A4
2066	DURR, Nathan	27	SWNW	1898-01-19		A4
2056	EAKINS, Louvenia	9	E½SE	1897-11-22		A4
2073	ECHOLS, Patrick	29	S½NE	1894-12-17		A4
2080	ECHOLS, Robert	29	NWSE	1897-11-01		A4
2004	EUBANKS, Isaac	18	W½NW	1846-09-01		A1
2005	" "	18	W½SW	1846-09-01		A1
1970	FAIR, Eliza	9	SWSE	1881-09-17		A4 G71
1991	FLETCHER, Finley	32	E½NW	1846-09-01		A1 G75
2107	FOSTER, William C	11	E½NW	1846-09-01		A1
2108	" "	11	W½NW	1846-09-01		A1
2123	GAINES, William M	1	NW	1847-04-01		A1
1951	GORDON, Charles	27	W½SE	1859-05-02		A1
2076	GORDON, Richmond T	23	NESW	1860-04-02		A1
2087	GORDON, Samuel	23	SWNW	1883-04-30		A4
2086	GRIFFITH, Samuel A	25	E½NW	1846-09-01		A1
2085	" "	23	SESW	1859-05-02		A1
2084	" "	23	E½SE	1860-04-02		A1
2027	HAMMOND, Jobe	34	E½NE	1846-09-01		A1 G96
2028	" "	34	W½NE	1846-09-01		A1 G96
2025	HAMMONDS, Job	27	E½SE	1846-09-01		A1 G97
2026	" "	35	W½NE	1846-09-01		A1 G98
2043	HARVARD, Celia	26	E½NE	1846-09-01		A1 G33
2094	HEARN, Sidney	23	N½NW	1916-01-07		A4
2016	HENDERSON, James	32	W½SE	1848-09-01		A1
2115	HENDERSON, William	32	E½SW	1841-01-05		A1 G107
2114	" "	32	W½NW	1841-01-05		A1 G108
1991	" "	32	E½NW	1846-09-01		A1 G75
2113	" "	32	NWSW	1848-09-01		A1
2130	HENDERSON, Wilson	19	E½NE	1846-09-01		A1
2131	" "	20	NWNW	1846-09-01		A1
2116	HUMPHREYS, William	36	W½SE	1841-01-05		A1
1965	HUSSEY, Edward G	7	E½SE	1846-09-01		A1
2081	HUSSEY, Robert	15	W½SE	1906-06-21		A4
2089	HUSSEY, Sarah E	5	N½SE	1859-11-10		A1
2088	" "	17	SESE	1860-04-02		A1
2055	INGRAM, Lemuel N	15	NESE	1859-11-10		A1
2117	JAMES, William	33	SWSW	1860-04-02		A1
2057	JOHNSON, Lula	15	SESE	1919-05-26		A4 G28
1968	JONES, Elijah	11	SWSE	1850/06/01		A2
1967	" "	11	SESE	1861-02-01		A1
2060	JONES, Martha	11	NWSE	1848-09-01		A1
2118	JONES, William	11	E½NE	1846-09-01		A1
2121	" "	11	SWNE	1848-09-01		A1
2119	" "	11	NESE	1850/06/01		A2
2122	" "	12	NWSW	1850/06/01		A2
2120	" "	11	NWNE	1859-11-10		A1
2024	KILLINGSWORTH, Jesse	4	E½NW	1841-01-05		A1 G129
2015	" "	4	E½SW	1841-01-05		A1 G3
2023	" "	4	W½NW	1846-09-01		A1 G128
2021	" "	4	NWNE	1849-12-01		A1

147

Family Maps of Lauderdale County, Mississippi

ID	Individual in Patent	Sec.	Sec. Part	Date Issued	Other Counties	For More Info...
2022	KILLINGSWORTH, Jesse (Cont'd)	4	NWSW	1849-12-01		A1
2067	LAMBERT, Nathan	23	W½SW	1841-01-05		A1 G131
2068	" "	36	W½NW	1841-01-05		A1 G131
2017	LOVE, James M	13	SWNE	1860-10-01		A1
2050	LOVE, Joseph W	5	NWSW	1848-09-01		A1
2051	" "	6	NESE	1848-09-01		A1
2035	LUCAS, Levi	25	W½SW	1841-01-05		A1 G65
1970	MAGRUE, Eliza	9	SWSE	1881-09-17		A4 G71
1945	MANN, William	31	W½NW	1841-01-05		A1 G16
2059	MCBRIDE, Marion	15	W½SW	1890-08-16		A4
2058	" "	15	NESW	1895-05-11		A4
1957	MCILWAIN, David	22	E½NW	1846-09-01		A1
2023	" "	4	W½NW	1846-09-01		A1 G128
1998	MCKINLEY, John	5	W½NW	1841-01-05		A1 G43
2045	MCLAURIN, John R	25	SWNE	1849-12-01		A1 R2046
2046	" "	25	SWNE	1849-12-01		A1 R2045
2069	MCLAURIN, Neal	25	SENE	1846-09-01		A1
2070	MCLAURIN, Neill	26	E½SW	1846-09-01		A1
2071	" "	26	W½SW	1846-09-01		A1
2112	MCLIN, William H	22	E½SE	1841-01-05		A1 G152
2067	" "	23	W½SW	1841-01-05		A1 G131
2068	" "	36	W½NW	1841-01-05		A1 G131
1941	MCNEILL, Alexander	26	W½SE	1846-09-01		A1
2102	MILLS, Willard C	3	N½NE	1861-05-01		A1
2103	" "	7	E½NE	1861-05-01		A1
2104	" "	7	E½NW	1861-05-01		A1
2052	MOODY, Josiah	9	E½NW	1846-09-01		A1
2024	MYRICK, Henry	4	E½NW	1841-01-05		A1 G129
2015	" "	4	E½SW	1841-01-05		A1 G3
2065	NEWELL, Nancy	3	SENE	1860-07-02		A1
2115	NICHOLS, Noah	32	E½SW	1841-01-05		A1 G107
1990	ODOM, Ephraim	25	E½SE	1841-01-05		A1
2124	OTT, William	17	NESE	1899-06-13		A4
1938	PAGE, Abraham	21	NENW	1846-09-01		A1
1939	" "	22	SWNE	1846-09-01		A1
2053	PARKER, Julia	13	E½NE	1909-01-21		A4
2075	PATHEA, Randle	15	W½NE	1885-06-12		A4
1942	PATTON, Amanda	29	NENW	1899-01-23		A4
1995	PATTON, George W	19	NWNE	1899-07-15		A4
2048	PEARSON, John W	33	NWSW	1861-02-01		A1
2042	PERCE, John	36	E½SE	1846-09-01		A1
1971	PIGFORD, Elizabeth	36	NE	1846-08-31		A1 G161
2129	PIGFORD, William W	35	E½SE	1846-09-01		A1
1962	PUGH, Edgeworth	35	E½NE	1846-09-01		A1
1963	" "	35	NENW	1846-09-01		A1
1964	" "	35	NWNE	1846-09-01		A1
2002	RAWSON, Henry H	33	SESE	1860-04-02		A1
1953	ROGERS, Daniel	36	E½NW	1846-09-01		A1
1954	" "	36	NESW	1846-09-01		A1
1947	RUSHING, Charles E	29	NESE	1859-06-01		A1
1948	" "	29	SWSE	1859-06-01		A1
1946	" "	29	NENE	1859-11-10		A1
1949	" "	33	NESW	1860-10-01		A1
1950	" "	7	NWNW	1860-10-01		A1
1958	RUSSELL, David	31	SE	1841-01-05		A1 G166
1992	RUSSELL, Francis	31	SENE	1875-07-01		A4
1958	RUSSELL, Isaac	31	SE	1841-01-05		A1 G166
2008	" "	29	SESW	1859-11-10		A1
2009	" "	31	NWNE	1859-11-10		A1
2010	" "	31	SWNE	1859-11-10		A1
2020	RUSSELL, Jane	31	NENE	1873-06-10		A4
2027	RUSSELL, William	34	E½NE	1846-09-01		A1 G96
2028	" "	34	W½NE	1846-09-01		A1 G96
2025	SCARBOROUGH, Absolam L	27	E½SE	1846-09-01		A1 G97
2026	SCARBOROUGH, David	35	W½NW	1846-09-01		A1 G98
1972	SCOTT, Ellen	29	SESE	1898-08-27		A4
2054	SECREST, Lawrence Wiley	31	NENW	1916-06-10		A4
2047	SHOWS, John	31	W½SW	1846-09-01		A1
2036	SMITH, John J	10	NESW	1852/05/01		A2
2037	" "	19	SWNE	1852/05/01		A2
2038	" "	28	NESW	1852/05/01		A2
2039	" "	3	NESW	1852/05/01		A2
2040	" "	32	SWSW	1852/05/01		A2

Township 7-N Range 17-E (Choctaw) - Map Group 10

ID	Individual in Patent	Sec.	Sec. Part	Date Issued	Other Counties	For More Info...
2041	SMITH, John J (Cont'd)	33	NWNW	1852/05/01		A2
2072	SMITH, Owen P	21	NWSW	1888-04-05		A4
2006	SUTTLE, Isaac G	1	E½SW	1846-09-01		A1
2007	" "	1	W½SE	1846-09-01		A1
1952	TILL, Claudius P	9	W½SW	1899-06-13		A4
1943	TUCKER, Arthur	18	E½NW	1846-09-01		A1
1944	" "	5	E½NW	1846-09-01		A1
1996	URY, George W	18	E½SE	1846-09-01		A1
1997	WALKER, George W	15	W½NW	1891-06-30		A4
2029	WALKER, Joel P	1	E½SE	1846-09-01		A1
2030	" "	1	NE	1846-09-01		A1
1973	WARBINGTON, Ellen	29	SENW	1895-02-21		A4
2061	WARLINGTON, Martha	31	SENW	1873-06-10		A4
1940	WEISSINGER, Alexander J	7	SWNW	1848-09-01		A1
1960	WHITE, David	12	SENW	1846-09-01		A1
1961	" "	12	W½NW	1846-09-01		A1
1959	" "	12	NENW	1849-12-01		A1
2112	WHITE, Elijah	22	E½SE	1841-01-05		A1 G152
1969	" "	22	W½SE	1846-09-01		A1
2014	" "	9	NE	1846-09-01		A1 G4
2018	WHITE, James	1	W½SW	1846-09-01		A1
2019	" "	2	NESE	1850/06/01		A2
2114	WILLIAMS, Joel	32	W½NW	1841-01-05		A1 G108
1971	WILLIAMSON, Francis	36	NE	1846-08-31		A1 G161
1971	WILLIAMSON, Wiley W	36	NE	1846-08-31		A1 G161
2012	WORBINGTON, Jacob B	30	E½SW	1846-09-01		A1 C
2095	YARRELL, Simon	27	NWSW	1896-07-11		A4
2011	YATES, Isaac	21	E½SE	1846-09-01		A1
2014	" "	9	NE	1846-09-01		A1 G4
1955	YEAGER, Daniel W	13	SW	1905-12-30		A4
2032	YOES, John C	33	NENE	1846-09-01		A1

Family Maps of Lauderdale County, Mississippi

Patent Map

T7-N R17-E
Choctaw Meridian

Map Group 10

Township Statistics

Parcels Mapped	:	196
Number of Patents	:	166
Number of Individuals	:	138
Patentees Identified	:	131
Number of Surnames	:	103
Multi-Patentee Parcels	:	24
Oldest Patent Date	:	1/5/1841
Most Recent Patent	:	5/26/1919
Block/Lot Parcels	:	0
Parcels Re-Issued	:	1
Parcels that Overlap	:	0
Cities and Towns	:	3
Cemeteries	:	4

Section 6
- BROWN, John A — 1841
- LOVE, Joseph W — 1848

Section 5
- CLAYTON [43], Henry A — 1841
- TUCKER, Arthur — 1846
- ANDING, David — 1846
- BATCHEE, Abraham — 1849
- LOVE, Joseph W — 1848
- HUSSEY, Sarah E — 1859

Section 4
- KILLINGSWORTH [128], Jesse — 1846
- KILLINGSWORTH, Jesse — 1849
- KILLINGSWORTH [129], Jesse — 1841
- KILLINGSWORTH, Jesse — 1849
- AGNEW [3], James — 1841
- AGNEW, James — 1841

Section 7
- RUSHING, Charles E — 1860
- MILLS, Willard C — 1861
- WEISSINGER, Alexander J — 1848
- MILLS, Willard C — 1861
- HUSSEY, Edward G — 1846

Section 8

Section 9
- BUTCHEE, Abraham — 1859
- AGNEW [4], James — 1846
- MOODY, Josiah — 1846
- TILL, Claudius P — 1899
- FAIR [71], Eliza — 1881
- EAKINS, Louvenia — 1897

Section 18
- EUBANKS, Isaac — 1846
- TUCKER, Arthur — 1846
- EUBANKS, Isaac — 1846

Section 17
- CAINS, Mucindine D — 1846
- DUPREE, William W — 1846
- BOYKIN, Rowell — 1846
- CAINS, Mucindine D — 1846
- URY, George W — 1846
- CAINS, Mucindine D — 1846
- OTT, William — 1899
- HUSSEY, Sarah E — 1860

Section 16

Section 19
- ANDREWS, William P — 1860
- PATTON, George W — 1899
- SMITH, John J — 1852
- HENDERSON, Wilson — 1846

Section 20
- HENDERSON, Wilson — 1846
- DURR, Emanuel A — 1852
- DURR, Emanuel A — 1852
- DURR, Emanuel A — 1852

Section 21
- DURR, Emanuel A — 1852
- PAGE, Abraham — 1846
- DURR, Emanuel A — 1852
- DURR, Frank — 1885
- SMITH, Owen P — 1888
- DURR, Emanuel A — 1859
- DURR, Emanuel A — 1852
- DURR, Emanuel A — 1852
- BUNYARD, George G — 1875
- DURR, Emanuel A — 1859
- CULBRAITH, John — 1846
- YATES, Isaac — 1846

Section 30
- DUBOSE, Wade H — 1846
- WORBINGTON, Jacob B — 1846
- BOSWELL, William S — 1846

Section 29
- DUBOSE, William W — 1861
- PATTON, Amanda — 1899
- DUN, Emanuel A — 1859
- RUSHING, Charles E — 1859
- WARBINGTON, Ellen — 1895
- ECHOLS, Patrick — 1894
- ECHOLS, Robert — 1897
- RUSHING, Charles E — 1859
- RUSSELL, Isaac — 1859
- RUSHING, Charles E — 1859
- SCOTT, Ellen — 1898

Section 28
- SMITH, John J — 1852

Section 31
- BALLARD [16], Charles — 1841
- SECREST, Lawrence Wiley — 1916
- RUSSELL, Isaac — 1859
- RUSSELL, Jane — 1873
- WARLINGTON, Martha — 1873
- RUSSELL, Isaac — 1859
- RUSSELL, Francis — 1875
- SHOWS, John — 1846
- RUSSELL [166], David — 1841

Section 32
- HENDERSON, William [108] — 1841
- FLETCHER [75], Finley — 1846
- HENDERSON, William — 1848
- HENDERSON, James — 1848
- HENDERSON [107], William — 1841
- SMITH, John J — 1852

Section 33
- SMITH, John J — 1852
- BROWN, Robert — 1860
- YOES, John C — 1846
- BROWN, Robert — 1860
- PEARSON, John W — 1861
- RUSHING, Charles E — 1860
- JAMES, William — 1860
- BROWN, William — 1859
- RAWSON, Henry H — 1860

Copyright 2006 Boyd IT, Inc. All Rights Reserved

150

Township 7-N Range 17-E (Choctaw) - Map Group 10

Section 1
- GAINES, William M 1847
- WALKER, Joel P 1846
- WHITE, James 1846
- SUTTLE, Isaac G 1846
- SUTTLE, Isaac G 1846
- WALKER, Joel P 1846

Section 2
- WHITE, James 1850

Section 3
- BUSBY, Stephen 1860
- MILLS, Willard C 1861
- BUSBY, Sheppard 1859
- NEWELL, Nancy 1860
- SMITH, John J 1852
- BURKHALTER, Elias 1846
- BUSBY, Sheppard 1859

Section 10
- BUSBY, Spencer 1846
- BUSBY, Shepherd 1846
- BUSBY, Shepherd 1846
- SMITH, John J 1852

Section 11
- FOSTER, William C 1846
- JONES, William 1859
- FOSTER, William C 1846
- JONES, William 1848
- JONES, William 1846
- JONES, Martha 1848
- JONES, William 1850
- CARMICHAEL, Hugh 1846
- JONES, Elijah 1850
- JONES, Elijah 1861

Section 12
- WHITE, David 1846
- WHITE, David 1849
- WHITE, David 1846
- JONES, William 1850

Section 13
- ASKEW, William D 1899
- ASKEW, William D 1899
- ASKEW, William D 1899
- LOVE, James M 1860
- PARKER, Julia 1909
- YEAGER, Daniel W 1905
- BROWN, Primus 1907

Section 14

Section 15
- WALKER, George W 1891
- PATHEA, Randle 1885
- DURR, Joseph A 1905
- BRASSIELL, Stephen 1895
- MCBRIDE, Marion 1890
- MCBRIDE, Marion 1895
- HUSSEY, Robert 1906
- INGRAM, Lemuel N 1859
- DURR, Emanuel A 1852
- BOZEMAN [28], Lula 1919

Section 22
- CULBRAITH, John 1846
- MCILWAIN, David 1846
- PAGE, Abraham 1846
- WHITE, Elijah 1846
- BRUNER, William 1846

Section 23
- HEARN, Sidney 1916
- GORDON, Samuel 1883
- CARPENTER, Robert L 1890
- LAMBERT [131], Nathan 1841
- GORDON, Richmond T 1860
- MCLIN [152], William H 1841
- GRIFFITH, Samuel A 1859
- GRIFFITH, Emanuel A 1852
- GRIFFITH, Samuel A 1860

Section 24
- DURR, Emanuel A 1852

Section 25
- BROWN [32], John R 1846
- BROWN [33], John R 1846
- GRIFFITH, Samuel A 1846
- MCLAURIN, John R 1849
- MCLAURIN, Neal 1846
- DOVE [65], John 1841
- ALEXANDER, Henry 1841
- ALEXANDER, Henry 1841
- ODOM, Ephraim 1841

Section 26
- BARNES, Thomas 1841
- GORDON, Charles 1859
- HAMMONDS [97], Job 1846
- MCLAURIN, Neill 1846
- MCLAURIN, Neill 1846
- MCNEILL, Alexander 1846
- ADAMS [2], Washington B 1841

Section 27
- DURR, Emanuel A 1852
- DURR, Nathan 1898
- YARRELL, Simon 1896
- BROWN, Robert 1860

Section 34
- BRANNAN, Henry 1846

Section 35
- HAMMOND [96], Jobe 1846
- HAMMOND [96], Jobe 1846
- HAMMONDS [98], Job 1846
- PUGH, Edgeworth 1846
- PUGH, Edgeworth 1846
- PUGH, Edgeworth 1846
- DURR, Emanuel A 1852
- PIGFORD, William W 1846

Section 36
- LAMBERT [131], Nathan 1841
- ROGERS, Daniel 1846
- PIGFORD [161], Elizabeth 1846
- ROGERS, Daniel 1846
- HUMPHREYS, William 1841
- PERCE, John 1846

Helpful Hints

1. This Map's INDEX can be found on the preceding pages.

2. Refer to Map "C" to see where this Township lies within Lauderdale County, Mississippi.

3. Numbers within square brackets [ ] denote a multi-patentee land parcel (multi-owner). Refer to Appendix "C" for a full list of members in this group.

4. Areas that look to be crowded with Patentees usually indicate multiple sales of the same parcel (Re-issues) or Overlapping parcels. See this Township's Index for an explanation of these and other circumstances that might explain "odd" groupings of Patentees on this map.

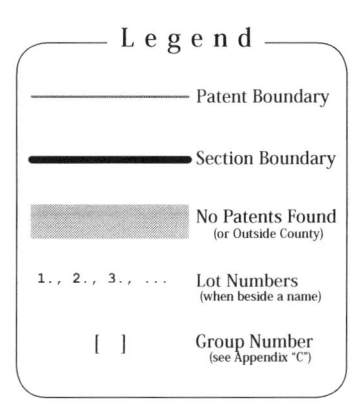

Copyright 2006 Boyd IT, Inc. All Rights Reserved

Legend

- ——— Patent Boundary
- ━━━ Section Boundary
- ▓▓▓ No Patents Found (or Outside County)
- 1., 2., 3., ... Lot Numbers (when beside a name)
- [ ] Group Number (see Appendix "C")

Scale: Section = 1 mile X 1 mile (generally, with some exceptions)

151

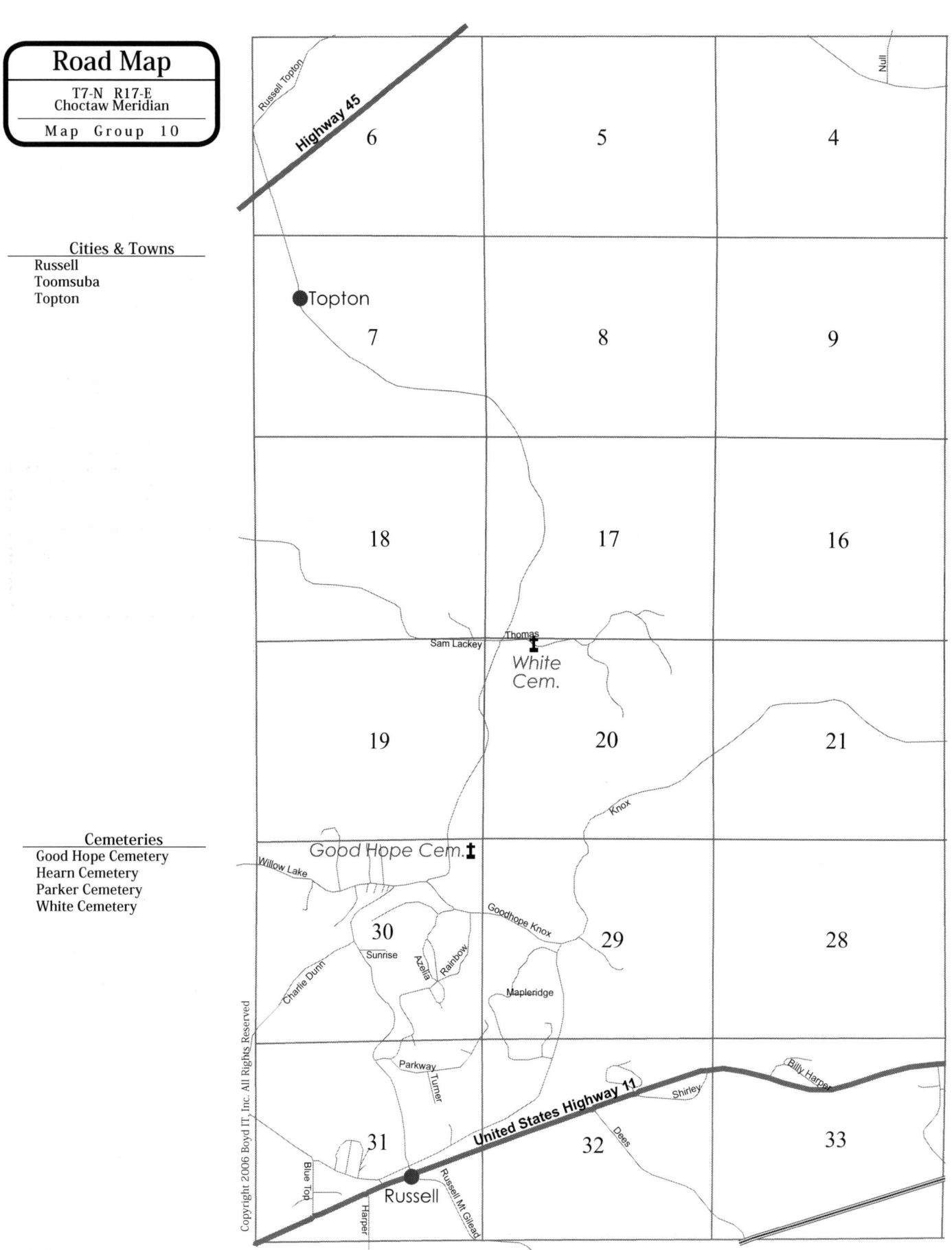

Family Maps of Lauderdale County, Mississippi

Road Map

T7-N R17-E
Choctaw Meridian

Map Group 10

Cities & Towns
Russell
Toomsuba
Topton

Cemeteries
Good Hope Cemetery
Hearn Cemetery
Parker Cemetery
White Cemetery

Township 7-N Range 17-E (Choctaw) - Map Group 10

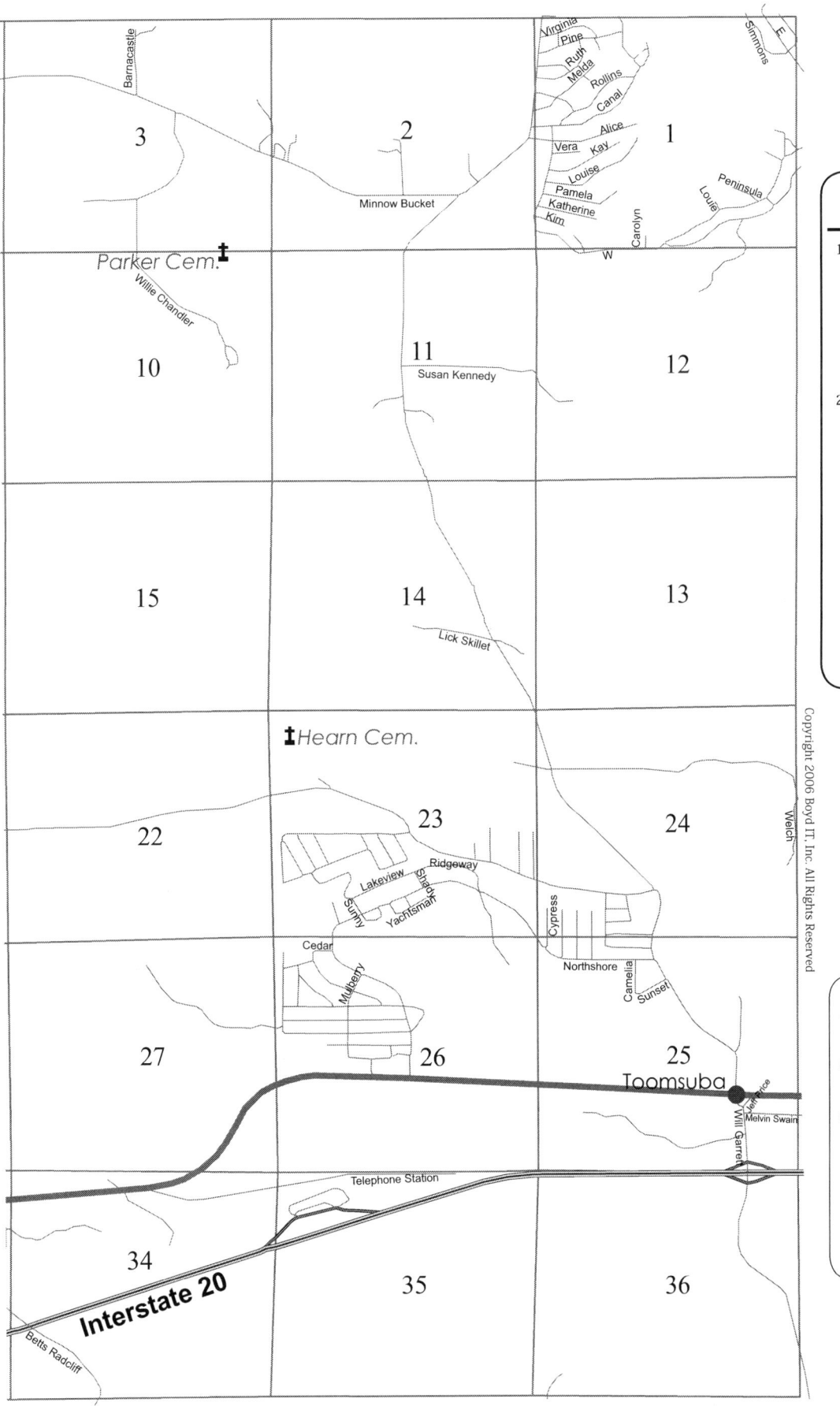

Helpful Hints

1. This road map has a number of uses, but primarily it is to help you: a) find the present location of land owned by your ancestors (at least the general area), b) find cemeteries and city-centers, and c) estimate the route/roads used by Census-takers & tax-assessors.

2. If you plan to travel to Lauderdale County to locate cemeteries or land parcels, please pick up a modern travel map for the area before you do. Mapping old land parcels on modern maps is not as exact a science as you might think. Just the slightest variations in public land survey coordinates, estimates of parcel boundaries, or road-map deviations can greatly alter a map's representation of how a road either does or doesn't cross a particular parcel of land.

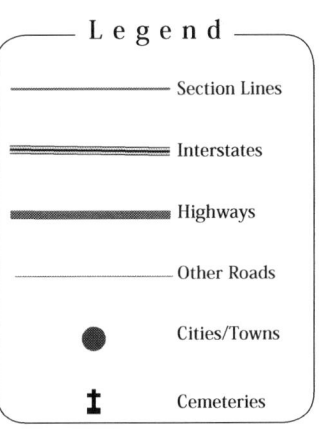

Legend

- Section Lines
- Interstates
- Highways
- Other Roads
- ● Cities/Towns
- ☩ Cemeteries

Scale: Section = 1 mile X 1 mile
(generally, with some exceptions)

153

Family Maps of Lauderdale County, Mississippi

Historical Map
T7-N R17-E
Choctaw Meridian
Map Group 10

Cities & Towns
Russell
Toomsuba
Topton

Cemeteries
Good Hope Cemetery
Hearn Cemetery
Parker Cemetery
White Cemetery

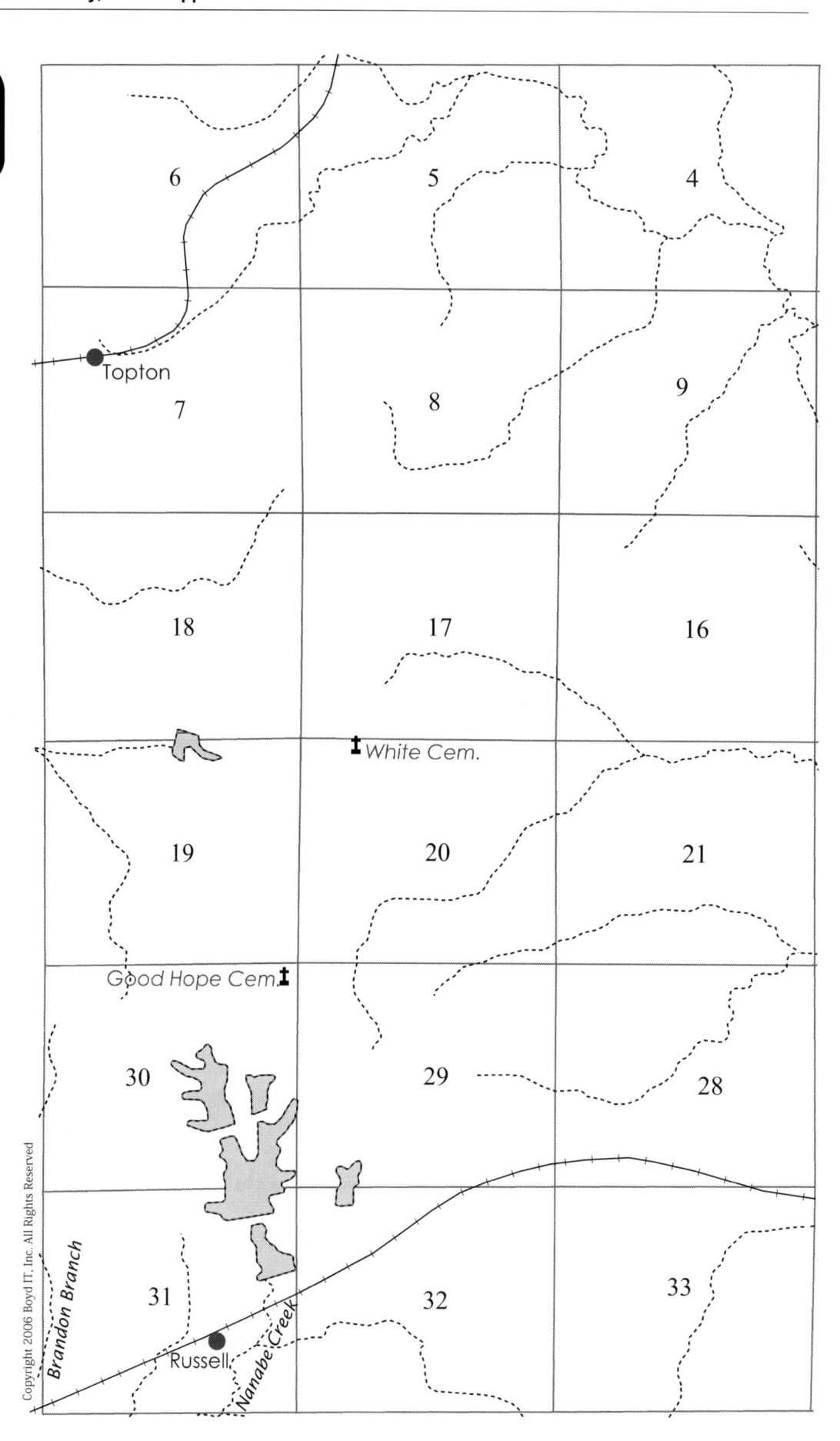

Township 7-N Range 17-E (Choctaw) - Map Group 10

Helpful Hints

1. This Map takes a different look at the same Congressional Township displayed in the preceding two maps. It presents features that can help you better envision the historical development of the area: a) Water-bodies (lakes & ponds), b) Water-courses (rivers, streams, etc.), c) Railroads, d) City/town center-points (where they were oftentimes located when first settled), and e) Cemeteries.

2. Using this "Historical" map in tandem with this Township's Patent Map and Road Map, may lead you to some interesting discoveries. You will often find roads, towns, cemeteries, and waterways are named after nearby landowners: sometimes those names will be the ones you are researching. See how many of these research gems you can find here in Lauderdale County.

Copyright 2006 Boyd IT, Inc. All Rights Reserved

Legend

— Section Lines
—+—+— Railroads
▨ Large Rivers & Bodies of Water
----- Streams/Creeks & Small Rivers
● Cities/Towns
† Cemeteries

Scale: Section = 1 mile X 1 mile
(there are some exceptions)

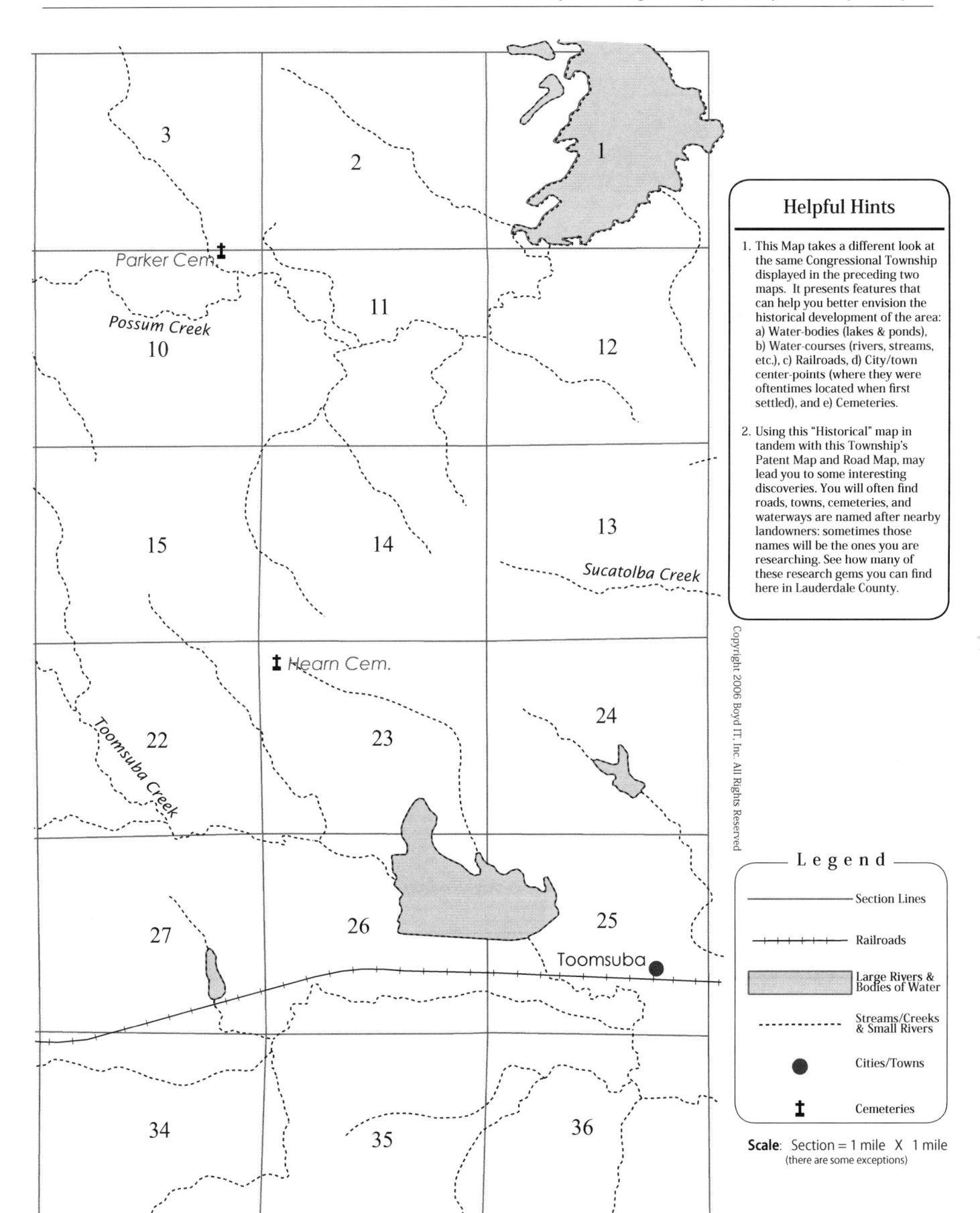

155

Map Group 11: Index to Land Patents
Township 7-North Range 18-East (Choctaw)

After you locate an individual in this Index, take note of the Section and Section Part then proceed to the Land Patent map on the pages immediately following. You should have no difficulty locating the corresponding parcel of land.

The "For More Info" Column will lead you to more information about the underlying Patents. See the *Legend* at right, and the "How to Use this Book" chapter, for more information.

```
LEGEND
         "For More Info . . ." column
A = Authority (Legislative Act, See Appendix "A")
B = Block or Lot (location in Section unknown)
C = Cancelled Patent
F = Fractional Section
G = Group (Multi-Patentee Patent, see Appendix "C")
V = Overlaps another Parcel
R = Re-Issued (Parcel patented more than once)

(A & G items require you to look in the Appendixes referred
to above. All other Letter-designations followed by a number
require you to locate line-items in this index that possess
the ID number found after the letter).
```

ID	Individual in Patent	Sec.	Sec. Part	Date Issued	Other Counties	For More Info . . .
2252	ABERNATHY, Martha	19	SWNE	1913-10-27		A4 G1
2191	ALEXANDER, Henry	28	W½NW	1846-09-01		A1 G5
2190	" "	21	E½NE	1848-09-01		A1
2189	" "	15	E½SW	1849-12-01		A1
2231	ALLEN, Joseph C	7	E½SE	1901-11-16		A4
2232	" "	7	SWSE	1901-11-16		A4
2204	BASTIN, James M	13	10	1849-12-01		A1 F
2205	" "	13	11	1849-12-01		A1 F
2206	" "	13	8	1849-12-01		A1 F
2192	BELL, Henry	7	E½NW	1906-06-30		A4 G20
2192	BELL, Mary	7	E½NW	1906-06-30		A4 G20
2267	BERDEAUX, Thomas	3	SWNW	1860-04-02		A1
2268	BOURDEAUX, Thomas D	10	E½NE	1846-09-01		A1
2269	" "	10	SENE	1849-12-01		A1 C
2270	" "	3	SESE	1849-12-01		A1
2271	" "	3	SWSE	1849-12-01		A1 C R2279
2278	BOURDEAUX, Thomas M	10	SWNE	1919-04-26		A1
2279	" "	3	SWSE	1919-04-26		A1 R2271
2256	BOZEMAN, Peter	31	SWSW	1859-11-10		A1
2257	BOZMAN, Peter H	31	E½SW	1846-09-01		A1
2258	" "	31	NWSW	1846-09-01		A1
2259	" "	32	W½NW	1846-09-01		A1
2147	BRAGG, Branch K	28	E½SE	1848-09-01		A1
2149	" "	26	E½SW	1859-05-02		A1 G29
2150	" "	26	NE	1859-05-02		A1 G29
2148	" "	34	SESW	1859-05-02		A1
2178	BROWER, Franklin P	14	SWSW	1846-09-01		A1
2179	" "	22	S½SE	1859-05-02		A1
2149	" "	26	E½SW	1859-05-02		A1 G29
2150	" "	26	NE	1859-05-02		A1 G29
2180	BROWN, Franklin P	22	E½NW	1846-09-01		A1
2181	" "	22	NESE	1846-09-01		A1
2244	BROWN, Little B	30	NWNE	1846-09-01		A1
2245	BROWN, Littleberry	30	E½SE	1846-09-01		A1
2247	" "	30	W½SE	1846-09-01		A1
2248	" "	31	E½NW	1846-09-01		A1
2249	" "	31	W½NW	1846-09-01		A1
2246	" "	30	SWSW	1849-12-01		A1
2233	BRUNSON, Lawrence M	25	7	1850/06/01		A2 C
2234	" "	25	8	1850/06/01		A2 C
2209	BUSTIN, James M	24	4	1859-05-02		A1
2207	" "	24	2	1859-06-01		A1
2208	" "	24	3	1859-06-01		A1
2140	BYNUM, Benjamin	29	SE	1841-01-05		A1 G36
2140	BYNUM, William	29	SE	1841-01-05		A1 G36
2230	CAMPBELL, John Thomas	7	NWNW	1924-06-18		A1

Township 7-N Range 18-E (Choctaw)   -   Map Group 11

ID	Individual in Patent	Sec.	Sec. Part	Date Issued	Other Counties	For More Info . . .
2254	CARTER, Norval R	17	SENE	1860-04-02		A1
2255	CARTER, Norvell R	17	SESE	1860-10-01		A1
2236	CLANTON, Lewis D	10	NWNW	1852/05/01		A2
2237	" "	4	SWNW	1852/05/01		A2
2238	" "	4	SWSE	1852/05/01		A2
2239	" "	8	NWNE	1852/05/01		A2
2240	" "	9	SENW	1852/05/01		A2
2241	" "	9	SWNE	1852/05/01		A2
2243	" "	9	W½SE	1852/05/01		A2
2242	" "	9	SWNW	1859-05-02		A1
2140	CLINTON, George W	29	SE	1841-01-05		A1 G36
2184	" "	29	W½SW	1841-01-05		A1 G44
2183	" "	29	SWNW	1846-09-01		A1
2199	CREWS, James D	14	W½NE	1846-09-01		A1
2198	" "	14	SENE	1848-09-01		A1
2282	CREWS, Thomas R	11	E½NW	1846-09-01		A1
2283	" "	11	W½NW	1846-09-01		A1
2284	" "	2	SW	1846-09-01		A1
2289	DEWITT, William	21	NESE	1849-12-01		A1
2197	DUNAGIN, James A	17	W½SE	1908-07-23		A4
2290	EAKENS, William	3	NWNW	1859-05-02		A1
2291	EAKIN, William	3	E½NW	1859-05-02		A1
2169	ELLIS, Elisha G	5	NESE	1875-07-01		A4
2288	ELLIS, Wiley	5	SESE	1849-12-01		A1
2299	ELLIS, William J	5	NESW	1879-05-06		A4
2300	" "	5	NWSE	1879-05-06		A4
2252	EVANS, Martha	19	SWNE	1913-10-27		A4 G1
2144	EZELL, Benjamin	9	E½SE	1846-09-01		A1
2145	" "	9	E½SW	1846-09-01		A1
2146	" "	9	W½SW	1846-09-01		A1
2139	FALLIN, Andrew O	22	E½NE	1846-09-01		A1
2272	GRAHAM, Thomas	21	NWNE	1849-12-01		A1
2275	" "	22	SWNW	1849-12-01		A1
2273	" "	21	NWSE	1859-11-10		A1
2274	" "	21	SWNE	1859-11-10		A1
2162	HALL, David	11	E½SW	1846-09-01		A1
2163	" "	11	SE	1846-09-01		A1
2164	" "	11	W½NE	1846-09-01		A1
2157	HAMSWORTH, Christopher C	15	E½NE	1914-10-27		A3
2253	HUNTER, Nicholas	33	E½SW	1846-09-01		A1
2292	JEMISON, William H	1	10	1846-09-01		A1 F
2293	" "	1	11	1846-09-01		A1 F
2294	" "	1	12	1846-09-01		A1 F
2295	" "	1	13	1846-09-01		A1 F
2296	" "	1	14	1846-09-01		A1 F
2297	" "	1	15	1846-09-01		A1 F
2298	" "	1	9	1846-09-01		A1 F
2167	KELLY, Duncan	29	NWNE	1849-12-01		A1 R2168
2168	" "	29	NWNE	1849-12-01		A1 R2167
2165	" "	21	E½SW	1859-11-10		A1
2166	" "	21	NWSW	1859-11-10		A1
2285	KELLY, Viney	19	NENW	1895-12-14		A4
2286	" "	19	NWNE	1895-12-14		A4
2266	LEWIS, Stephen	32	E½NE	1841-01-05		A1
2225	LINZEY, John	5	NWNE	1850/06/01		A2
2226	" "	5	SENE	1850/06/01		A2
2260	MARSH, Peter	26	W½SW	1846-09-01		A1 G139
2261	" "	35	E½NW	1846-09-01		A1 G141
2228	MARTIN, John P	28	SW	1841-01-05		A1 G143
2193	MCCOMB, Hugh H	22	W½NE	1846-09-01		A1
2194	" "	23	W½NW	1846-09-01		A1
2229	MCLARRIN, John R	15	W½NE	1860-04-02		A1
2151	MILLER, Charles G	12	1	1846-09-01		A1 F
2153	" "	12	12	1846-09-01		A1 F
2154	" "	12	2	1846-09-01		A1 F
2155	" "	12	5	1846-09-01		A1 F
2156	" "	12	7	1846-09-01		A1 F
2152	" "	12	11	1848-09-01		A1 F
2276	MILLER, Thomas J	14	SESE	1859-05-02		A1
2277	" "	14	W½SE	1859-05-02		A1
2306	MILLER, William L	11	W½SW	1918-06-05		A4
2227	MIXON, John	3	E½SW	1852/05/01		A2
2191	NUTT, James M	28	W½NW	1846-09-01		A1 G5

157

Family Maps of Lauderdale County, Mississippi

ID	Individual in Patent	Sec.	Sec. Part	Date Issued	Other Counties	For More Info...
2177	ODOM, Ephraim	31	NE	1846-09-01		A1
2200	PATON, James F	1	1	1846-09-01		A1 F
2201	" "	1	2	1846-09-01		A1 F
2202	" "	1	7	1846-09-01		A1 F
2203	" "	1	8	1846-09-01		A1 F
2307	PAYNE, William P	2	SWNW	1849-12-01		A1
2170	PEAVY, Elizabeth	13	12	1849-12-01		A1 F
2171	" "	13	2	1849-12-01		A1 F
2172	" "	13	3	1849-12-01		A1 F
2173	" "	13	7	1849-12-01		A1 F
2280	PHILIPS, Thomas	28	E½NW	1846-09-01		A1
2281	" "	28	W½NE	1846-09-01		A1
2228	PHILLIPS, Allinson	28	SW	1841-01-05		A1 G143
2160	PHILLIPS, Cyndrilia	21	SWSW	1849-12-01		A1
2250	PHILLIPS, Lorenzo J	21	NWNW	1860-04-02		A1
2251	" "	21	SWNW	1860-04-02		A1
2141	PRICE, Benjamin D	17	NESW	1860-04-02		A1
2142	" "	17	NWNW	1860-04-02		A1
2185	PRICE, George W	17	E½NW	1859-11-10		A1
2186	" "	17	N½NE	1859-11-10		A1
2187	" "	17	NWSW	1860-04-02		A1
2188	" "	17	SWNE	1860-04-02		A1
2235	PRICE, Leonidas L	17	S½SW	1882-12-30		A4
2195	PRINGLE, Isham K	32	SWSE	1852/05/01		A2
2196	RAMSEY, Jacob	19	S½SE	1901-10-01		A4
2158	REA, Constantine	21	S½SE	1848-09-01		A1
2143	RICE, Benjamin D	17	NESE	1859-11-10		A1
2134	ROGERS, Alexander T	32	E½NW	1846-09-01		A1
2135	" "	32	W½NE	1846-09-01		A1
2159	SHOWS, Cornelius	30	E½SW	1846-09-01		A1
2136	SMITH, Allen	29	NWNW	1849-12-01		A1 R2137
2137	" "	29	NWNW	1849-12-01		A1 R2136
2138	" "	29	SWNE	1849-12-01		A1
2182	SMITH, George	19	S½NW	1913-02-14		A4
2222	SMITH, John J	29	E½NW	1852/05/01		A2
2223	" "	3	NWSE	1852/05/01		A2
2224	" "	3	W½NE	1852/09/10		A2
2228	SPINKS, Rolley	28	SW	1841-01-05		A1 G143
2263	" "	29	E½NE	1841-01-05		A1 G171
2174	STEWART, Elizabeth	29	E½SW	1846-09-01		A1
2263	TAYLOR, Richard	29	E½NE	1841-01-05		A1 G171
2260	TEMPLETON, George	26	W½SW	1846-09-01		A1 G139
2161	THORNTON, Daniel	3	W½SW	1906-06-21		A4
2184	TOUCHSTONE, Daniel	29	W½SW	1841-01-05		A1 G44
2261	WALKER, James	35	E½NW	1846-09-01		A1 G141
2214	WALKER, Joel P	6	NESW	1846-09-01		A1
2217	" "	6	W½NE	1850/06/01		A2
2215	" "	6	NWSW	1852/05/01		A2
2216	" "	6	SWNW	1852/05/01		A2
2220	" "	9	NWNW	1859-05-02		A1
2213	" "	5	W½SW	1859-11-10		A1
2218	" "	7	E½NE	1859-11-10		A1
2211	" "	5	SESW	1862/01/20		A2
2212	" "	5	SWSE	1862/01/20		A2
2219	" "	7	W½NE	1862/01/20		A2
2262	WALKER, Richard T	5	NENE	1859-11-10		A1
2264	WALKER, Samuel	5	NW	1859-05-02		A1
2287	WALKER, Walter J	5	SWNE	1859-11-10		A1
2221	WATSON, John B	17	SWNW	1859-06-01		A1
2265	WILSON, Sheerwood	27	E½NE	1846-09-01		A1
2175	WITHERSPOON, Ellen	7	NWSW	1904-11-15		A4 G182
2176	" "	7	SWNW	1904-11-15		A4 G182
2175	WITHERSPOON, Wat	7	NWSW	1904-11-15		A4 G182
2176	" "	7	SWNW	1904-11-15		A4 G182
2210	WOOD, James	15	W½SW	1846-09-01		A1
2132	WRIGHT, Aleck	19	NESE	1905-12-30		A4 G183
2133	" "	19	SENE	1905-12-30		A4 G183
2132	WRIGHT, Lucinda	19	NESE	1905-12-30		A4 G183
2133	" "	19	SENE	1905-12-30		A4 G183
2302	WRIGHT, William J	11	E½NE	1846-09-01		A1
2304	" "	2	NE	1846-09-01		A1
2305	" "	2	SE	1846-09-01		A1
2301	" "	10	NWNE	1849-12-01		A1

Township 7-N Range 18-E (Choctaw)   -   Map Group 11

ID	Individual in Patent	Sec.	Sec. Part	Date Issued	Other Counties	For More Info . . .
2303	WRIGHT, William J (Cont'd)	2	E½NW	1849-12-01		A1

Family Maps of Lauderdale County, Mississippi

Patent Map

T7-N R18-E
Choctaw Meridian

Map Group 11

Township Statistics

Parcels Mapped	:	176
Number of Patents	:	138
Number of Individuals	:	94
Patentees Identified	:	87
Number of Surnames	:	69
Multi-Patentee Parcels	:	15
Oldest Patent Date	:	1/5/1841
Most Recent Patent	:	6/18/1924
Block/Lot Parcels	:	29
Parcels Re-Issued	:	3
Parcels that Overlap	:	0
Cities and Towns	:	2
Cemeteries	:	4

Copyright 2006 Boyd IT, Inc. All Rights Reserved

160

Township 7-N Range 18-E (Choctaw) - Map Group 11

Section 3
- EAKENS William 1859
- BERDEAUX Thomas 1860
- EAKIN William 1859
- SMITH John J 1852
- THORNTON Daniel 1906
- SMITH John J 1852
- MIXON John 1852
- BOURDEAUX Thomas M 1919
- BOURDEAUX Thomas D 1849
- BOURDEAUX Thomas D 1849

Section 2
- PAYNE William P 1849
- WRIGHT William J 1849
- WRIGHT William J 1846
- CREWS Thomas R 1846
- WRIGHT William J 1846

Section 1 — Lots-Sec. 1
Lot	Patentee	Year
1	PATON, James F	1846
2	PATON, James F	1846
7	PATON, James F	1846
8	PATON, James F	1846
9	JEMISON, William H	1846
10	JEMISON, William H	1846
11	JEMISON, William H	1846
12	JEMISON, William H	1846
13	JEMISON, William H	1846
14	JEMISON, William H	1846
15	JEMISON, William H	1846

Section 10
- CLANTON Lewis D 1852
- WRIGHT William J 1849
- BOURDEAUX Thomas M 1919
- BOURDEAUX Thomas D 1846

Section 11
- CREWS Thomas R 1846
- CREWS Thomas R 1846
- HALL David 1846
- WRIGHT William J 1846
- MILLER William L 1918
- HALL David 1846
- HALL David 1846

Section 12 — Lots-Sec. 12
Lot	Patentee	Year
1	MILLER, Charles G	1846
2	MILLER, Charles G	1846
5	MILLER, Charles G	1846
7	MILLER, Charles G	1846
11	MILLER, Charles G	1848
12	MILLER, Charles G	1846

Section 15
- MCLARRIN John R 1860
- HAMSWORTH Christopher C 1914
- WOOD James 1846
- ALEXANDER Henry 1849

Section 14
- CREWS James D 1846
- CREWS James D 1848
- MILLER Thomas J 1859
- BROWER Franklin P 1846
- MILLER Thomas J 1859

Section 13 — Lots-Sec. 13
Lot	Patentee	Year
2	PEAVY, Elizabeth	1849
3	PEAVY, Elizabeth	1849
7	PEAVY, Elizabeth	1849
8	BASTIN, James M	1849
10	BASTIN, James M	1849
11	BASTIN, James M	1849
12	PEAVY, Elizabeth	1849

Section 22
- MCCOMB Hugh H 1846
- GRAHAM Thomas 1849
- BROWN Franklin P 1846
- FALLIN Andrew O 1846
- BROWN Franklin P 1846
- BROWER Franklin P 1859

Section 23
- MCCOMB Hugh H 1846

Section 24 — Lots-Sec. 24
Lot	Patentee	Year
2	BUSTIN, James M	1859
3	BUSTIN, James M	1859
4	BUSTIN, James M	1859

Section 27
- WILSON Sheerwood 1846

Section 26
- BRAGG [29] Branch K 1859
- MARSH [139] Peter 1846
- BRAGG [29] Branch K 1859

Section 25 — Lots-Sec. 25
Lot	Patentee	Year
7	BRUNSON, Lawrence M	1850
8	BRUNSON, Lawrence M	1850

Section 34
- BRAGG Branch K 1859

Section 35
- MARSH [141] Peter 1846

Section 36

Helpful Hints

1. This Map's INDEX can be found on the preceding pages.
2. Refer to Map "C" to see where this Township lies within Lauderdale County, Mississippi.
3. Numbers within square brackets [ ] denote a multi-patentee land parcel (multi-owner). Refer to Appendix "C" for a full list of members in this group.
4. Areas that look to be crowded with Patentees usually indicate multiple sales of the same parcel (Re-issues) or Overlapping parcels. See this Township's Index for an explanation of these and other circumstances that might explain "odd" groupings of Patentees on this map.

Copyright 2006 Boyd IT, Inc. All Rights Reserved

Legend
- Patent Boundary
- Section Boundary
- No Patents Found (or Outside County)
- 1., 2., 3., ... Lot Numbers (when beside a name)
- [ ] Group Number (see Appendix "C")

Scale: Section = 1 mile X 1 mile (generally, with some exceptions)

Family Maps of Lauderdale County, Mississippi

Road Map
T7-N R18-E
Choctaw Meridian
Map Group 11

Cities & Towns
Kewanee
Smith

Cemeteries
Bethel Cemetery
Kelly Cemetery
Pack Cemetery
Payne Cemetery

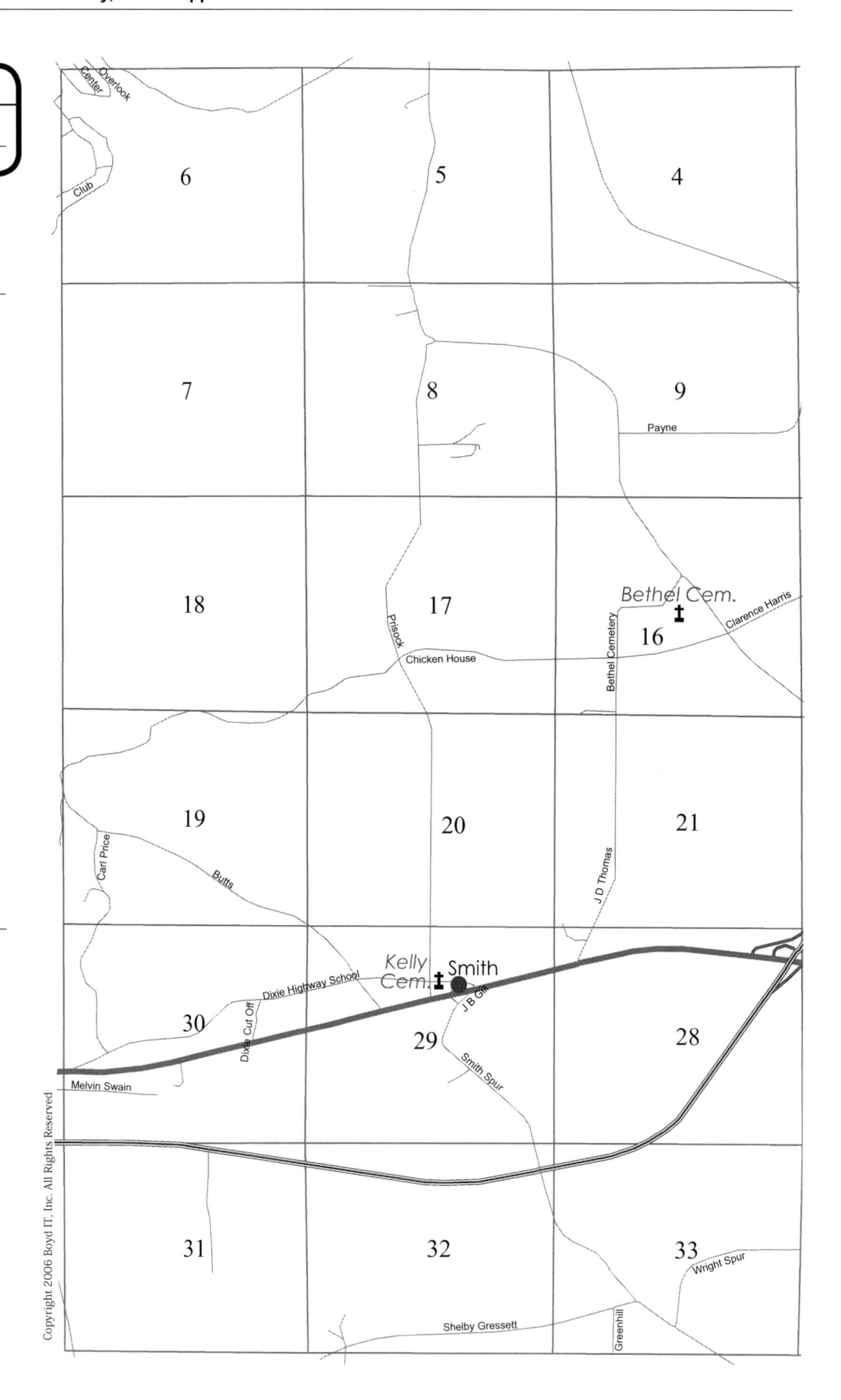

Township 7-N Range 18-E (Choctaw) - Map Group 11

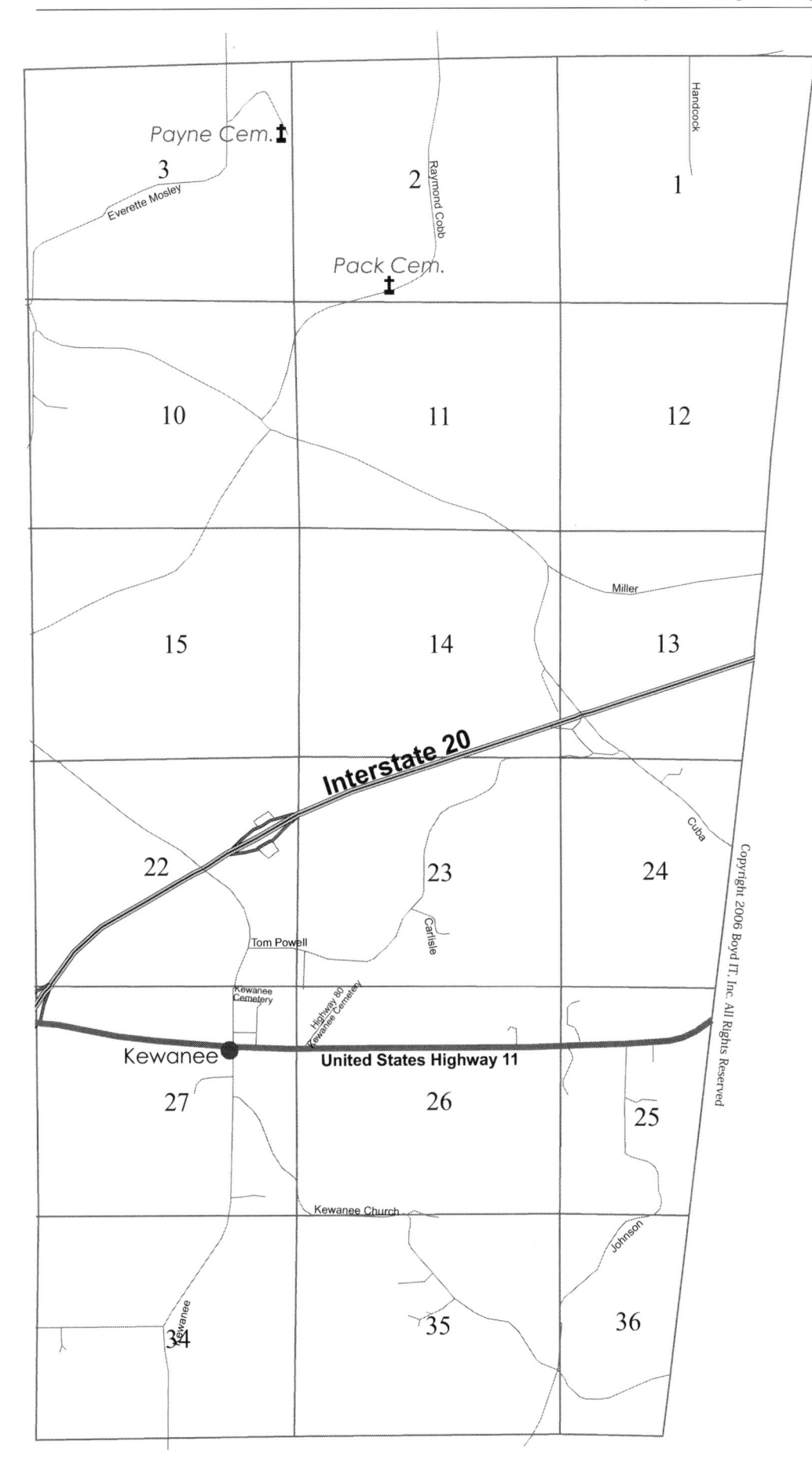

Helpful Hints

1. This road map has a number of uses, but primarily it is to help you: a) find the present location of land owned by your ancestors (at least the general area), b) find cemeteries and city-centers, and c) estimate the route/roads used by Census-takers & tax-assessors.

2. If you plan to travel to Lauderdale County to locate cemeteries or land parcels, please pick up a modern travel map for the area before you do. Mapping old land parcels on modern maps is not as exact a science as you might think. Just the slightest variations in public land survey coordinates, estimates of parcel boundaries, or road-map deviations can greatly alter a map's representation of how a road either does or doesn't cross a particular parcel of land.

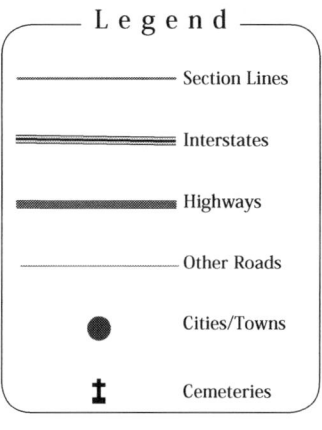

Legend
- Section Lines
- Interstates
- Highways
- Other Roads
- ● Cities/Towns
- ✝ Cemeteries

Scale: Section = 1 mile X 1 mile
(generally, with some exceptions)

163

Family Maps of Lauderdale County, Mississippi

Historical Map
T7-N R18-E
Choctaw Meridian
Map Group 11

Cities & Towns
Kewanee
Smith

Cemeteries
Bethel Cemetery
Kelly Cemetery
Pack Cemetery
Payne Cemetery

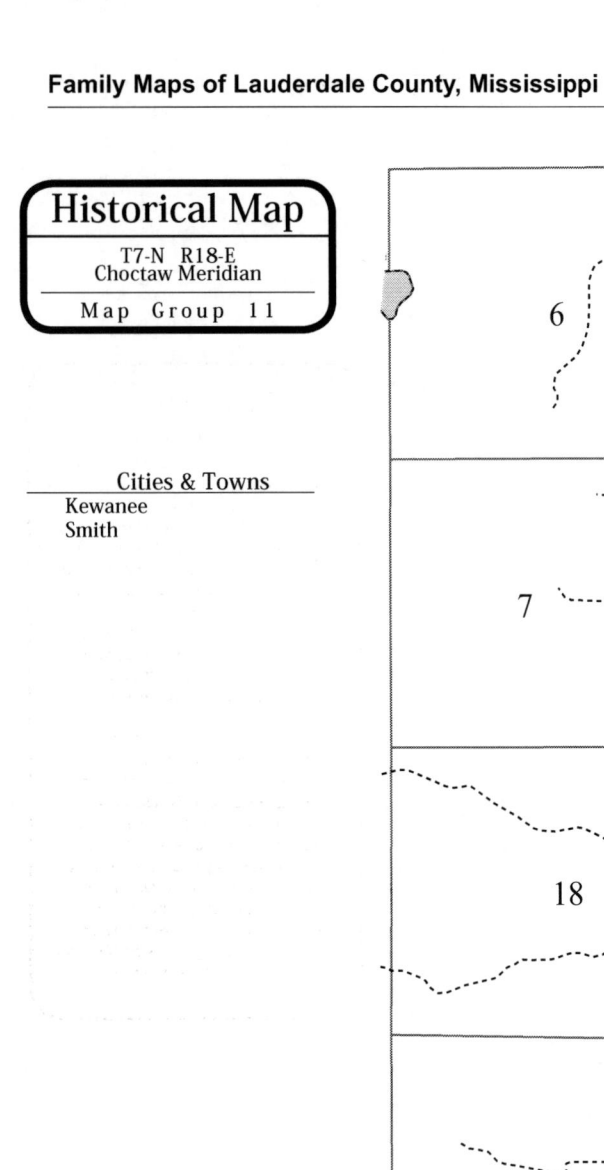

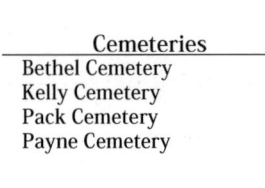

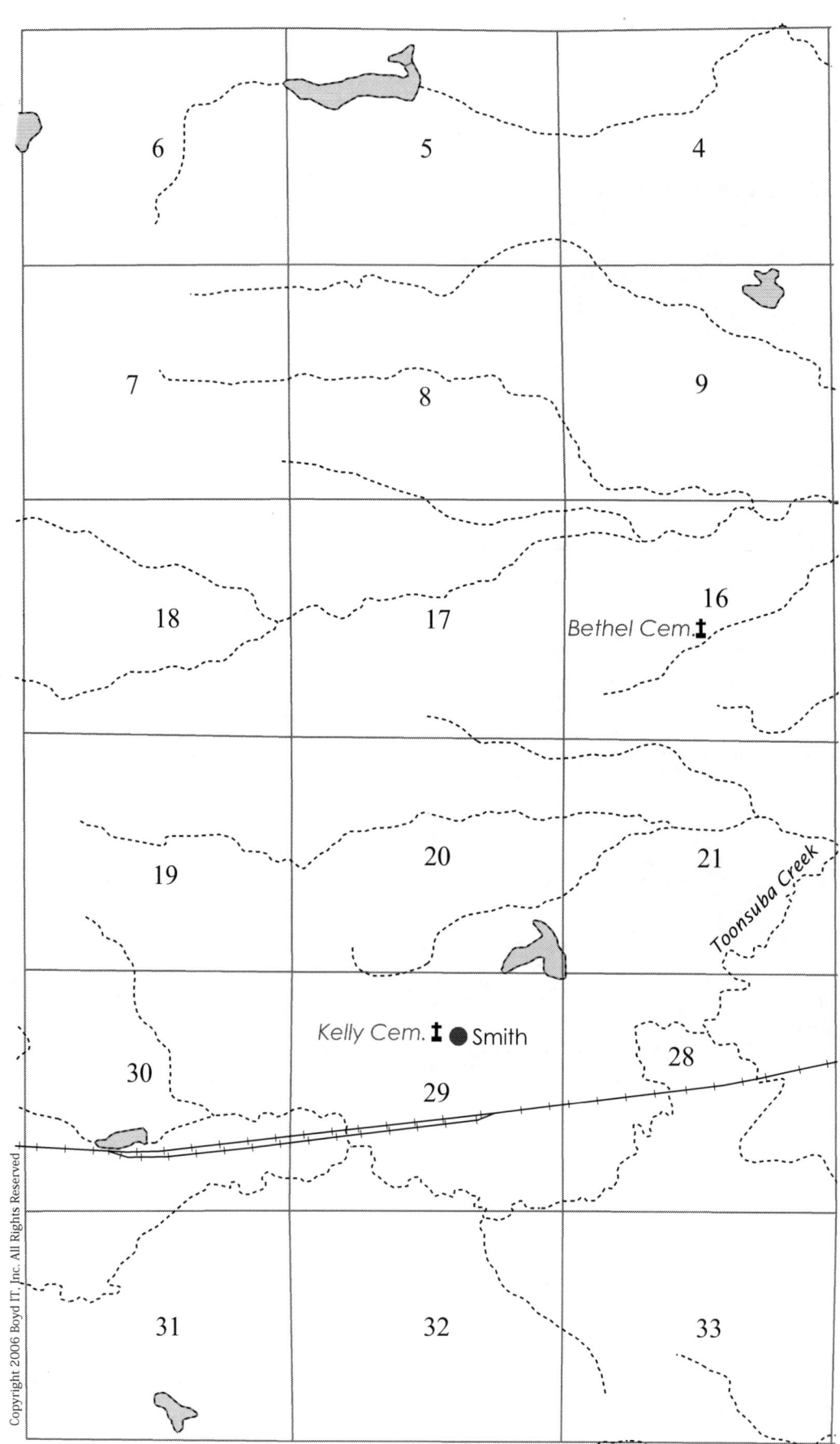

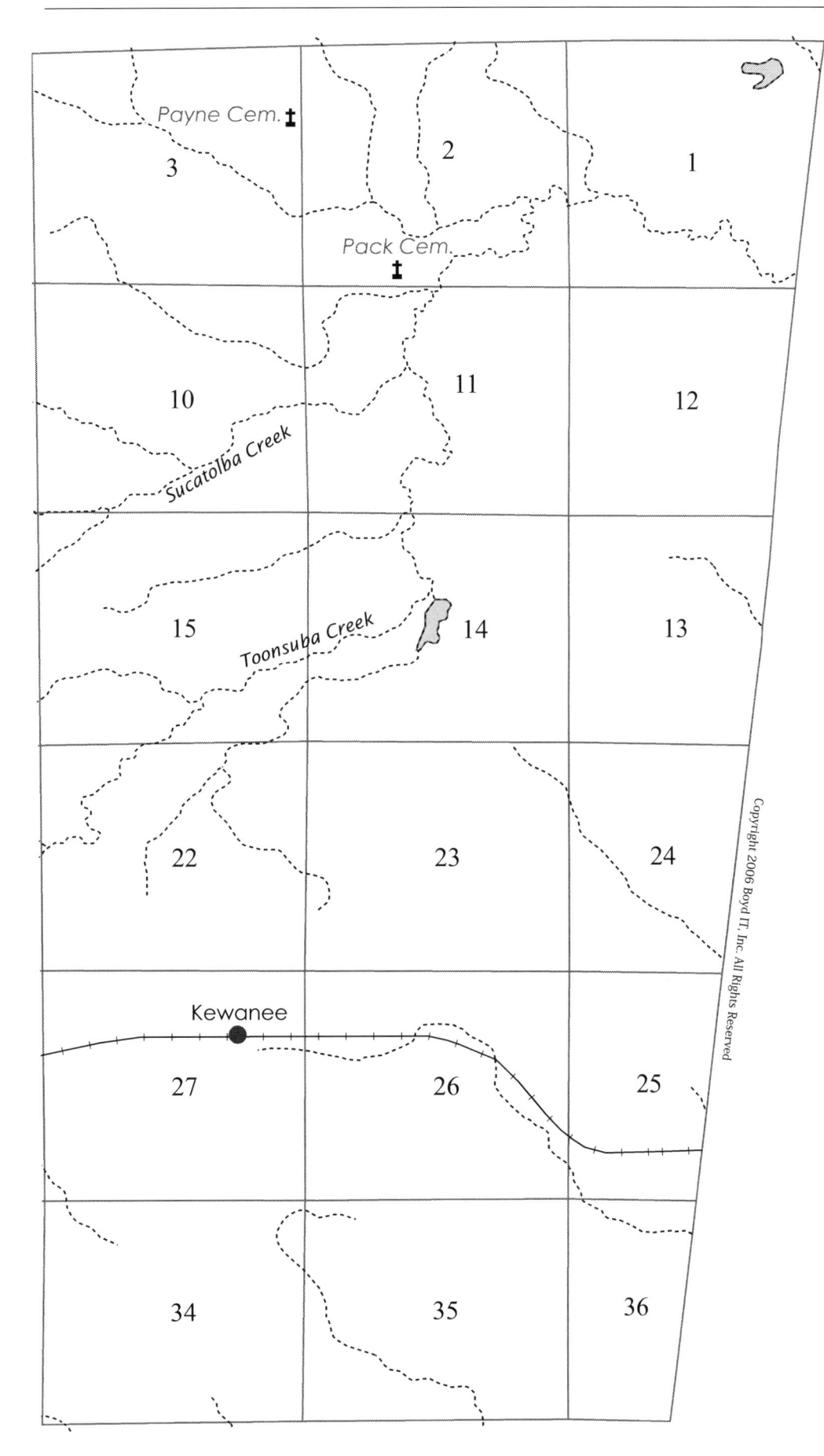

Township 7-N Range 18-E (Choctaw) - Map Group 11

Helpful Hints

1. This Map takes a different look at the same Congressional Township displayed in the preceding two maps. It presents features that can help you better envision the historical development of the area: a) Water-bodies (lakes & ponds), b) Water-courses (rivers, streams, etc.), c) Railroads, d) City/town center-points (where they were oftentimes located when first settled), and e) Cemeteries.

2. Using this "Historical" map in tandem with this Township's Patent Map and Road Map, may lead you to some interesting discoveries. You will often find roads, towns, cemeteries, and waterways are named after nearby landowners: sometimes those names will be the ones you are researching. See how many of these research gems you can find here in Lauderdale County.

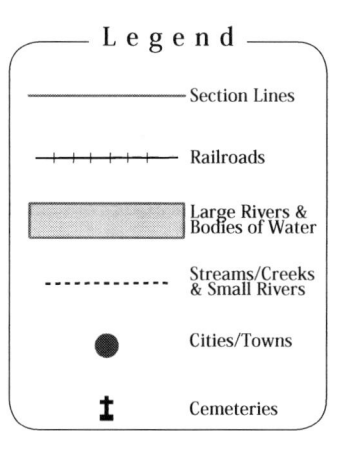

Legend

— Section Lines
+++++ Railroads
▨ Large Rivers & Bodies of Water
----- Streams/Creeks & Small Rivers
● Cities/Towns
✝ Cemeteries

Scale: Section = 1 mile X 1 mile
(there are some exceptions)

Features on map: Payne Cem., Pack Cem., Sucatolba Creek, Toonsuba Creek, Kewanee

Sections: 3, 2, 1, 10, 11, 12, 15, 14, 13, 22, 23, 24, 27, 26, 25, 34, 35, 36

Copyright 2006 Boyd IT, Inc. All Rights Reserved

165

Family Maps of Lauderdale County, Mississippi

Map Group 12: Index to Land Patents
Township 6-North Range 14-East (Choctaw)

After you locate an individual in this Index, take note of the Section and Section Part then proceed to the Land Patent map on the pages immediately following. You should have no difficulty locating the corresponding parcel of land.

The "For More Info" Column will lead you to more information about the underlying Patents. See the *Legend* at right, and the "How to Use this Book" chapter, for more information.

```
                    LEGEND
            "For More Info . . ." column
A = Authority (Legislative Act, See Appendix "A")
B = Block or Lot (location in Section unknown)
C = Cancelled Patent
F = Fractional Section
G = Group (Multi-Patentee Patent, see Appendix "C")
V = Overlaps another Parcel
R = Re-Issued (Parcel patented more than once)

(A & G items require you to look in the Appendixes referred
to above. All other Letter-designations followed by a number
require you to locate line-items in this index that possess
the ID number found after the letter).
```

ID	Individual in Patent	Sec.	Sec. Part	Date Issued	Other Counties	For More Info . . .
2362	ADAMS, Spencer	7	NWNW	1841-01-05		A1
2314	BOUNDS, Daniel	13	E½SW	1891-05-20		A4
2315	" "	13	W½SE	1891-05-20		A4
2354	BOUNDS, Richard D	13	E½NW	1891-05-20		A4
2355	" "	13	W½NE	1891-05-20		A4
2311	BROWN, Bill	11	E½SE	1899-04-17		A4
2320	COOLEY, Edward Fedory	1	E½SE	1923-08-20		A4
2384	COON, Willis	23	NWSE	1882-03-30		A4
2326	COX, Henry E	23	NESE	1860-10-01		A1 G54
2334	CURRIE, Jane C	23	N½SW	1895-12-14		A4 G56
2373	CURRIE, William	27	E½NE	1882-08-03		A4
2334	" "	23	N½SW	1895-12-14		A4 G56
2317	DEAR, Darling	18	NENE	1859-11-10		A1
2318	" "	18	NWNE	1859-11-10		A1
2329	DRUMOND, James	27	SESW	1892-03-23		A4
2330	" "	27	W½SE	1892-03-23		A4
2359	FOWLER, Samuel	31	E½NE	1841-01-05		A1
2360	" "	33	NE	1841-01-05		A1
2361	" "	33	W½SE	1841-01-05		A1
2337	GADDIS, Jimmie	25	NENE	1916-11-28		A4
2341	GADDIS, John	23	N½NE	1913-08-19		A4
2340	GARDNER, John G	22	S½SE	1848-09-01		A1
2352	GILBERT, Nathan	8	E½NE	1841-01-05		A1
2353	GOODWIN, Peter J	27	E½SE	1892-04-16		A1
2374	GRESSETT, William	20	SESW	1859-11-10		A1
2375	GRISSETT, William	17	SWSE	1848-09-01		A1
2376	" "	20	NENE	1848-09-01		A1
2377	" "	30	NENE	1859-05-02		A1
2312	HOUSTON, Columbus W	13	E½NE	0012-00-00		A4
2313	" "	13	E½SE	0012-00-00		A4
2336	HUEY, Jesse	33	NENW	1846-09-01		A1
2356	JONES, Richard	4	E½NW	1841-01-05		A1
2357	" "	4	SW	1841-01-05		A1
2358	" "	4	W½NE	1841-01-05		A1
2370	KEATH, William B	14	SWNW	1848-09-01		A1
2308	KEITH, Allen J	2	NESW	1859-05-02		A1
2309	" "	2	W½SE	1859-05-02		A1
2371	KEITH, William B	14	NENW	1896-06-06		A1
2372	" "	14	NWSW	1896-06-06		A1
2379	LACY, William H	32	E½SW	1848-09-01		A1
2380	" "	32	SWNE	1859-05-02		A1
2378	" "	32	E½NE	1859-11-10		A1
2364	LATTNEY, Thomas	23	S½SE	1882-03-30		A4
2316	MABRY, Daniel	8	N½SE	1859-05-02		A1
2331	MAXEY, James	35	SE	1895-08-30		A4
2344	MAXWELL, Joshua	28	SESE	1859-05-02		A1

Township 6-N Range 14-E (Choctaw)   -   Map Group 12

ID	Individual in Patent	Sec.	Sec. Part	Date Issued	Other Counties	For More Info . . .
2345	MAXWELL, Joshua (Cont'd)	34	W½NW	1859-11-10		A1
2368	MAYBERRY, Walter	17	NESW	1848-09-01		A1
2369	" "	17	NWNE	1848-09-01		A1
2332	MCLAREN, James	32	W½NW	1841-01-05		A1
2322	MOFFETT, Gabriel	32	E½SE	1841-01-05		A1
2323	" "	33	E½SE	1841-01-05		A1
2324	" "	34	NWSW	1841-01-05		A1
2321	NEW, Elijah	35	SW	1892-08-08		A4
2342	RUSH, John H	7	NWSE	1848-09-01		A1
2385	SAVELL, Willoughby	22	NENE	1848-09-01		A1
2343	SIMPSON, John	35	NE	1892-06-06		A1
2346	SNOWDEN, Ludie B	1	SESW	1916-05-31		A4
2347	" "	1	W½SE	1916-05-31		A4
2335	STALLINGS, Jeff	35	NW	1892-04-16		A1
2310	STEDIVANT, Allen	2	NENW	1859-06-01		A1
2381	STEPHENS, William L	2	SENW	1859-05-02		A1
2382	" "	2	SWNE	1859-05-02		A1
2348	SUMMERS, Lydia	25	N½SE	1909-01-25		A3 G175
2326	SWILLEY, James J	23	NESE	1860-10-01		A1 G54
2325	TILLMAN, Gideon	25	SESE	1859-11-10		A1
2327	TILLMAN, Henry	25	NESW	1894-11-17		A1
2328	" "	25	SENE	1894-11-17		A1
2348	" "	25	N½SE	1909-01-25		A3 G175
2349	TILLMAN, Martha	25	SESW	1892-06-15		A4
2350	" "	25	SWSE	1892-06-15		A4
2351	" "	25	W½SW	1892-06-15		A4
2319	TWILLEY, Ed	23	S½NE	1912-05-16		A4
2363	WADE, Susanna	23	N½NW	1894-12-17		A4
2333	WALKER, James P	23	SWSW	1860-04-02		A1
2338	WALKER, Joel P	23	SESW	1860-04-02		A1
2339	" "	27	NWNE	1860-10-01		A1
2365	WALLACE, Thomas	23	S½NW	1911-04-05		A4
2366	WILSON, Thomas	25	E½NW	1892-09-02		A4
2367	" "	25	W½NE	1892-09-02		A4
2383	YARBROUGH, William	25	W½NW	1896-04-23		A4

Family Maps of Lauderdale County, Mississippi

Patent Map

**T6-N R14-E
Choctaw Meridian**

Map Group 12

Township Statistics

Parcels Mapped	:	78
Number of Patents	:	62
Number of Individuals	:	53
Patentees Identified	:	52
Number of Surnames	:	46
Multi-Patentee Parcels	:	3
Oldest Patent Date	:	1/5/1841
Most Recent Patent	:	8/20/1923
Block/Lot Parcels	:	0
Parcels Re-Issued	:	0
Parcels that Overlap	:	0
Cities and Towns	:	2
Cemeteries	:	1

Section 6

Section 5

Section 4
- JONES, Richard 1841
- JONES, Richard 1841
- JONES, Richard 1841

Section 7
- ADAMS, Spencer 1841
- RUSH, John H 1848

Section 8
- GILBERT, Nathan 1841
- MABRY, Daniel 1859

Section 9

Section 18
- DEAR, Darling 1859
- DEAR, Darling 1859

Section 17
- MAYBERRY, Walter 1848
- MAYBERRY, Walter 1848
- GRISSETT, William 1848

Section 16

Section 19

Section 20
- GRISSETT, William 1848
- GRESSETT, William 1859

Section 21

Section 30
- GRISSETT, William 1859

Section 29

Section 28

Section 31
- FOWLER, Samuel 1841

Section 32
- MCLAREN, James 1841
- LACY, William H 1859
- LACY, William H 1859
- LACY, William H 1848
- MOFFETT, Gabriel 1841

Section 33
- HUEY, Jesse 1846
- FOWLER, Samuel 1841
- MAXWELL, Joshua 1859
- FOWLER, Samuel 1841
- MOFFETT, Gabriel 1841

Copyright 2006 Boyd IT, Inc. All Rights Reserved

168

Township 6-N Range 14-E (Choctaw) - Map Group 12

Section 1
- SNOWDEN, Ludie B — 1916
- SNOWDEN, Ludie B — 1916
- COOLEY, Edward Fedory — 1923

Section 2
- STEDIVANT, Allen — 1859
- STEPHENS, William L — 1859
- STEPHENS, William L — 1859
- KEITH, Allen J — 1859
- KEITH, Allen J — 1859

Section 11
- BROWN, Bill — 1899

Section 13
- BOUNDS, Richard D — 1891
- BOUNDS, Richard D — 1891
- HOUSTON, Columbus W — 0012
- BOUNDS, Daniel — 1891
- BOUNDS, Daniel — 1891
- HOUSTON, Columbus W — 0012

Section 14
- KEITH, William B — 1896
- KEATH, William B — 1848
- KEITH, William B — 1896

Section 23
- SAVELL, Willoughby — 1848
- WADE, Susanna — 1894
- GADDIS, John — 1913
- WALLACE, Thomas — 1911
- TWILLEY, Ed — 1912
- CURRIE [56], Jane C — 1895
- COON, Willis — 1882
- COX [54], Henry E — 1860
- GARDNER, John G — 1848
- WALKER, James P — 1860
- WALKER, Joel P — 1860
- LATTNEY, Thomas — 1882

Section 25
- YARBROUGH, William — 1896
- WILSON, Thomas — 1892
- GADDIS, Jimmie — 1916
- WILSON, Thomas — 1892
- TILLMAN, Henry — 1894
- TILLMAN, Martha — 1892
- TILLMAN, Henry — 1894
- SUMMERS [175], Lydia — 1909
- TILLMAN, Martha — 1892
- TILLMAN, Martha — 1892
- TILLMAN, Gideon — 1859

Section 27
- WALKER, Joel P — 1860
- CURRIE, William — 1882
- DRUMOND, James — 1892
- DRUMOND, James — 1892
- GOODWIN, Peter J — 1892

Section 34
- MAXWELL, Joshua — 1859
- MOFFETT, Gabriel — 1841

Section 35
- STALLINGS, Jeff — 1892
- SIMPSON, John — 1892
- NEW, Elijah — 1892
- MAXEY, James — 1895

Helpful Hints

1. This Map's INDEX can be found on the preceding pages.
2. Refer to Map "C" to see where this Township lies within Lauderdale County, Mississippi.
3. Numbers within square brackets [ ] denote a multi-patentee land parcel (multi-owner). Refer to Appendix "C" for a full list of members in this group.
4. Areas that look to be crowded with Patentees usually indicate multiple sales of the same parcel (Re-issues) or Overlapping parcels. See this Township's Index for an explanation of these and other circumstances that might explain "odd" groupings of Patentees on this map.

Copyright 2006 Boyd IT, Inc. All Rights Reserved

Legend
- ——— Patent Boundary
- ━━━ Section Boundary
- (shaded) No Patents Found (or Outside County)
- 1., 2., 3., ... Lot Numbers (when beside a name)
- [ ] Group Number (see Appendix "C")

Scale: Section = 1 mile X 1 mile (generally, with some exceptions)

169

Family Maps of Lauderdale County, Mississippi

Road Map

T6-N R14-E
Choctaw Meridian
Map Group 12

Cities & Towns
Graham
Meehan

Cemeteries
Tunnel Hill Cemetery

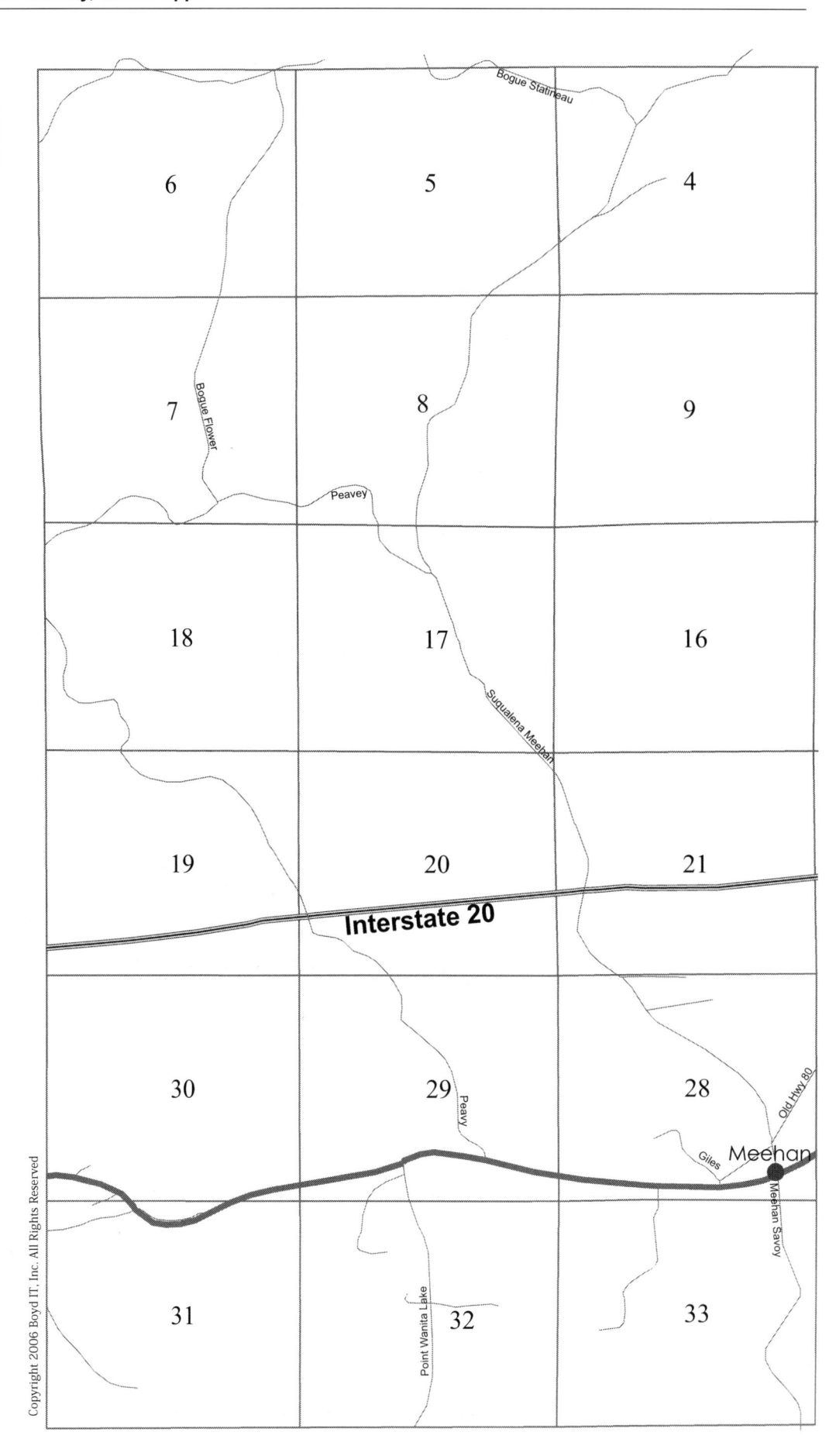

Township 6-N Range 14-E (Choctaw) - Map Group 12

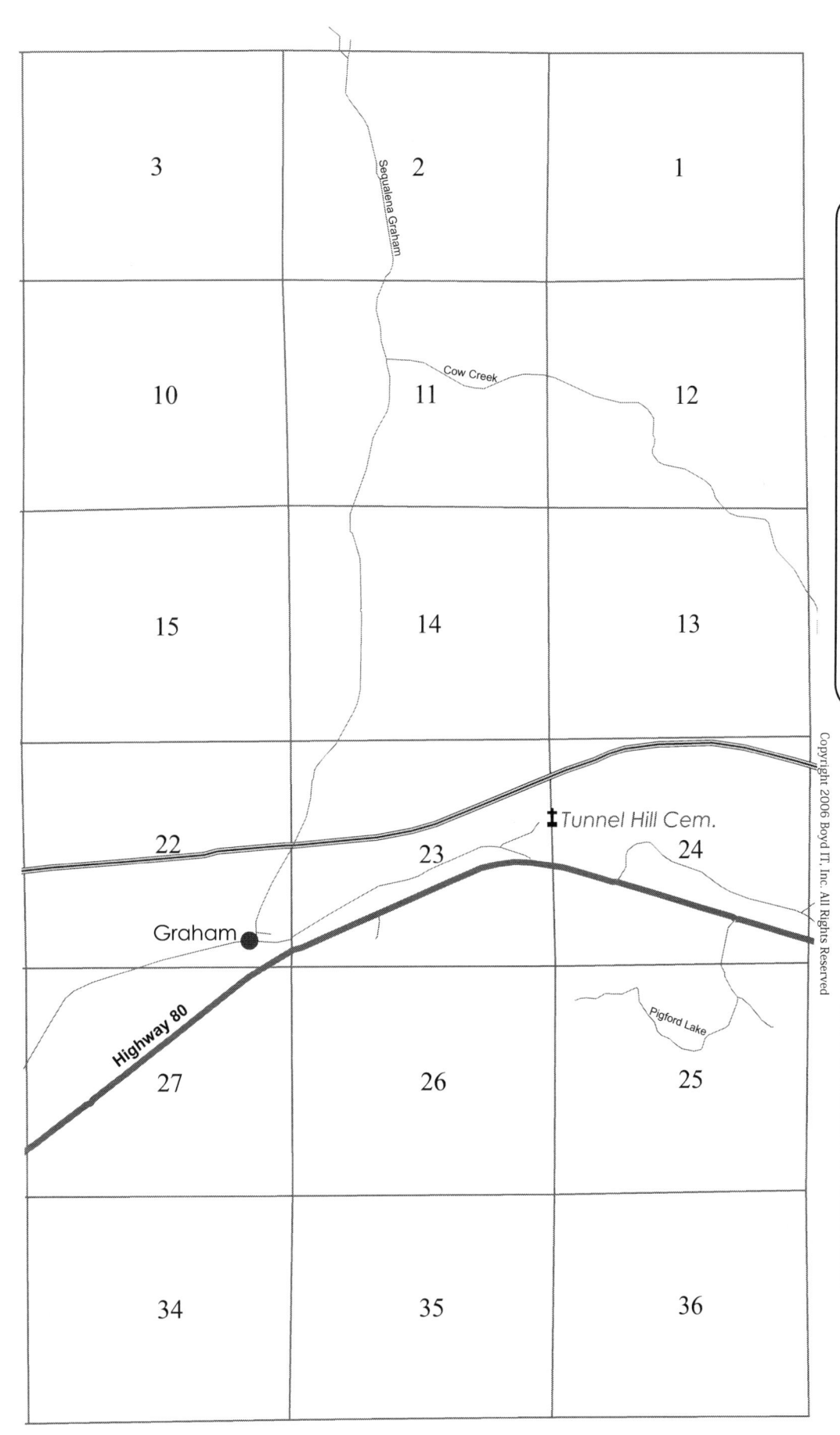

Helpful Hints

1. This road map has a number of uses, but primarily it is to help you: a) find the present location of land owned by your ancestors (at least the general area), b) find cemeteries and city-centers, and c) estimate the route/roads used by Census-takers & tax-assessors.

2. If you plan to travel to Lauderdale County to locate cemeteries or land parcels, please pick up a modern travel map for the area before you do. Mapping old land parcels on modern maps is not as exact a science as you might think. Just the slightest variations in public land survey coordinates, estimates of parcel boundaries, or road-map deviations can greatly alter a map's representation of how a road either does or doesn't cross a particular parcel of land.

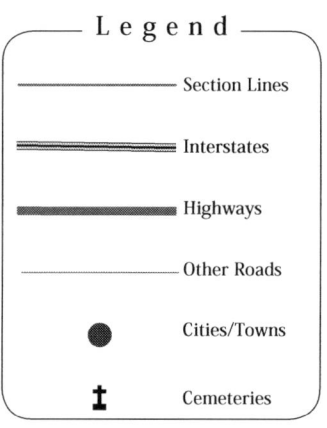

Copyright 2006 Boyd IT, Inc. All Rights Reserved

Legend

— Section Lines
═ Interstates
▬ Highways
— Other Roads
● Cities/Towns
☦ Cemeteries

Scale: Section = 1 mile X 1 mile
(generally, with some exceptions)

Family Maps of Lauderdale County, Mississippi

Historical Map
T6-N R14-E
Choctaw Meridian
Map Group 12

Cities & Towns
Graham
Meehan

Cemeteries
Tunnel Hill Cemetery

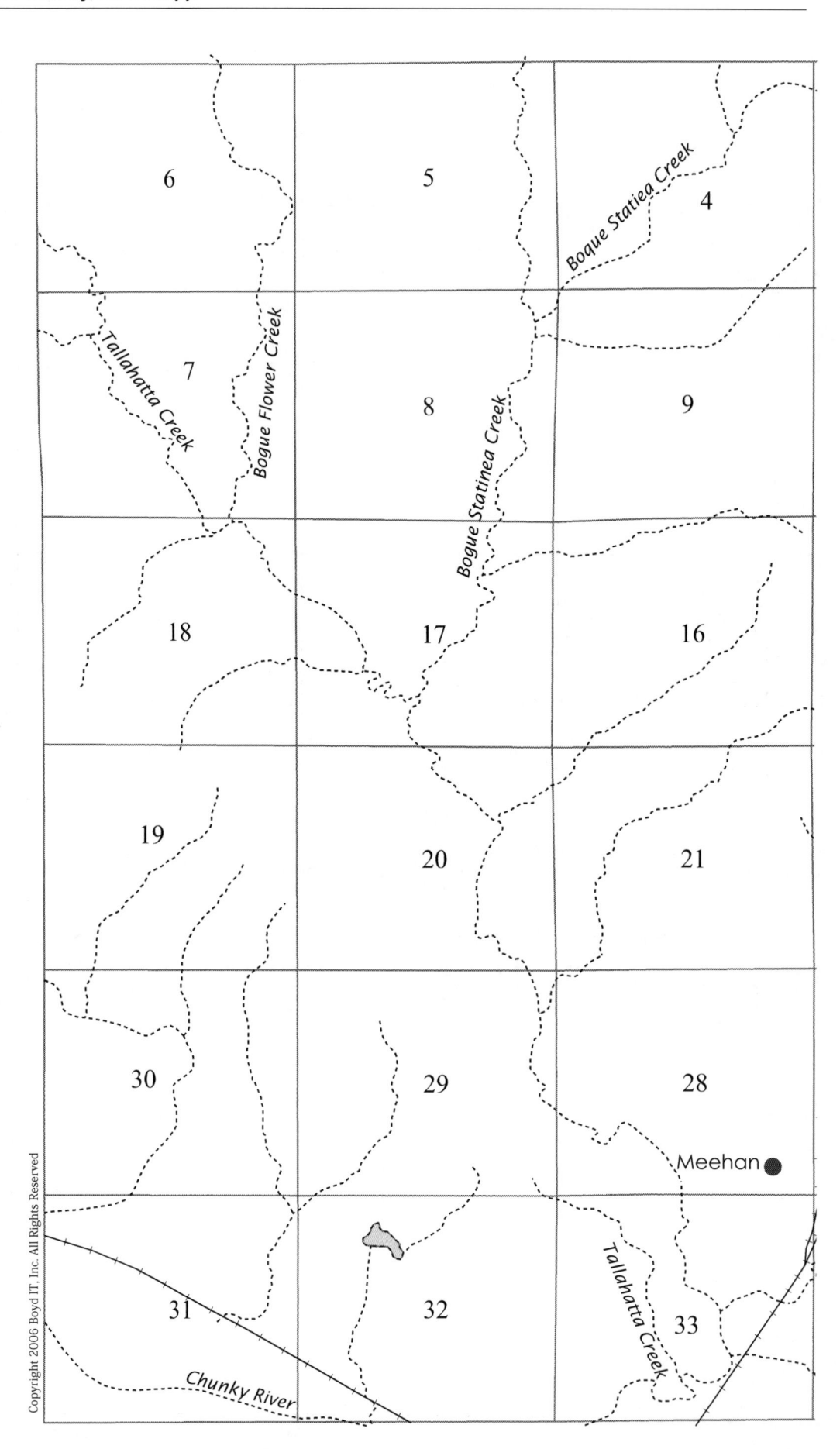

Township 6-N Range 14-E (Choctaw) - Map Group 12

Helpful Hints

1. This Map takes a different look at the same Congressional Township displayed in the preceding two maps. It presents features that can help you better envision the historical development of the area: a) Water-bodies (lakes & ponds), b) Water-courses (rivers, streams, etc.), c) Railroads, d) City/town center-points (where they were oftentimes located when first settled), and e) Cemeteries.

2. Using this "Historical" map in tandem with this Township's Patent Map and Road Map, may lead you to some interesting discoveries. You will often find roads, towns, cemeteries, and waterways are named after nearby landowners: sometimes those names will be the ones you are researching. See how many of these research gems you can find here in Lauderdale County.

Copyright 2006 Boyd IT, Inc. All Rights Reserved

Legend

— Section Lines
+++++ Railroads
▬ Large Rivers & Bodies of Water
- - - - Streams/Creeks & Small Rivers
● Cities/Towns
✝ Cemeteries

Scale: Section = 1 mile X 1 mile
(there are some exceptions)

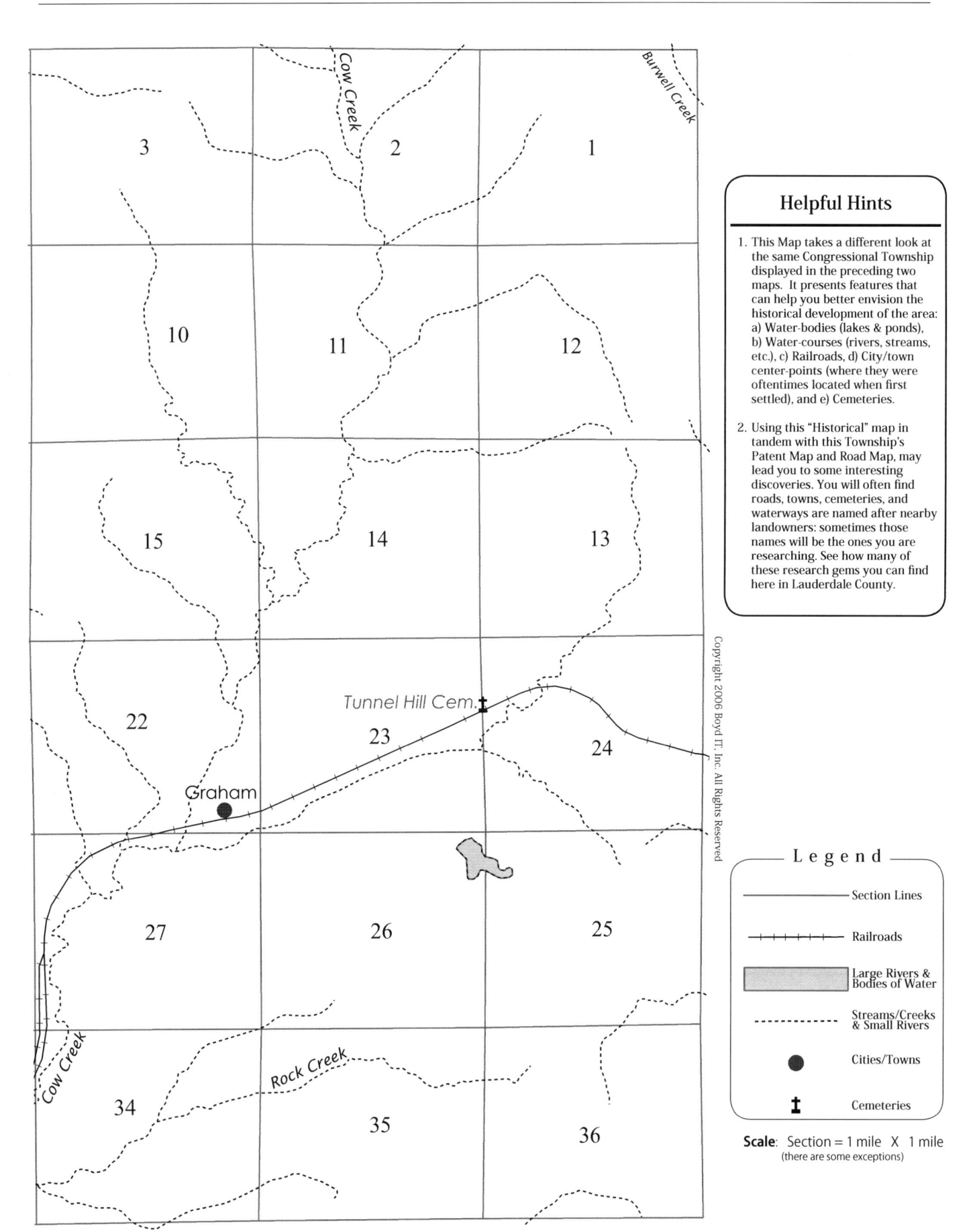

173

Family Maps of Lauderdale County, Mississippi

Map Group 13: Index to Land Patents
Township 6-North Range 15-East (Choctaw)

After you locate an individual in this Index, take note of the Section and Section Part then proceed to the Land Patent map on the pages immediately following. You should have no difficulty locating the corresponding parcel of land.

The "For More Info" Column will lead you to more information about the underlying Patents. See the *Legend* at right, and the "How to Use this Book" chapter, for more information.

LEGEND
"For More Info . . . " column

- **A** = Authority (Legislative Act, See Appendix "A")
- **B** = Block or Lot (location in Section unknown)
- **C** = Cancelled Patent
- **F** = Fractional Section
- **G** = Group (Multi-Patentee Patent, see Appendix "C")
- **V** = Overlaps another Parcel
- **R** = Re-Issued (Parcel patented more than once)

(A & G items require you to look in the Appendixes referred to above. All other Letter-designations followed by a number require you to locate line-items in this index that possess the ID number found after the letter).

ID	Individual in Patent	Sec.	Sec. Part	Date Issued	Other Counties	For More Info . . .
2497	BALL, John T	1	SESE	1860-04-02		A1 G14
2498	" "	11	SESW	1860-04-02		A1 G15
2509	BARNETT, Joseph	1	NWSE	1841-01-05		A1 G18
2505	" "	1	W½NE	1841-01-05		A1
2506	" "	12	E½NW	1841-01-05		A1
2508	" "	13	E½NW	1841-01-05		A1
2507	" "	12	NWNE	1846-09-01		A1
2593	BERRY, William J	21	E½NE	1859-05-02		A1
2570	BISHOP, Stanmore	26	SE	1841-01-05		A1
2405	BOON, Alen	10	W½SE	1841-01-05		A1
2596	BREWER, William M	17	SESE	1895-06-19		A4
2426	BRIDGES, Daniel B	25	NWSE	1860-04-02		A1
2433	BROWN, Edward	25	NWSE	1896-10-31		A4
2512	BUCHANAN, Joseph S	25	NENE	1904-11-15		A4
2396	BURWELL, Abram L	21	SESE	1853/01/22		A2
2397	" "	21	SESW	1853/01/22		A2
2399	" "	28	NENE	1853/01/22		A2
2400	" "	28	NWNE	1853/01/22		A2
2394	" "	21	N½SE	1859-05-02		A1
2395	" "	21	N½SW	1859-05-02		A1
2390	" "	17	NESW	1860-04-02		A1
2391	" "	17	SWNE	1860-04-02		A1
2393	" "	19	SENE	1860-04-02		A1
2392	" "	19	NESW	1860-10-01		A1
2398	" "	21	SWSW	1875-07-01		A4
2487	BURWELL, John	31	SWNE	1894-03-12		A4
2456	CLAY, George W	19	E½SE	1919-06-03		A4
2500	COATES, John W	35	NWSE	1859-06-01		A1
2571	COLE, Stephen M	13	W½NW	1859-05-02		A1 G47
2461	CONNER, Isaac	35	E½NE	1882-03-30		A4
2403	CREAL, Albert	24	SWNE	1841-01-05		A1
2423	CREEL, Colin W	35	W½NE	1841-01-05		A1
2459	CREEL, Henry	35	E½SW	1841-01-05		A1
2477	DANIEL, James W	15	W½NW	1841-01-05		A1
2479	" "	3	W½NW	1841-01-05		A1
2480	" "	4	E½SW	1841-01-05		A1
2478	" "	21	W½NW	1859-05-02		A1
2468	DANIELS, James	29	NWNW	1915-08-11		A4
2522	DANIELS, Louis	29	NENW	1917-08-11		A4 R2445
2420	DAVID, Charles S	23	NENE	1859-05-02		A1
2421	" "	23	NWSE	1859-05-02		A1
2422	" "	23	SWSE	1859-05-02		A1
2488	DAVIS, John	9	W½SW	1901-06-25		A4
2489	DEARMAN, John	10	E½SE	1841-01-05		A1
2567	DEARMAN, Solomon	10	E½SW	1841-01-05		A1
2568	" "	15	E½NW	1841-01-05		A1

174

Township 6-N Range 15-E (Choctaw) - Map Group 13

ID	Individual in Patent	Sec.	Sec. Part	Date Issued	Other Counties	For More Info . . .
2444	DEERMAN, Elisha	22	E½NW	1841-01-05		A1
2490	DEERMAN, John	10	W½SW	1841-01-05		A1
2497	DUNCAN, L A	1	SESE	1860-04-02		A1 G14
2514	" "	11	NENE	1860-04-02		A1
2498	" "	11	SESW	1860-04-02		A1 G15
2515	" "	11	SWSW	1860-04-02		A1 G66
2513	" "	1	NWSE	1860-10-01		A1
2590	DUNCAN, William C	1	SWSW	1860-04-02		A1 R2599
2498	" "	11	SESW	1860-04-02		A1 G15
2515	" "	11	SWSW	1860-04-02		A1 G66
2455	DYASS, George	28	SENE	1841-01-05		A1
2491	EARLONG, John	33	W½NW	1896-10-10		A4
2401	EUBANKS, Adam	7	NWNE	1859-06-01		A1
2462	EUBANKS, Isaac	3	NENE	1848-09-01		A1
2530	EUBANKS, Nancy	7	NENW	1875-07-01		A4
2520	FAIRCHILD, Lofton B	17	SENW	1841-01-05		A1
2521	FAIRCHILD, Lofton E	17	SESW	1882-03-30		A4
2552	FAIRCHILD, Robert	20	NENE	1841-01-05		A1
2592	FARRELL, William H	17	NWSW	1906-12-17		A4
2481	FAULKNER, James W	33	SENE	1860-04-02		A1
2402	FENDALL, Adeline	31	SW	1894-09-28		A4 G73
2402	FENDALL, George	31	SW	1894-09-28		A4 G73
2546	FOUNTAIN, Robert A	3	E½SW	1841-01-05		A1
2547	" "	3	W½SW	1841-01-05		A1
2548	" "	9	W½NE	1841-01-05		A1
2549	FOUNTAINE, Robert A	4	E½SE	1841-01-05		A1
2424	GADDIS, Cornelius M	33	SESW	1859-05-02		A1
2425	" "	33	SWSW	1860-04-02		A1
2535	GADDIS, Neal	31	NENW	1860-07-02		A1
2536	" "	33	NWSW	1860-07-02		A1
2524	GARDNER, Mahala	25	NWNE	1875-07-01		A4
2438	GARRETT, Edward T	12	E½SW	1841-01-05		A1
2439	" "	35	E½NW	1841-01-05		A1
2597	GILBERT, William R	17	NESE	1895-06-19		A4
2386	GOODSON, Aaron	33	NENE	1841-01-05		A1
2387	" "	34	NWNW	1841-01-05		A1
2412	GRAHAM, Archibald	2	NWSW	1841-01-05		A1
2492	GRAHAM, John	3	W½SE	1841-01-05		A1
2469	HAMRICK, James	1	E½NE	1841-01-05		A1
2418	HARRISON, Charles R	7	W½NW	1900-11-28		A4
2419	" "	7	W½SW	1900-11-28		A4
2485	HARRISON, Joel	7	E½SW	1906-06-26		A4
2486	" "	7	NWSE	1906-06-26		A4
2493	HARRISON, John	7	S½SE	1905-05-09		A4
2550	HARRISON, Robert C	7	SENW	1900-11-28		A4
2551	" "	7	SWNE	1900-11-28		A4
2529	HOGANS, Moses	17	NWNE	1899-06-28		A4
2538	JACKSON, Perry	5	W½NW	1897-06-07		A4
2539	" "	5	W½SW	1897-06-07		A4
2571	JACKSON, William H	13	W½NW	1859-05-02		A1 G47
2523	JAMES, Louisa	17	E½SW	1895-06-27		A4
2510	JOHNSON, Joseph	5	W½NE	1891-05-20		A4
2511	" "	5	W½SE	1891-05-20		A4
2457	JONES, Green	5	E½NW	1894-09-28		A4
2458	" "	5	E½SW	1894-09-28		A4
2517	JONES, Lenos	19	N½NE	1905-05-02		A4
2518	" "	19	SWNE	1905-05-02		A4
2435	JOSEPH, Edward	5	NESE	1881-09-17		A4
2436	" "	9	E½SW	1884-12-30		A4
2484	JOSEPH, Joe	5	SENE	1898-12-27		A4
2572	KEYS, Tanner	29	SWSE	1908-10-26		A4
2542	KILLENS, Richard	19	NENW	1901-12-17		A4
2470	KILLILEA, James	25	SESE	1893-09-23		A1
2583	LANG, William A	13	W½SW	1841-01-05		A1 G132
2575	" "	15	W½NE	1841-01-05		A1
2576	" "	15	W½SE	1841-01-05		A1
2577	" "	22	E½NE	1841-01-05		A1
2578	" "	22	E½SE	1841-01-05		A1
2584	" "	23	E½SE	1841-01-05		A1 G132
2585	" "	23	SW	1841-01-05		A1 G132
2579	" "	23	W½NW	1841-01-05		A1
2586	" "	24	W½NW	1841-01-05		A1 G132
2587	" "	25	W½NW	1841-01-05		A1 G132

Family Maps of Lauderdale County, Mississippi

ID	Individual in Patent	Sec.	Sec. Part	Date Issued	Other Counties	For More Info . . .
2588	LANG, William A (Cont'd)	26	E½NE	1841-01-05		A1 G132
2580	" "	26	E½SW	1841-01-05		A1
2581	" "	26	W½NE	1841-01-05		A1
2582	" "	34	E½SE	1841-01-05		A1
2589	" "	35	W½SW	1841-01-05		A1 G132
2431	LEWIS, Debreaux	33	NENW	1894-08-04		A4
2432	" "	33	NWNE	1894-08-04		A4
2503	LEWIS, Jonas	33	SWNE	1912-08-26		A4
2573	LITTLETON, Thomas	31	SENE	1874-04-10		A4
2463	LOPER, Isaac	3	E½SE	1841-01-05		A1
2464	" "	3	SENE	1841-01-05		A1
2465	" "	3	SWNE	1841-01-05		A1
2504	LOPER, Joseph A	3	NWNE	1841-01-05		A1
2531	LOTT, Nancy	25	SESW	1894-07-24		A4
2460	MARCY, Henry	29	SESE	1893-09-01		A4
2416	MARTIN, Carter	24	NWNE	1841-01-05		A1
2574	MASSENGALL, Warren	1	NESW	1860-07-02		A1
2404	MATHEWS, Albert	5	NENE	1875-08-20		A4
2543	MCLEMORE, Richard	13	W½NE	1859-05-02		A1
2525	MESSLEY, Malinda	29	NESW	1916-07-31		A4
2519	MILLER, Lewis	32	SESE	1841-01-05		A1
2545	MITCHELL, Rilla	25	NESW	1895-02-21		A4
2526	MONTAGUE, Mickelborough S	19	W½SE	1859-05-02		A1
2527	" "	29	W½SW	1859-05-02		A1
2553	MONTAGUE, Robert V	33	SWSE	1859-05-02		A1
2413	MOORE, Bartholomew F	33	NESW	1859-11-10		A1
2414	" "	33	NWSE	1859-11-10		A1
2415	" "	33	SENW	1860-04-02		A1
2443	NEWSOM, Elbert	24	E½NE	1841-01-05		A1
2569	NEWSOM, Solomon	15	E½SW	1841-01-05		A1
2471	PARKER, James	28	E½SE	1841-01-05		A1
2406	PEW, Alexander	27	SWNW	1841-01-05		A1
2445	REAVES, Elizabeth	29	NENW	1875-07-01		A4 C R2522
2537	REED, Nora	31	W½NW	1898-02-24		A4
2446	REEVES, Elizabeth	29	SWNW	1914-06-25		A4
2466	REEVES, Isaah	31	SE	1895-02-21		A4
2532	REEVES, Nancy	29	SWNE	1860-04-02		A1
2533	REIVES, Nancy	29	SENW	1859-11-10		A1
2534	" "	29	SESW	1859-11-10		A1
2410	RHODES, Anderson W	25	NESE	1882-12-30		A4
2411	" "	25	SENE	1882-12-30		A4
2417	RUSHING, Charles E	3	E½NW	1859-05-02		A1
2407	RUSSELL, Allen	4	E½NE	1841-01-05		A1
2408	" "	4	W½NE	1841-01-05		A1
2409	" "	4	W½SE	1841-01-05		A1
2472	RUSSELL, James	10	E½NE	1841-01-05		A1
2473	" "	10	E½NW	1841-01-05		A1
2474	" "	10	W½NE	1841-01-05		A1
2475	" "	11	W½NW	1841-01-05		A1
2494	RUSSELL, John	10	SWNW	1841-01-05		A1
2495	" "	4	E½NW	1841-01-05		A1
2496	" "	9	SENE	1841-01-05		A1
2565	RUSSELL, Simeon	10	NWNW	1841-01-05		A1
2566	" "	9	NENE	1841-01-05		A1
2598	RUSSELL, William	1	SESW	1841-01-05		A1
2599	" "	1	SWSW	1841-01-05		A1 R2590
2453	SEMMES, Francis C	1	NESE	1859-05-02		A1
2454	" "	11	NWSW	1859-11-10		A1
2516	SEMMES, Lem	9	N½NW	1891-06-30		A4
2447	SHACKELFORD, Eveline	23	SENE	1859-11-10		A1
2583	SHAW, Dougald C	13	W½NW	1841-01-05		A1 G132
2584	" "	23	E½SE	1841-01-05		A1 G132
2585	" "	23	SW	1841-01-05		A1 G132
2586	" "	24	W½NW	1841-01-05		A1 G132
2587	" "	25	W½NW	1841-01-05		A1 G132
2588	" "	26	E½NE	1841-01-05		A1 G132
2589	" "	35	W½SW	1841-01-05		A1 G132
2388	SIKES, Aaron M	21	SWSE	1841-01-05		A1
2389	STAFFORD, Abraham	13	SESE	1841-01-05		A1
2437	STAFFORD, Edward	25	SWSW	1859-06-01		A1
2448	STAFFORD, Ezekiel	13	E½SW	1841-01-05		A1
2449	" "	13	W½SE	1841-01-05		A1
2450	" "	24	E½NW	1841-01-05		A1

Township 6-N Range 15-E (Choctaw) - Map Group 13

ID	Individual in Patent	Sec.	Sec. Part	Date Issued	Other Counties	For More Info...
2451	STAFFORD, Ezekiel (Cont'd)	24	E½SW	1841-01-05		A1
2452	" "	24	W½SW	1841-01-05		A1
2554	TABB, Samuel G	27	E½NW	1841-01-05		A1
2555	" "	27	NE	1841-01-05		A1
2556	" "	27	NWNW	1841-01-05		A1
2557	" "	27	S½	1841-01-05		A1
2558	" "	33	E½SE	1841-01-05		A1
2559	" "	34	E½NW	1841-01-05		A1
2560	" "	34	NE	1841-01-05		A1
2561	" "	34	SW	1841-01-05		A1
2562	" "	34	SWNW	1841-01-05		A1
2563	" "	34	W½SE	1841-01-05		A1
2564	" "	35	W½NW	1841-01-05		A1
2440	TAYLOR, Edward	19	NWNW	1892-06-15		A4
2441	" "	19	NWSW	1892-06-15		A4
2442	" "	19	S½NW	1892-06-15		A4
2476	THOMAS, James	9	S½NW	1895-06-27		A4
2540	THOMAS, Phillis	29	N½NE	1906-06-30		A4
2541	" "	29	SENE	1906-06-30		A4
2499	TURNER, John	9	E½SE	1841-01-05		A1
2544	ULMER, Richard	1	E½NW	1860-04-02		A1
2428	VANCE, David	15	E½SE	1859-05-02		A1
2429	" "	15	SENE	1859-05-02		A1
2430	" "	15	W½SW	1859-05-02		A1
2591	WALTHALL, William C	11	E½SE	1859-11-10		A1
2501	WASHINGTON, John	25	SWSE	1881-05-10		A4
2502	WATSON, John	31	SENW	1875-07-01		A4
2427	WEBSTER, Daniel	29	N½SE	1874-04-10		A4
2467	WELCH, Jacob P	11	W½SE	1859-11-10		A1
2482	WHITEHEAD, James	15	NENE	1841-01-05		A1
2483	" "	35	SWSE	1841-01-05		A1
2528	WILLIAMS, Monroe	5	SESE	1895-02-21		A4
2434	WILSON, Edward H	25	SWNE	1882-12-30		A4
2509	WILSON, William J	1	NWSE	1841-01-05		A1 G18
2594	WOLF, William L	21	W½NE	1848-09-01		A1
2595	" "	9	W½SE	1859-05-02		A1

Family Maps of Lauderdale County, Mississippi

Patent Map

T6-N R15-E
Choctaw Meridian

Map Group 13

Township Statistics

Parcels Mapped	:	214
Number of Patents	:	166
Number of Individuals	:	128
Patentees Identified	:	128
Number of Surnames	:	94
Multi-Patentee Parcels	:	13
Oldest Patent Date	:	1/5/1841
Most Recent Patent	:	6/3/1919
Block/Lot Parcels	:	0
Parcels Re-Issued	:	2
Parcels that Overlap	:	0
Cities and Towns	:	3
Cemeteries	:	7

Copyright 2006 Boyd IT, Inc. All Rights Reserved

178

Township 6-N Range 15-E (Choctaw) - Map Group 13

Helpful Hints

1. This Map's INDEX can be found on the preceding pages.

2. Refer to Map "C" to see where this Township lies within Lauderdale County, Mississippi.

3. Numbers within square brackets [ ] denote a multi-patentee land parcel (multi-owner). Refer to Appendix "C" for a full list of members in this group.

4. Areas that look to be crowded with Patentees usually indicate multiple sales of the same parcel (Re-issues) or Overlapping parcels. See this Township's Index for an explanation of these and other circumstances that might explain "odd" groupings of Patentees on this map.

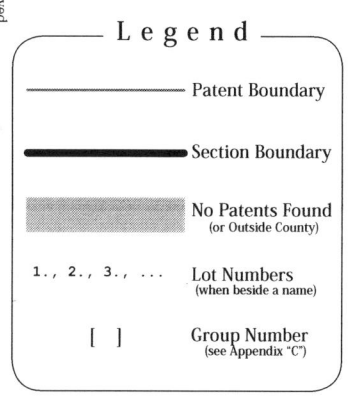

Legend

———	Patent Boundary
▬▬▬	Section Boundary
▓▓▓	No Patents Found (or Outside County)
1., 2., 3., ...	Lot Numbers (when beside a name)
[ ]	Group Number (see Appendix "C")

Scale: Section = 1 mile X 1 mile (generally, with some exceptions)

Copyright 2006 Boyd IT, Inc. All Rights Reserved

Family Maps of Lauderdale County, Mississippi

Road Map
T6-N R15-E Choctaw Meridian
Map Group 13

Cities & Towns
- Complete
- Lost Gap
- Savannah Grove

Cemeteries
- Burwell Cemetery
- Fairchild Cemetery
- Gordon Cemetery
- Rose Hill Cemetery
- Saint Patricks Cemetery
- Semmes Cemetery
- West Mount Moriah Cemetery

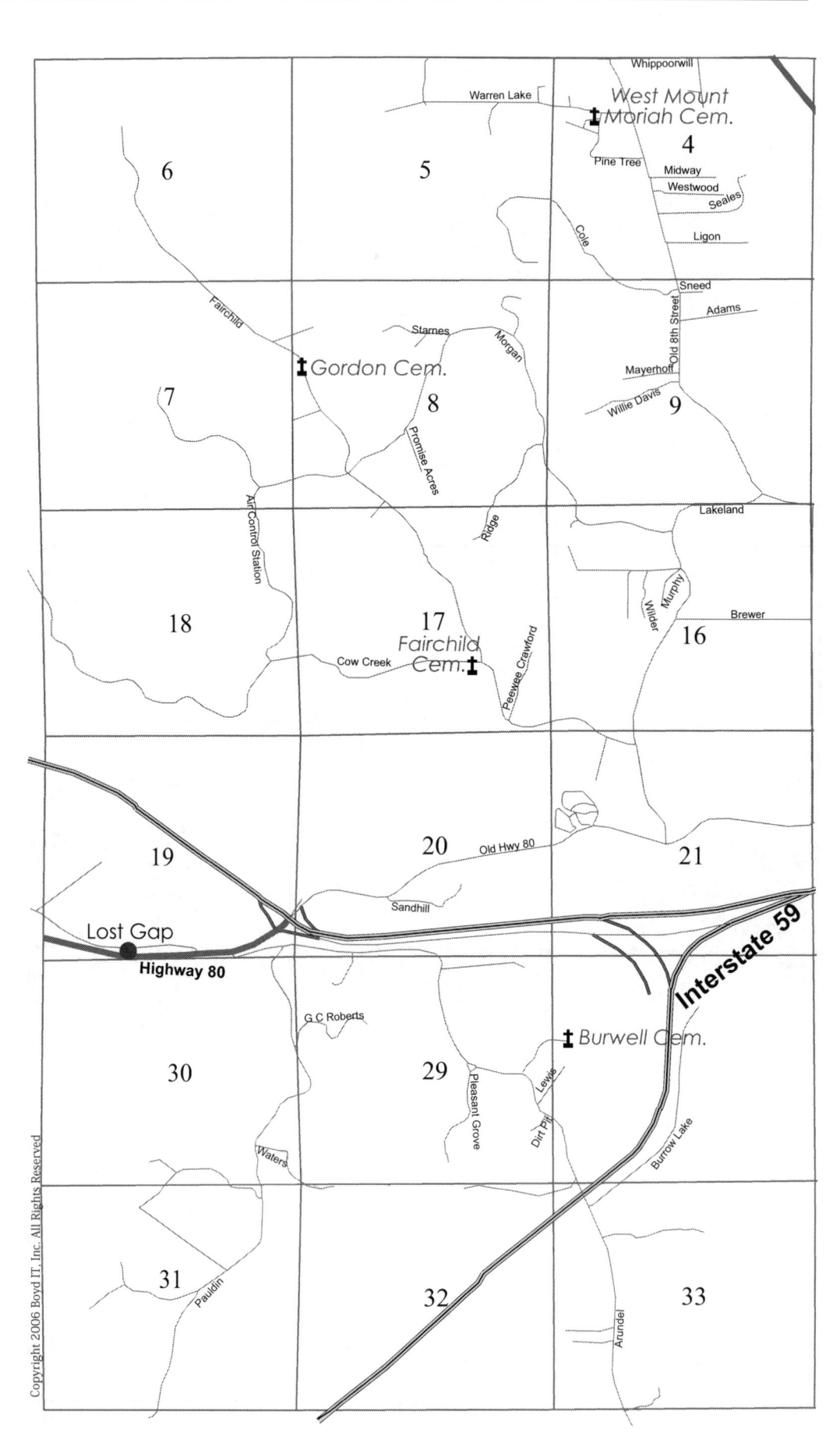

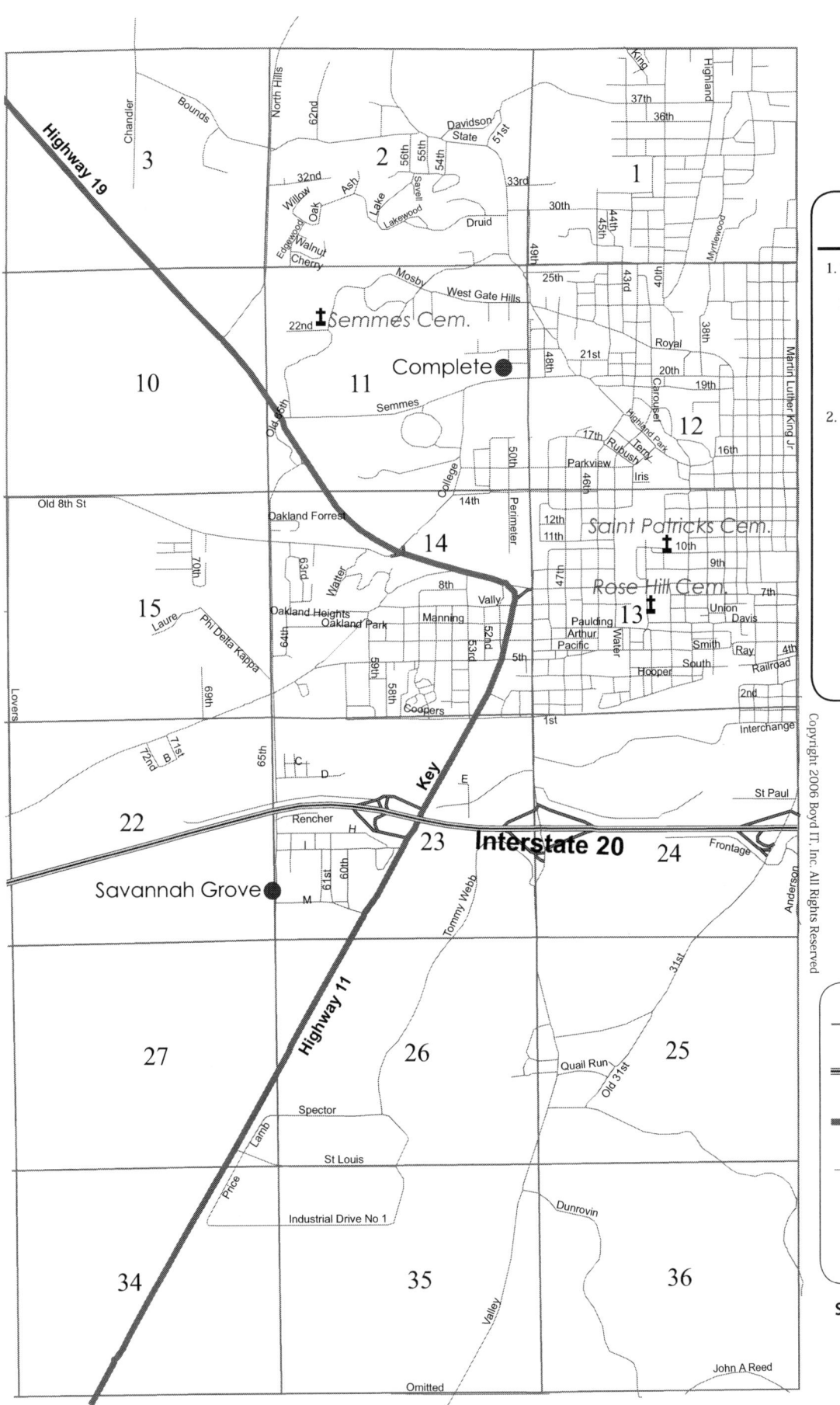

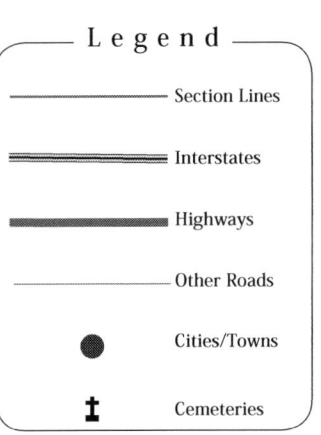

Township 6-N Range 15-E (Choctaw) - Map Group 13

Family Maps of Lauderdale County, Mississippi

Historical Map
T6-N R15-E
Choctaw Meridian
Map Group 13

Cities & Towns
Complete
Lost Gap
Savannah Grove

Cemeteries
Burwell Cemetery
Fairchild Cemetery
Gordon Cemetery
Rose Hill Cemetery
Saint Patricks Cemetery
Semmes Cemetery
West Mount Moriah Cemetery

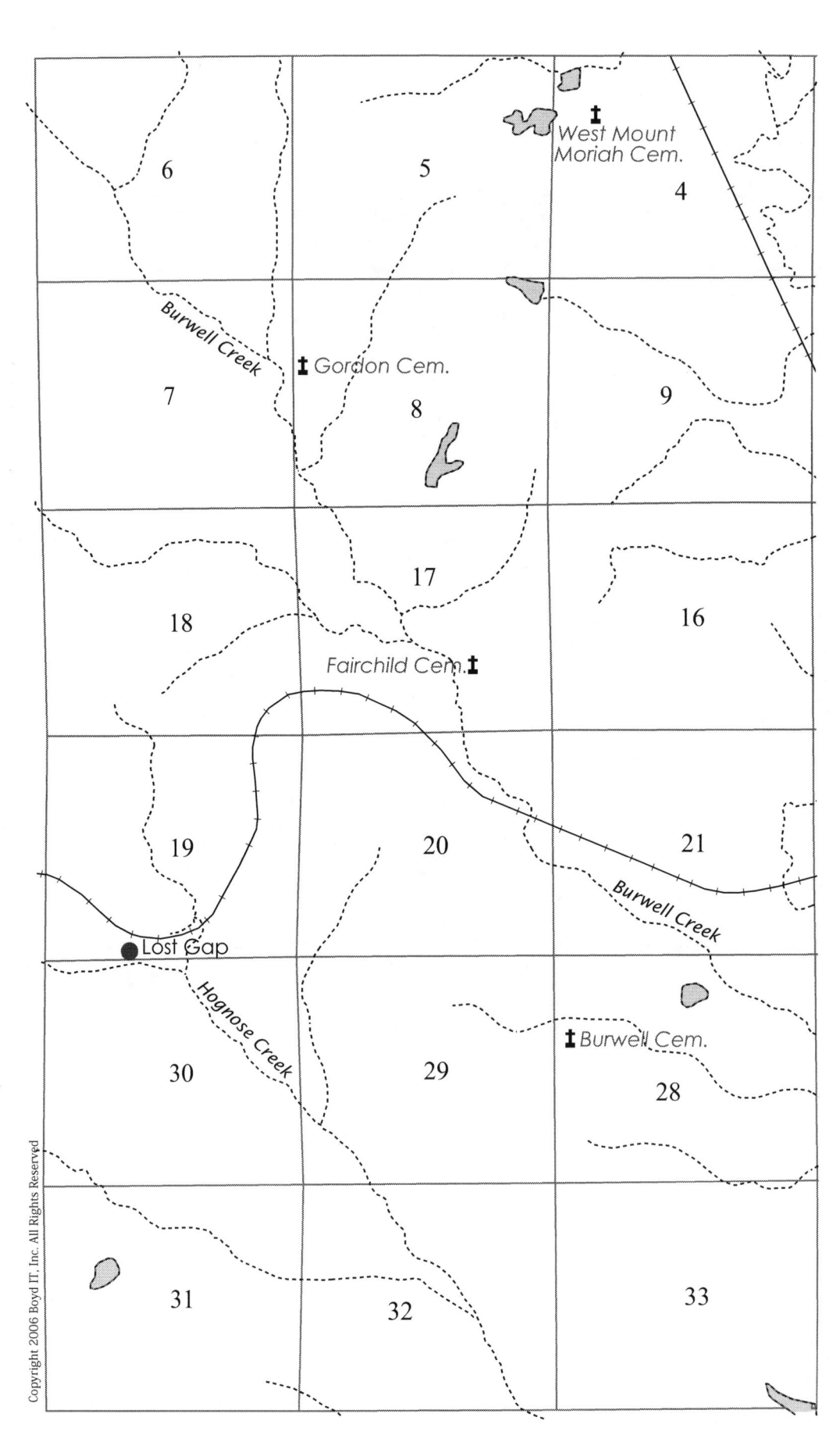

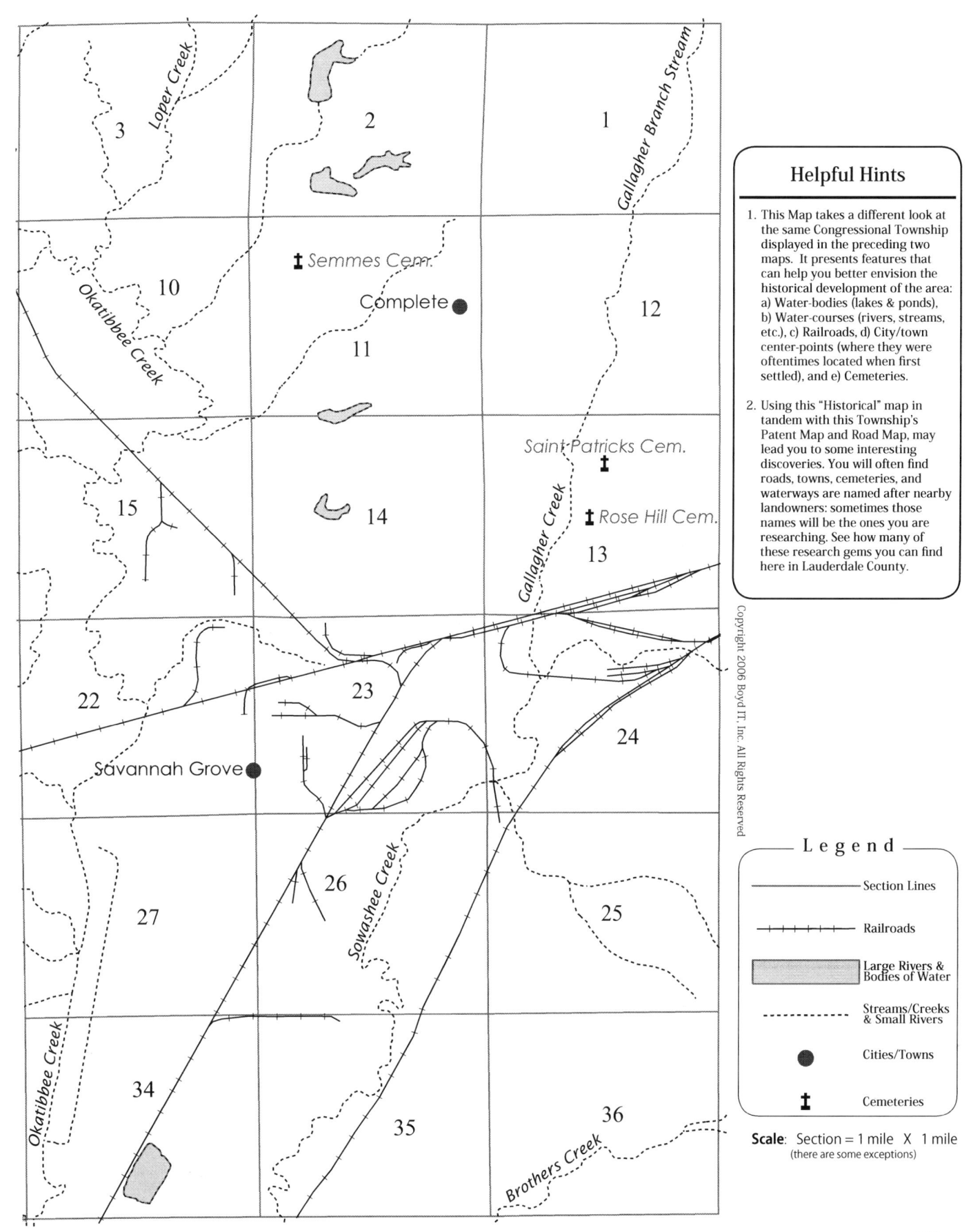

Township 6-N Range 15-E (Choctaw) - Map Group 13

Helpful Hints

1. This Map takes a different look at the same Congressional Township displayed in the preceding two maps. It presents features that can help you better envision the historical development of the area: a) Water-bodies (lakes & ponds), b) Water-courses (rivers, streams, etc.), c) Railroads, d) City/town center-points (where they were oftentimes located when first settled), and e) Cemeteries.

2. Using this "Historical" map in tandem with this Township's Patent Map and Road Map, may lead you to some interesting discoveries. You will often find roads, towns, cemeteries, and waterways are named after nearby landowners: sometimes those names will be the ones you are researching. See how many of these research gems you can find here in Lauderdale County.

Legend

- Section Lines
- Railroads
- Large Rivers & Bodies of Water
- Streams/Creeks & Small Rivers
- Cities/Towns
- Cemeteries

Scale: Section = 1 mile X 1 mile
(there are some exceptions)

Family Maps of Lauderdale County, Mississippi

Map Group 14: Index to Land Patents
Township 6-North Range 16-East (Choctaw)

After you locate an individual in this Index, take note of the Section and Section Part then proceed to the Land Patent map on the pages immediately following. You should have no difficulty locating the corresponding parcel of land.

The "For More Info" Column will lead you to more information about the underlying Patents. See the *Legend* at right, and the "How to Use this Book" chapter, for more information.

```
LEGEND
          "For More Info . . . " column
A = Authority (Legislative Act, See Appendix "A")
B = Block or Lot (location in Section unknown)
C = Cancelled Patent
F = Fractional Section
G = Group  (Multi-Patentee Patent, see Appendix "C")
V = Overlaps another Parcel
R = Re-Issued (Parcel patented more than once)

(A & G items require you to look in the Appendixes referred
to above. All other Letter-designations followed by a number
require you to locate line-items in this index that possess
the ID number found after the letter).
```

ID	Individual in Patent	Sec.	Sec. Part	Date Issued	Other Counties	For More Info . . .
2817	ADKINSON, Solomon	23	SWNE	1893-09-28		A4
2718	ALEXANDER, John	4	NESW	1841-01-05		A1
2719	" "	4	W½SW	1841-01-05		A1
2837	ALFORD, Warren F	1	SWSE	1860-04-02		A1
2836	" "	1	NESE	1861-05-01		A1
2693	ALLBROOKS, James F	25	N½NE	1892-07-11		A4
2737	BALL, John T	19	SESE	1860-04-02		A1
2739	" "	19	SWSW	1860-04-02		A1
2740	" "	29	SWNE	1860-04-02		A1
2741	" "	29	SWSE	1860-04-02		A1
2738	" "	19	SESW	1860-07-02		A1
2700	BERRIEN, James W	1	SWNW	1840-06-10		A1 G21
2701	" "	2	E½NW	1840-06-10		A1 G21
2702	" "	2	SENE	1840-06-10		A1 G21
2703	" "	2	SWNW	1840-06-10		A1 G21
2704	" "	2	W½NE	1840-06-10		A1 G21
2705	" "	3	E½NW	1840-06-10		A1 G21 R2657
2706	" "	3	NE	1840-06-10		A1 G21
2707	" "	3	SWNW	1840-06-10		A1 G21
2708	" "	4	NW	1840-06-10		A1 G21
2709	" "	4	SESE	1840-06-10		A1 G21
2710	" "	4	SESW	1840-06-10		A1 G21
2711	" "	4	W½NE	1840-06-10		A1 G21
2712	" "	4	W½SE	1840-06-10		A1 G21
2776	BICKART, Mary	11	SESW	1884-11-20		A4
2827	BISHOP, Tilmon	17	NENW	1841-01-05		A1
2828	" "	17	W½NE	1841-01-05		A1
2717	BLACKMAN, John A	15	NWNE	1850/06/01		A2
2674	BORDERS, Henry	27	SENE	1896-10-10		A4
2700	BOTHWELL, David E	1	SWNW	1840-06-10		A1 G21
2701	" "	2	E½NW	1840-06-10		A1 G21
2702	" "	2	SENE	1840-06-10		A1 G21
2703	" "	2	SWNW	1840-06-10		A1 G21
2704	" "	2	W½NE	1840-06-10		A1 G21
2705	" "	3	E½NW	1840-06-10		A1 G21 R2657
2706	" "	3	NE	1840-06-10		A1 G21
2707	" "	3	SWNW	1840-06-10		A1 G21
2708	" "	4	NW	1840-06-10		A1 G21
2709	" "	4	SESE	1840-06-10		A1 G21
2710	" "	4	SESW	1840-06-10		A1 G21
2711	" "	4	W½NE	1840-06-10		A1 G21
2712	" "	4	W½SE	1840-06-10		A1 G21
2807	BOZONE, Samuel	20	E½NE	1841-01-05		A1
2657	BROWER, Franklin P	3	E½NW	1848-09-01		A1 R2705
2667	BUNTYN, Geraldres E	31	SW	1899-04-28		A4
2689	BURWELL, James	29	NWSE	1895-06-27		A4

Township 6-N Range 16-E (Choctaw) - Map Group 14

ID	Individual in Patent	Sec.	Sec. Part	Date Issued	Other Counties	For More Info...
2675	BYNUM, Henry	33	S½SW	1882-05-10		A1
2751	CLINTON, Joseph	9	NENE	1846-09-01		A1
2752	" "	9	NENW	1846-09-01		A1
2640	COLE, Daniel	15	E½SE	1883-05-25		A4
2793	COLEMAN, Remus	33	N½SW	1892-03-17		A4
2639	CREAL, Colden W	17	NESE	1841-01-05		A1
2826	CREAL, Thomas D	19	E½NW	1841-01-05		A1
2691	CREEL, James	17	NWSE	1841-01-05		A1
2692	" "	17	SENE	1841-01-05		A1
2728	CREEL, John H	15	SWNE	1892-06-06		A1
2824	CREEL, Thomas B	17	NESW	1841-01-05		A1
2825	" "	17	SENW	1841-01-05		A1
2690	CRORY, James C	18	W½NE	1841-01-05		A1
2656	DANIELS, Ephraim C	31	S½SE	1905-03-30		A4
2731	DAVENPORT, John M	29	NWSW	1898-01-19		A4
2644	DAVIS, Ed	3	SESE	1913-02-14		A4
2676	DEAN, Henry H	25	NWSW	1875-06-01		A4
2750	DEAN, Joseph A	25	S½NE	1894-12-17		A4
2721	DEARMAN, John	7	N½NW	1859-11-10		A1 G63
2722	" "	7	NWNE	1859-11-10		A1 G63
2723	" "	7	SENW	1859-11-10		A1 G63
2724	" "	7	SWNE	1859-11-10		A1 G63
2720	" "	5	SENW	1860-10-01		A1
2795	DEARMAN, Richard	17	SWSE	1841-01-05		A1
2838	DEARMAN, Wiley	17	NWSW	1841-01-05		A1
2647	DEEN, Edmund	9	SWNW	1841-01-05		A1
2782	DEEN, Nathan P	8	NENW	1841-01-05		A1
2816	DEEN, Sermon	8	NENE	1841-01-05		A1
2820	DEENS, Surmon	8	NWNE	1841-01-05		A1
2821	" "	8	SENE	1841-01-05		A1
2855	DUBOSE, William W	11	SESE	1860-04-02		A1
2794	DUNGEN, Reuben	23	SWNW	1882-08-03		A4
2620	DURR, Annie	23	NESE	1904-11-15		A4
2654	DURR, Emanuel A	11	E½NE	1852/05/01		A2
2617	EDWARDS, Andrew J	21	W½NW	1892-04-16		A1
2774	EDWARDS, Marion	21	NWNE	1907-05-13		A4
2789	EDWARDS, Peter	21	SENE	1882-03-30		A4
2725	EVANS, John	19	NWNW	1841-01-05		A1
2726	" "	19	W½NE	1841-01-05		A1
2641	FAGIN, Daniel	23	E½SW	1892-06-15		A4
2771	FAIRCHILD, Loftin B	21	W½SW	1841-01-05		A1
2600	FARCHILES, Abram	35	NENE	1841-01-05		A1
2732	FARISS, John M	13	N½SW	1893-09-01		A4
2733	" "	13	NWSE	1893-09-01		A4
2734	" "	13	SESW	1893-09-01		A4
2694	FENNER, James	11	W½SW	1884-12-30		A4
2753	FORTSON, Joseph G	15	S½SW	1859-05-02		A1
2638	FUNK, Christian	27	W½SE	1885-12-19		A4
2648	GARRETT, Edward T	1	E½SW	1841-01-05		A1
2649	" "	2	NENE	1841-01-05		A1
2650	" "	22	W½NW	1841-01-05		A1
2713	GARRETT, James W	11	NWNE	1841-01-05		A1
2714	" "	21	NENE	1841-01-05		A1
2777	GARY, Mathias E	27	NESE	1881-05-10		A4
2684	GILLINDER, Hugh	35	W½NW	1841-01-05		A1
2727	GILLINDER, John	34	SENE	1841-01-05		A1
2610	GORDON, Alexander	8	W½NW	1841-01-05		A1
2669	GRANT, Green W	3	W½SW	1841-01-05		A1 G84
2846	GREEN, William	21	SENW	1901-07-09		A4
2685	GREER, Isaac S	29	NWNW	1860-04-02		A1
2643	HALL, Darling	11	NWNW	1860-04-02		A1
2642	" "	11	NENW	1860-10-01		A1
2849	HALL, William	5	E½SW	1841-01-05		A1
2621	HARPER, Benjamin F	25	S½NW	1898-12-01		A4
2637	HENDERSON, Charles J	35	NENW	1893-09-01		A4
2651	HENDERSON, Eliza	35	NESW	1875-07-01		A4
2755	HENDERSON, Joseph T	35	S½SESW	1918-05-03		A4
2756	" "	35	S½SESW	1918-05-03		A4
2757	" "	35	S½SWSE	1918-05-03		A4
2758	" "	35	S½SWSE	1918-05-03		A4
2829	HENDERSON, Tyree	35	NWNE	1860-04-02		A1
2830	" "	35	SENE	1860-10-01		A1
2831	" "	35	SENW	1860-10-01		A1

Family Maps of Lauderdale County, Mississippi

ID	Individual in Patent	Sec.	Sec. Part	Date Issued	Other Counties	For More Info . . .
2832	HENDERSON, Tyrone	35	SESE	1860-04-02		A1
2833	" "	35	SWNE	1860-04-02		A1
2850	HENDERSON, William J	35	NESE	1918-04-11		A4
2669	HERNDON, Thomas H	3	W½SW	1841-01-05		A1 G84
2623	HILL, Caleb	34	SENW	1841-01-05		A1
2658	HILL, George M	8	SWNE	1841-01-05		A1
2636	HOBBS, Charles	1	NWNE	1841-01-05		A1
2812	HOLDEN, Sarah A	27	N½NW	1884-12-30		A4
2677	IRBY, Henry	17	NWNW	1841-01-05		A1
2678	" "	20	NWNW	1841-01-05		A1
2679	" "	8	E½SW	1841-01-05		A1
2602	JOHNSON, Albert P	5	NENW	1860-04-02		A1
2695	JOHNSON, James	29	SWNW	1891-06-30		A4
2749	JOINER, Jordan W	2	SE	1848-09-01		A1
2772	JONES, Louis J	25	E½SE	1894-12-17		A4
2851	JONES, William	28	NWNW	1841-01-05		A1
2790	KIELY, Phillip	29	NESW	1884-12-30		A4
2791	" "	29	SENW	1884-12-30		A4
2818	KNOX, Stephen	23	NWSW	1882-03-30		A4
2819	" "	23	SESE	1921-04-27		A4
2839	LANG, William A	18	W½SW	1841-01-05		A1 G132
2786	LEE, Owen	1	SENW	1841-01-05		A1
2787	LEE, Owin	1	NWNW	1841-01-05		A1
2788	" "	1	SWNE	1841-01-05		A1
2808	LEHMAN, Samuel	29	E½SE	1884-12-30		A4
2611	LINDLEY, Ammond	5	W½SE	1859-11-10		A1
2603	MAASS, Albert W	29	NENW	1882-04-20		A1
2604	MAASS, Alex W	29	NWNE	1890-07-18		A1
2716	MANNING, Jesse B	31	E½NE	1893-08-23		A4
2770	MCCAREY, Lindsay	1	W½SW	1841-01-05		A1
2730	MCCARY, John L	1	NWSE	1841-01-05		A1
2601	MCDANIEL, Abram	23	SENW	1895-05-11		A4 G148
2601	MCDANIEL, Betsy	23	SENW	1895-05-11		A4 G148
2815	MCDANIEL, Scott	21	SWSE	1905-12-13		A4
2764	MCINNIS, Lafayett J	13	NENW	1860-10-01		A1
2765	MCINNIS, Lafayette J	13	NWNW	1881-06-30		A1
2759	MCLAMORE, Josiah	17	S½SW	1841-01-05		A1
2796	MCLAMORE, Richard	18	NWSE	1841-01-05		A1
2797	" "	4	E½NE	1841-01-05		A1
2612	MCLEMORE, Amos	11	N½SE	1860-07-02		A1
2614	" "	11	SENW	1860-07-02		A1
2615	" "	11	SWNE	1860-07-02		A1
2613	" "	11	NESW	1860-10-01		A1
2760	MCLEMORE, Josiah	34	E½SW	1841-01-05		A1
2763	MCLEMORE, Kesiann	22	SENW	1841-01-05		A1
2801	MCLEMORE, Richard	18	E½NE	1841-01-05		A1
2804	" "	18	SESE	1841-01-05		A1
2805	" "	34	SESE	1841-01-05		A1
2802	" "	18	E½NW	1848-09-01		A1
2803	" "	18	NESE	1848-09-01		A1
2798	" "	15	NWSW	1859-05-02		A1
2799	" "	15	SWNW	1859-05-02		A1
2800	" "	17	SESE	1859-05-02		A1
2835	MCMULLIN, Walter J	21	E½SW	1890-12-20		A1
2619	MCSHANN, Ann	33	NENW	1894-09-28		A4 G153
2619	MCSHANN, Burnell	33	NENW	1894-09-28		A4 G153
2735	MCSHANN, John	33	W½NE	1908-07-23		A4
2841	MILLS, William D	35	N½SESW	1917-05-25		A1
2842	" "	35	N½SESW	1917-05-25		A1
2843	" "	35	N½SWSE	1917-05-25		A1
2844	" "	35	N½SWSE	1917-05-25		A1
2845	" "	35	NWSE	1917-05-25		A1
2646	MOFFET, Edmond	21	E½SE	1892-07-11		A4
2754	MORGAN, Joseph J	15	NESW	1891-12-26		A1
2682	MORRIS, Hezekiah	17	SWNW	1841-01-05		A1
2834	NEAL, Vander	23	NWSE	1904-11-15		A4
2622	NEWSOM, Byrd	8	SENW	1841-01-05		A1
2823	NEWSOM, Taylor	20	NENW	1841-01-05		A1
2618	NICHOL, Andrew	23	SENE	1895-06-27		A4
2852	NICHOLS, William	23	NENE	1859-06-01		A1
2696	ODOM, James	5	W½NW	1841-01-05		A1
2792	ODOM, Randol	4	NESE	1841-01-05		A1
2659	PACE, George S	25	SWSW	1841-01-05		A1

Township 6-N Range 16-E (Choctaw) - Map Group 14

ID	Individual in Patent	Sec.	Sec. Part	Date Issued	Other Counties	For More Info . . .
2660	PACE, George S (Cont'd)	26	SESE	1841-01-05		A1
2785	PACE, Nicholas M	5	SWNE	1860-04-02		A1
2697	PETAGREW, James	27	SW	1841-01-05		A1
2605	PICKARD, Alexander A	22	NESE	1848-09-01		A1
2609	" "	9	E½SW	1848-09-01		A1
2606	" "	22	NWSE	1849-12-01		A1 R2607
2607	" "	22	NWSE	1849-12-01		A1 R2606
2608	" "	29	SWSW	1859-11-10		A1
2854	PRESTON, William	33	NWNW	1892-12-15		A4
2813	RAGSDALE, Sarah A	7	E½NE	1859-11-10		A1
2814	" "	7	SWNW	1859-11-10		A1
2766	RAGSDALL, Lewis A	19	NESE	1859-05-02		A1
2767	" "	19	SWNW	1859-05-02		A1
2768	" "	7	E½SW	1859-05-02		A1
2822	RANDALL, Susan M	23	NWNW	1891-06-30		A4
2775	ROBERTS, Mary A	15	SENW	1873-09-20		A4
2632	RUSHING, Charles E	27	NWNE	1859-11-10		A1
2628	" "	13	NESE	1860-04-02		A1
2629	" "	13	SWSE	1860-04-02		A1
2630	" "	23	SWSE	1860-04-02		A1
2631	" "	27	NENE	1860-10-01		A1
2633	" "	27	SWNE	1860-10-01		A1
2634	RUSSING, Charles E	13	W½NE	1860-04-02		A1
2686	RUTLEDGE, James A	25	NENW	1895-05-11		A4
2687	" "	25	NWNW	1895-05-11		A4
2673	SANFORD, Harriet	23	SWSW	1901-12-04		A4
2624	SCARBOROUGH, Calvin	10	E½NW	1841-01-05		A1 V2635
2625	" "	10	NWNE	1841-01-05		A1
2626	" "	3	E½SW	1841-01-05		A1
2627	" "	3	SWSE	1841-01-05		A1
2688	SCARBOROUGH, James B	9	NWNW	1841-01-05		A1
2698	SCARBOROUGH, James R	5	NESE	1841-01-05		A1
2699	" "	5	SESE	1841-01-05		A1 G167
2668	SCHANROCK, Gottliab	27	S½NW	1884-12-30		A4
2769	SCOTT, Lewis	29	SENE	1875-07-01		A4
2773	SCOTT, Malinda	29	NENE	1882-12-30		A4
2653	SHACKELFORD, Emaline	33	W½SE	1900-11-28		A4 G168
2655	SHACKELFORD, Ennis	33	E½SE	1893-09-28		A4
2653	SHACKELFORD, William	33	W½SE	1900-11-28		A4 G168
2839	SHAW, Dougald C	18	W½SW	1841-01-05		A1 G132
2736	SIMS, John	8	W½SE	1841-01-05		A1
2670	SMITH, Hampton S	19	N½SW	1859-05-02		A1
2671	" "	19	W½SE	1859-05-02		A1
2672	" "	7	W½SW	1859-05-02		A1
2784	STAFFORD, Navey	20	SWNW	1841-01-05		A1
2778	STONE, Miles	35	NWSW	1860-04-02		A1
2781	STONE, Moses	35	SWSW	1859-06-01		A1
2783	STONE, Nathan Y	27	SESE	1896-11-04		A4
2853	STONE, William P	11	SWSE	1861-02-01		A1
2652	STROUD, Elvira	9	SESE	1875-07-01		A4
2662	STROUD, George	9	N½SE	1859-05-02		A1
2665	" "	9	SWSE	1859-05-02		A1
2661	" "	15	NWNW	1882-05-10		A1
2663	" "	9	SENE	1882-05-10		A1
2664	" "	9	SENW	1882-05-10		A1
2666	" "	9	W½NE	1882-05-10		A1
2721	SWILLEY, James J	7	N½NW	1859-11-10		A1 G63
2722	" "	7	NWNE	1859-11-10		A1 G63
2723	" "	7	SENW	1859-11-10		A1 G63
2724	" "	7	SWNE	1859-11-10		A1 G63
2729	THOMPSON, John H	15	SENE	1882-03-30		A4
2806	THOMPSON, Sallie	15	NENW	1873-09-20		A4
2681	TIMBREL, Henry	15	W½SE	1885-04-04		A4
2742	TIMMS, John	17	NENE	1841-01-05		A1
2743	" "	8	E½SE	1841-01-05		A1
2744	TIMS, John	9	W½SW	1841-01-05		A1
2840	WABINGTON, William B	1	E½NE	1841-01-05		A1
2809	WAITS, Samuel W	13	SESE	1896-03-04		A4
2761	WALTHALL, Julian B	31	N½SE	1891-06-30		A4
2762	" "	31	W½NE	1891-06-30		A4
2683	WARBINGTON, Horatio B	13	E½NW	1841-01-05		A1
2745	WEST, John	3	NWNW	1841-01-05		A1
2746	WHITE, John	21	NWSE	1884-12-30		A4

Family Maps of Lauderdale County, Mississippi

ID	Individual in Patent	Sec.	Sec. Part	Date Issued	Other Counties	For More Info . . .
2747	WHITE, John (Cont'd)	21	SWNE	1884-12-30		A4
2847	WHITE, William H	10	NENE	1850/06/01		A2
2848	" "	11	SWNW	1850/06/01		A2
2635	WILLARD, Charles F	10	NW	1891-05-20		A4 V2624
2616	WILLIAMS, Anderson	15	NENE	1892-07-20		A4
2779	WILLIAMS, Milly	21	NENW	1885-05-25		A4
2780	WILLIAMS, Monroe Alexander	29	SESW	1912-09-16		A4
2810	WILLIAMS, Samuel	23	NWNE	1910-01-20		A4
2680	WILSON, Henry R	1	SESE	1861-02-01		A1
2699	WILSON, William J	5	SESE	1841-01-05		A1 G167
2856	WOOD, William	19	E½NE	1841-01-05		A1
2811	WORBINGTON, Samuel	1	NENW	1841-01-05		A1
2715	WRIGHT, Jane	33	S½NW	1891-11-03		A4
2645	YOUNG, Ed	13	SWSW	1901-10-08		A4
2748	YOUNG, John	23	NENW	1891-05-20		A4

Family Maps of Lauderdale County, Mississippi

Patent Map

T6-N R16-E
Choctaw Meridian

Map Group 14

Township Statistics

Parcels Mapped	:	257
Number of Patents	:	213
Number of Individuals	:	176
Patentees Identified	:	170
Number of Surnames	:	124
Multi-Patentee Parcels	:	23
Oldest Patent Date		6/10/1840
Most Recent Patent		4/27/1921
Block/Lot Parcels	:	0
Parcels Re - Issued	:	2
Parcels that Overlap	:	2
Cities and Towns	:	2
Cemeteries	:	6

Section 6

Section 5
- ODOM James 1841
- JOHNSON Albert P 1860
- DEARMAN John 1860
- PACE Nicholas M 1860
- SCARBOROUGH James R 1841
- LINDLEY Ammond 1859
- HALL William 1841
- SCARBOROUGH [167] James R 1841

Section 4
- BERRIEN [21] James W 1840
- BERRIEN [21] James W 1840
- MCLAMORE Richard 1841
- ALEXANDER John 1841
- BERRIEN [21] James W 1840
- ODOM Randol 1841
- ALEXANDER John 1841
- BERRIEN [21] James W 1840
- BERRIEN [21] James W 1840

Section 7
- DEARMAN [63] John 1859
- DEARMAN [63] John 1859
- RAGSDALE Sarah A 1859
- RAGSDALE Sarah A 1859
- DEARMAN [63] John 1859
- DEARMAN [63] John 1859
- SMITH Hampton S 1859
- RAGSDALL Lewis A 1859

Section 8
- GORDON Alexander 1841
- DEEN Nathan P 1841
- DEENS Surmon 1841
- DEEN Sermon 1841
- NEWSOM Byrd 1841
- HILL George M 1841
- DEENS Surmon 1841
- SIMS John 1841
- IRBY Henry 1841
- TIMMS John 1841

Section 9
- SCARBOROUGH James B 1841
- CLINTON Joseph 1846
- STROUD George 1882
- CLINTON Joseph 1846
- DEEN Edmund 1841
- STROUD George 1882
- STROUD George 1882
- TIMS John 1841
- PICKARD Alexander A 1848
- STROUD George 1859
- STROUD George 1859
- STROUD Elvira 1875

Section 18
- MCLEMORE Richard 1848
- CRORY James C 1841
- MCLEMORE Richard 1841
- LANG [132] William A 1841
- MCLAMORE Richard 1841
- MCLEMORE Richard 1848
- MCLEMORE Richard 1841

Section 17
- IRBY Henry 1841
- BISHOP Tilmon 1841
- BISHOP Tilmon 1841
- TIMMS John 1841
- MORRIS Hezekiah 1841
- CREEL Thomas B 1841
- CREEL James 1841
- DEARMAN Wiley 1841
- CREEL Thomas B 1841
- CREEL James 1841
- CREAL Colden W 1841
- MCLAMORE Josiah 1841
- DEARMAN Richard 1841
- MCLEMORE Richard 1859

Section 16

Section 19
- EVANS John 1841
- EVANS John 1841
- RAGSDALL Lewis A 1859
- CREAL Thomas D 1841
- WOOD William 1841
- SMITH Hampton S 1859
- SMITH Hampton S 1859
- RAGSDALL Lewis A 1859
- BALL John T 1860
- BALL John T 1860
- BALL John T 1860

Section 20
- IRBY Henry 1841
- NEWSOM Taylor 1841
- STAFFORD Navey 1841
- BOZONE Samuel 1841

Section 21
- EDWARDS Andrew J 1892
- WILLIAMS Milly 1885
- EDWARDS Marion 1907
- GARRETT James W 1841
- GREEN William 1901
- WHITE John 1884
- EDWARDS Peter 1882
- FAIRCHILD Loftin B 1841
- WHITE John 1884
- MCMULLIN Walter J 1890
- MCDANIEL Scott 1905
- MOFFET Edmond 1892

Section 30

Section 29
- GREER Isaac S 1860
- MAASS Albert W 1882
- MAASS Alex W 1890
- SCOTT Malinda 1882
- JOHNSON James 1891
- KIELY Phillip 1884
- BALL John T 1860
- SCOTT Lewis 1875
- DAVENPORT John M 1898
- KIELY Phillip 1884
- BURWELL James 1895
- PICKARD Alexander A 1859
- WILLIAMS Monroe Alexander 1912
- BALL John T 1860
- LEHMAN Samuel 1884

Section 28
- JONES William 1841

Section 31
- WALTHALL Julian B 1891
- MANNING Jesse B 1893
- WALTHALL Julian B 1891
- BUNTYN Geraldres E 1899
- DANIELS Ephraim C 1905

Section 32

Section 33
- PRESTON William 1892
- MCSHANN [153] Ann 1894
- MCSHANN John 1908
- WRIGHT Jane 1891
- COLEMAN Remus 1892
- SHACKELFORD [168] Emaline 1900
- BYNUM Henry 1882
- SHACKELFORD Ennis 1893

Copyright 2006 Boyd IT, Inc. All Rights Reserved

190

Township 6-N Range 16-E (Choctaw) - Map Group 14

Family Maps of Lauderdale County, Mississippi

Road Map

T6-N R16-E
Choctaw Meridian

Map Group 14

Cities & Towns
Bonita
Meridian

Cemeteries
Beth Israel Cemetery
Magnolia Cemetery
McLemore Cemetery
Oak Grove Cemetery
Saint Luke Cemetery
Tabernacle Cemetery

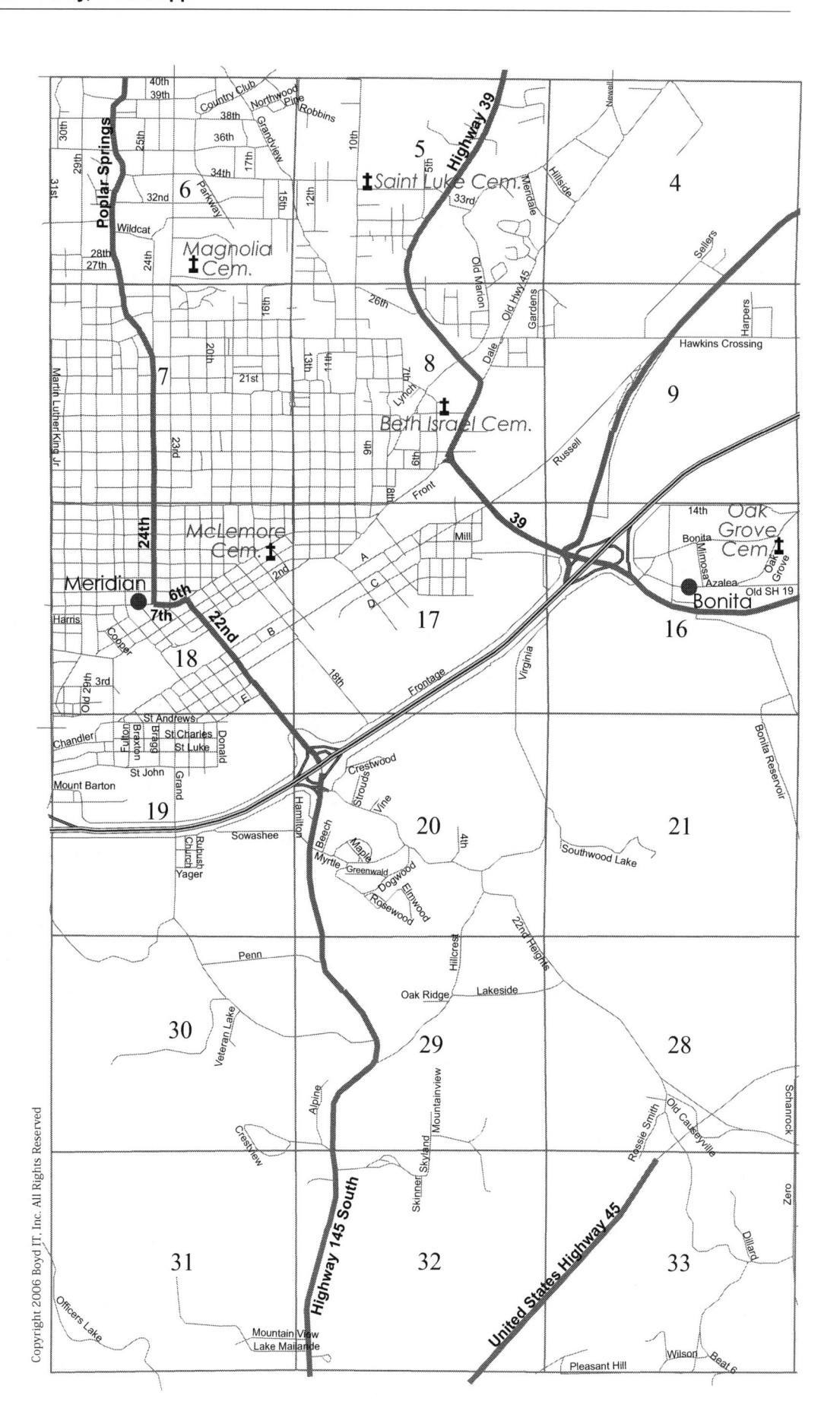

Township 6-N Range 16-E (Choctaw) - Map Group 14

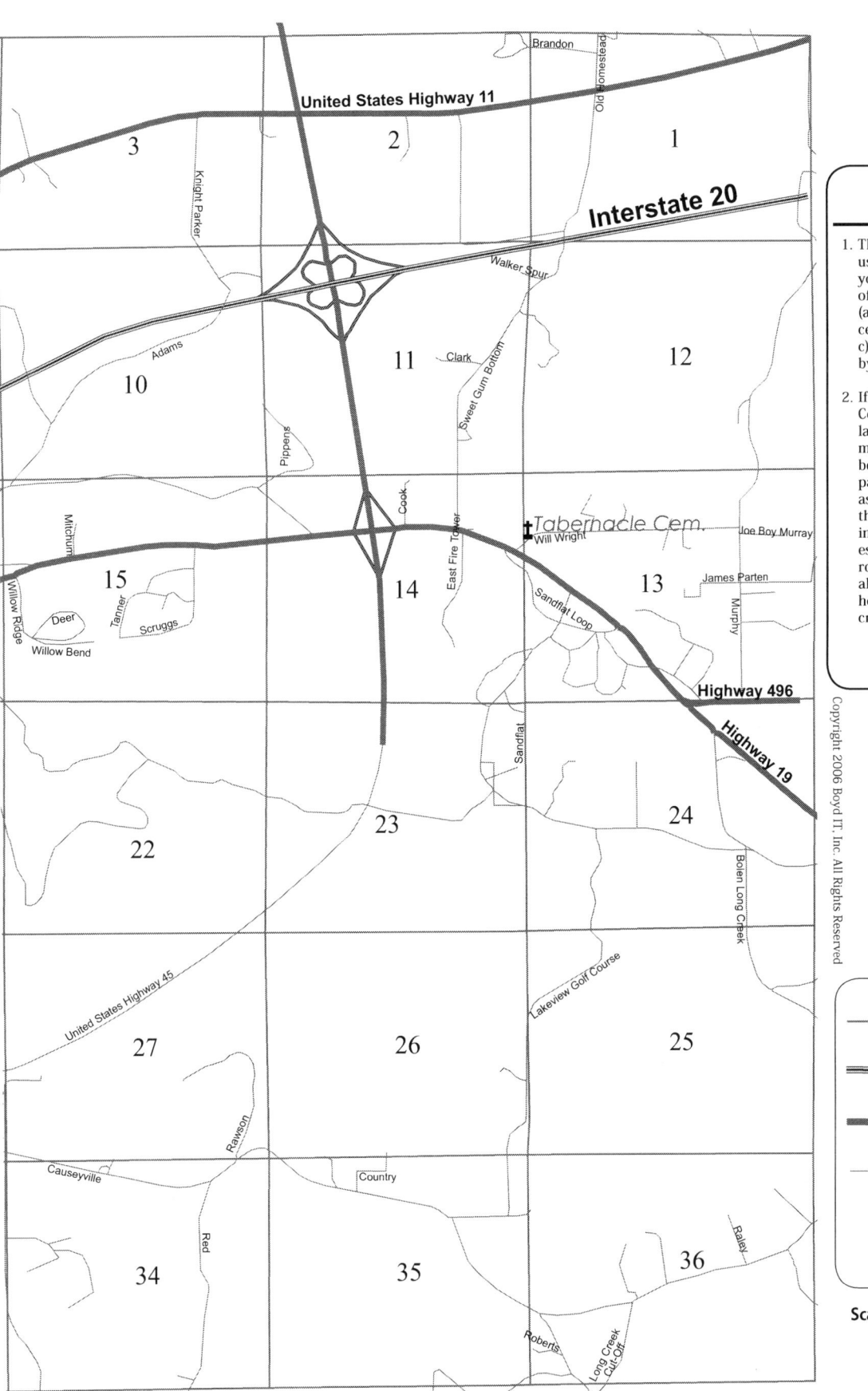

Helpful Hints

1. This road map has a number of uses, but primarily it is to help you: a) find the present location of land owned by your ancestors (at least the general area), b) find cemeteries and city-centers, and c) estimate the route/roads used by Census-takers & tax-assessors.

2. If you plan to travel to Lauderdale County to locate cemeteries or land parcels, please pick up a modern travel map for the area before you do. Mapping old land parcels on modern maps is not as exact a science as you might think. Just the slightest variations in public land survey coordinates, estimates of parcel boundaries, or road-map deviations can greatly alter a map's representation of how a road either does or doesn't cross a particular parcel of land.

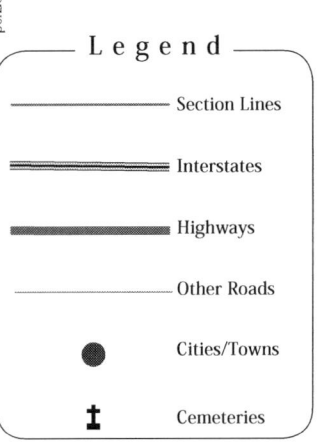

Copyright 2006 Boyd IT, Inc. All Rights Reserved

Legend

- Section Lines
- Interstates
- Highways
- Other Roads
- ● Cities/Towns
- ✝ Cemeteries

Scale: Section = 1 mile X 1 mile
(generally, with some exceptions)

193

Family Maps of Lauderdale County, Mississippi

Historical Map
T6-N R16-E
Choctaw Meridian
Map Group 14

Cities & Towns
Bonita
Meridian

Cemeteries
Beth Israel Cemetery
Magnolia Cemetery
McLemore Cemetery
Oak Grove Cemetery
Saint Luke Cemetery
Tabernacle Cemetery

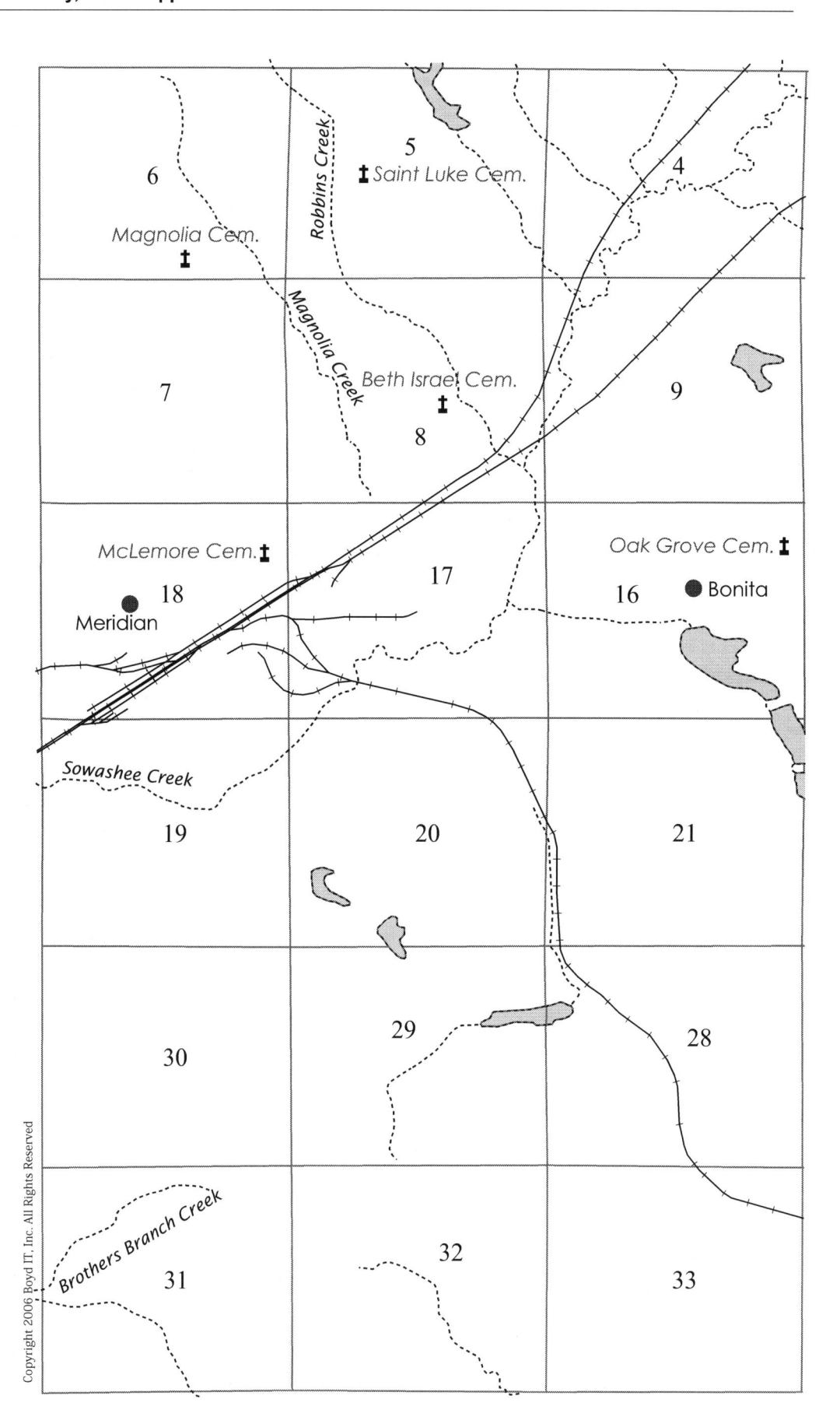

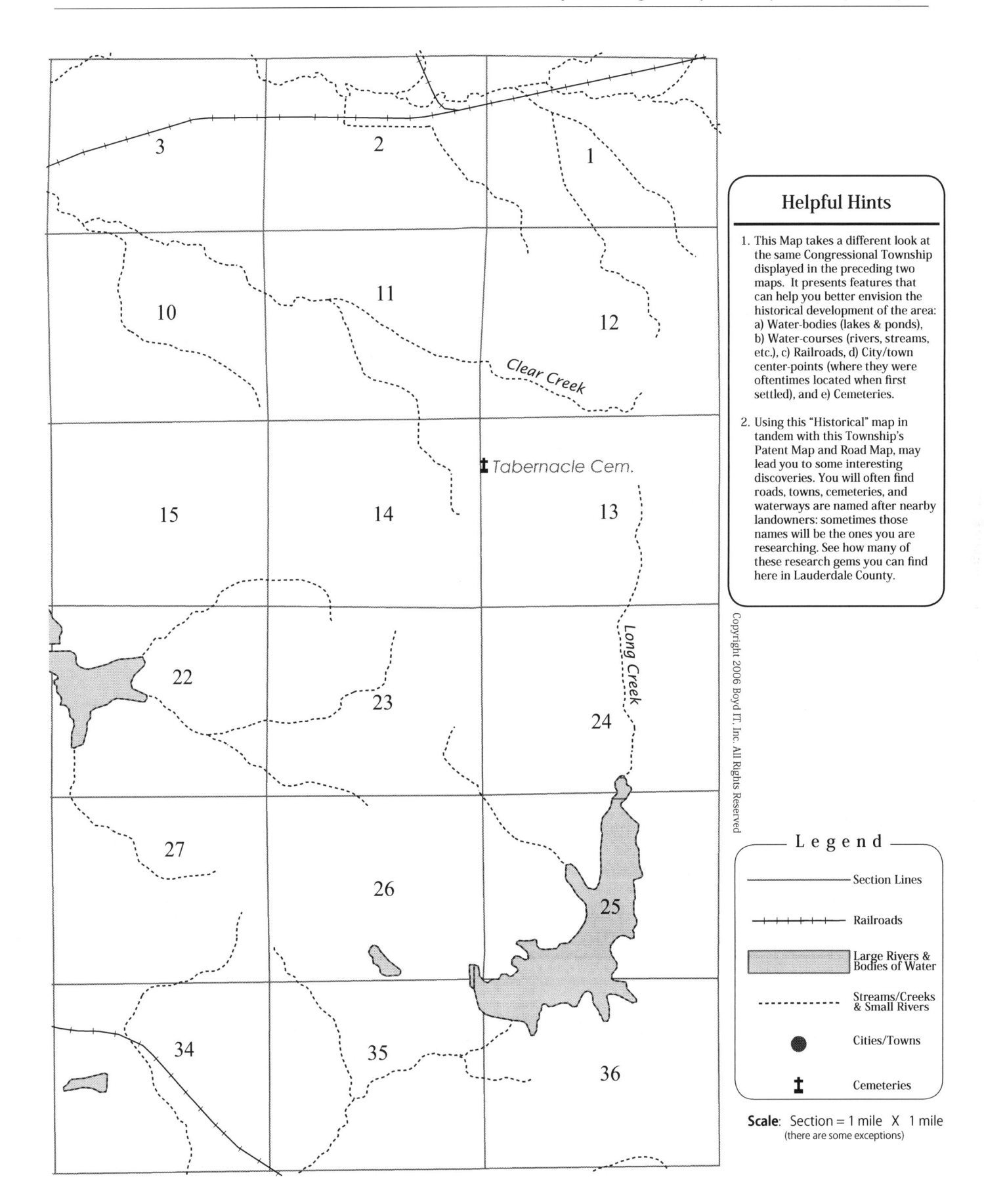

Township 6-N Range 16-E (Choctaw) - Map Group 14

Helpful Hints

1. This Map takes a different look at the same Congressional Township displayed in the preceding two maps. It presents features that can help you better envision the historical development of the area: a) Water-bodies (lakes & ponds), b) Water-courses (rivers, streams, etc.), c) Railroads, d) City/town center-points (where they were oftentimes located when first settled), and e) Cemeteries.

2. Using this "Historical" map in tandem with this Township's Patent Map and Road Map, may lead you to some interesting discoveries. You will often find roads, towns, cemeteries, and waterways are named after nearby landowners: sometimes those names will be the ones you are researching. See how many of these research gems you can find here in Lauderdale County.

Clear Creek

Tabernacle Cem.

Long Creek

Legend

— Section Lines
+++++ Railroads
▓ Large Rivers & Bodies of Water
---- Streams/Creeks & Small Rivers
● Cities/Towns
✝ Cemeteries

Scale: Section = 1 mile X 1 mile
(there are some exceptions)

Copyright 2006 Boyd IT, Inc. All Rights Reserved

195

Map Group 15: Index to Land Patents
Township 6-North Range 17-East (Choctaw)

After you locate an individual in this Index, take note of the Section and Section Part then proceed to the Land Patent map on the pages immediately following. You should have no difficulty locating the corresponding parcel of land.

The "For More Info" Column will lead you to more information about the underlying Patents. See the *Legend* at right, and the "How to Use this Book" chapter, for more information.

```
LEGEND
     "For More Info . . ." column
A = Authority (Legislative Act, See Appendix "A")
B = Block or Lot (location in Section unknown)
C = Cancelled Patent
F = Fractional Section
G = Group (Multi-Patentee Patent, see Appendix "C")
V = Overlaps another Parcel
R = Re-Issued (Parcel patented more than once)

(A & G items require you to look in the Appendixes referred
to above. All other Letter-designations followed by a number
require you to locate line-items in this index that possess
the ID number found after the letter).
```

ID	Individual in Patent	Sec.	Sec. Part	Date Issued	Other Counties	For More Info . . .
2888	BLACKWELL, Elemander	1	W½SE	1841-01-05		A1 G23
2888	BLACKWELL, Wright	1	W½SE	1841-01-05		A1 G23
2926	BOLDEN, John	5	SE	1911-04-05		A4
2941	BONEY, Kinsey	12	NWNE	1848-09-01		A1
2992	BOWLING, William F	28	NE	1846-09-01		A1 G25
2864	BRIANT, Benjamin	15	E½SW	1841-01-05		A1 G31
2918	BRUISTER, James W	33	NESW	1848-09-01		A1
2919	" "	33	SENW	1848-09-01		A1
2920	" "	34	NESE	1859-11-10		A1
2921	" "	34	SWSE	1859-11-10		A1
2983	BURKHALTER, Thomas	3	SWSW	1860-04-02		A1
2972	BUSON, Samuel A	7	NESW	1859-06-01		A1
2955	COAKER, Peyton	28	E½SW	1846-09-01		A1 G45
2942	COKER, Levi	28	NENW	1859-05-02		A1
2943	" "	28	SWNW	1859-05-02		A1
2956	COKER, Peyton	33	NWNE	1848-09-01		A1
2945	COOK, Lewis	25	NE	1846-09-01		A1 G48
2984	CRENSHAW, Thomas G	31	NWNE	1905-02-13		A4
2882	DAVIS, Christopher J	9	SWSW	1846-09-01		A1
2995	DEARMAN, William R	29	NWSW	1896-07-11		A4
2889	DURR, Emanuel A	2	NENW	1852/05/01		A2
2890	" "	2	SENE	1852/05/01		A2
2891	" "	3	SESW	1852/05/01		A2
2991	EMBRY, Warren	31	NWNW	1859-05-02		A1
2900	FREEMAN, Gideon A	7	NWSE	1859-06-01		A1
2902	" "	7	SWNE	1859-06-01		A1
2899	" "	7	NWNE	1860-04-02		A1
2903	" "	7	SWSE	1860-04-02		A1
2901	" "	7	SESE	1860-07-02		A1
2878	GORDON, Charles	15	W½NE	1846-09-01		A1
2879	" "	15	W½NW	1846-09-01		A1
2880	" "	9	SESE	1848-09-01		A1
2910	GORDON, James	10	NWSW	1859-05-02		A1
2961	GORDON, Richmond T	10	E½NW	1859-05-02		A1
2962	" "	10	E½SW	1859-05-02		A1
2963	" "	14	NW	1859-05-02		A1
2964	" "	14	W½NE	1859-05-02		A1
2965	" "	2	S½	1859-05-02		A1
2966	" "	22	NW	1859-05-02		A1
2967	" "	9	NESE	1859-11-10		A1
2968	" "	9	SENE	1860-10-01		A1
2969	GORDON, Robert B	3	NWSW	1896-12-14		A4
2863	GRANT, Andrew J	19	SWSW	1906-06-16		A4
2928	GRIFFITH, John	9	NWSW	1901-04-22		A4
2929	" "	9	SWNW	1901-04-22		A4
2970	HALL, Sabra	19	SENW	1895-02-21		A4

Township 6-N Range 17-E (Choctaw) - Map Group 15

ID	Individual in Patent	Sec.	Sec. Part	Date Issued	Other Counties	For More Info . . .
2971	HALL, Sabra (Cont'd)	19	SWNE	1895-02-21		A4
2860	HARPER, Alexander	1	E½SE	1846-09-01		A1 G102
2871	HARPER, Burrell W	9	SESW	1911-01-09		A4
2914	HARPER, James M	29	N½NW	1898-12-01		A4
2915	" "	29	SWNW	1898-12-01		A4
2951	HARPER, Nancy L	5	NWNE	1897-11-01		A4
2952	" "	5	S½NE	1897-11-01		A4
2996	HARPER, William R	5	SESW	1906-09-14		A4
2868	HENDERSON, Blank	17	SWNE	1860-04-02		A1 G106
2947	HENDERSON, Michael P	20	E½NE	1846-09-01		A1
2948	" "	21	SWNW	1846-09-01		A1
2993	HILL, William	19	NWSE	1905-02-13		A4
2887	HOLDEN, Ebenezer	14	W½SW	1841-01-05		A1 G113
2864	" "	15	E½SW	1841-01-05		A1 G31
2886	" "	15	SE	1846-09-01		A1 G112
2945	" "	25	NE	1846-09-01		A1 G48
2860	HUMPHREYS, William	1	E½SE	1846-09-01		A1 G102
2906	HUNT, Henry J	32	S½SE	1859-06-01		A1
2907	" "	32	SWSW	1859-06-01		A1
2978	HUTTON, Sonny	5	SWSW	1902-02-12		A4
2930	IRBY, John H	31	NESW	1906-06-16		A4
2931	" "	31	SENW	1906-06-16		A4
2932	" "	31	SWNE	1906-06-16		A4
2954	IRBY, Patrick	19	SWNW	1895-02-21		A4
2904	JACKSON, Green	7	NENE	1906-06-16		A4 G126
2904	JACKSON, Jane	7	NENE	1906-06-16		A4 G126
2883	JOHNSON, Darius G	19	SESW	1891-06-30		A4
2884	" "	19	SWSE	1891-06-30		A4
2955	JOHNSON, Jacob	28	E½SW	1846-09-01		A1 G45
2905	JONES, Henrietta	17	SENW	1894-12-17		A4
2886	JONES, James	15	SE	1846-09-01		A1 G112
2976	JONES, Sandy	17	SWNW	1895-11-11		A4
2992	JORDAN, Henry G	28	NE	1846-09-01		A1 G25
2896	KIMBRELL, George	5	SWNW	1909-11-08		A4
2862	KNIGHT, Allen E	9	NWNW	1906-06-30		A1
2979	KNOX, Stephen	19	NENW	1889-11-23		A4
2980	" "	19	NWNE	1889-11-23		A4
2973	LACKEY, Samuel L	9	W½NE	1898-08-27		A4
2945	LANE, Shepherd	25	NE	1846-09-01		A1 G48
2953	LARKIN, Oliver	19	NESW	1910-10-24		A4
2911	LASHLEY, James	9	SENW	1860-07-02		A1 R2913
2912	LASLEY, James	9	NESW	1860-10-01		A1
2913	" "	9	SENW	1897-08-09		A1 R2911
2893	LEE, Felix	5	NESW	1895-02-21		A4
2894	" "	5	SENW	1913-02-25		A4
2887	LEE, Zachariah	14	W½SW	1841-01-05		A1 G113
2934	LEWIS, John	5	NWNW	1846-09-01		A1
2933	" "	5	NENW	1859-11-10		A1
2935	" "	5	NWSW	1859-11-10		A1
2988	MARSHALL, Virinda E	19	NWSW	1905-02-13		A4
2908	MASTERS, Jack	19	E½SE	1895-02-21		A4
2957	MCCRANEY, Philip	24	E½SE	1841-05-18		A1 G146
2922	MCELROY, Joel	9	W½SE	1890-06-25		A4
2872	MCLEMORE, Caleb S	31	SWSW	1901-03-23		A4
2997	MCMORY, William S	24	SWSW	1846-09-01		A1
2998	" "	25	NWNW	1846-09-01		A1
2924	MCMULLEN, John A	17	W½SE	1890-06-25		A4
2923	" "	17	E½SW	1895-02-21		A4
2989	MCSHAN, Walter J	7	NWSW	1895-12-14		A4
2990	" "	7	W½NW	1895-12-14		A4
3000	MERRITT, Willis	23	E½NW	1846-09-01		A1
3001	" "	23	W½NE	1846-09-01		A1
3002	" "	23	W½NW	1846-09-01		A1
2999	" "	22	E½SE	1859-05-02		A1
2892	MOORE, Felix E	12	NENE	1860-04-02		A1
2960	MOTEN, Rederick	17	W½SW	1890-02-21		A4
2994	NEWTON, William	13	W½NE	1846-09-01		A1
3003	PIGFORD, Wright	2	NENE	1846-09-01		A1
3004	" "	2	W½NE	1860-04-10		A1
2916	POGUE, James P	34	SESE	1848-09-01		A1
2857	PRINGLE, Abraham	21	SWSE	1852/05/01		A2
2858	" "	28	SENW	1852/05/01		A2
2859	" "	28	W½SE	1852/05/01		A2

Family Maps of Lauderdale County, Mississippi

ID	Individual in Patent	Sec.	Sec. Part	Date Issued	Other Counties	For More Info...
2925	PRINGLE, John A	22	NWSW	1859-05-02		A1
2974	PRINGLE, Samuel T	21	SESE	1848-09-01		A1
2975	" "	22	SWSW	1848-09-01		A1
2985	PRINGLE, Thomas K	22	E½SW	1846-09-01		A1
2986	" "	22	W½SE	1846-09-01		A1
2987	" "	28	E½SE	1846-09-01		A1
2881	RAWSON, Charles	10	W½NE	1846-09-01		A1
2898	RAWSON, George W	9	NENE	1899-09-30		A4
2940	RICHEY, Joseph W	31	NENW	1894-04-10		A4
2869	RODGERS, Bonapart	13	SESE	1846-09-01		A1
2870	" "	13	SWSE	1846-09-01		A1
2877	RUSHING, Charles E	9	NENW	1859-05-02		A1
2875	" "	5	NENE	1859-06-01		A1
2876	" "	7	SESW	1859-11-10		A1
2873	" "	3	NENE	1860-04-02		A1
2874	" "	3	SWNE	1860-04-02		A1
2885	RUSSELL, David	6	E½NE	1841-01-05		A1
2977	SCARBROUGH, Silas	24	W½NE	1846-09-01		A1
2897	SIMS, George	24	SESW	1846-09-01		A1
2958	SIMS, Primus A	31	NWSW	1895-01-17		A4
2959	" "	31	SWNW	1895-01-17		A4
2861	SPINKS, Alfred	14	SESW	1846-09-01		A1
2927	SPINKS, John C	35	NESW	1852/05/01		A2
2957	SUGGS, Joshua	24	E½SE	1841-05-18		A1 G146
2937	TERRELL, Joseph	10	NENE	1859-06-01		A1
2938	" "	10	SE	1859-11-10		A1
2939	" "	10	SENE	1859-11-10		A1
2917	THOMPSON, James	19	NENE	1897-02-04		A4
2944	WAITS, Levi	6	E½SW	1846-09-01		A1
2950	WALKER, Moses A	7	SWSW	1911-02-23		A4
2909	WARBINGTON, Jacob B	6	NW	1841-01-05		A1
2946	WATES, Martin	7	NENW	1846-09-01		A1
2865	WATTS, Benjamin F	20	E½SE	1859-05-02		A1
2949	WELCH, Mina H	3	NESW	1860-04-02		A1
2981	WHITE, Thadeus C	14	NESW	1846-09-01		A1
2982	" "	14	NWSE	1846-09-01		A1
2866	WHITEHEAD, Benjamin F	17	N½NW	1892-04-29		A4
2867	" "	17	NWNE	1892-04-29		A4
2868	WILLIAMS, Charles	17	SWNE	1860-04-02		A1 G106
2936	WILLIAMS, John	19	SENE	1892-07-25		A4
2886	WILSON, Henry	15	SE	1846-09-01		A1 G112
2895	YOUNG, Fred	19	NWNW	1905-08-05		A4

Family Maps of Lauderdale County, Mississippi

Patent Map
T6-N R17-E Choctaw Meridian
Map Group 15

Township Statistics

Parcels Mapped	:	148
Number of Patents	:	120
Number of Individuals	:	103
Patentees Identified	:	93
Number of Surnames	:	80
Multi-Patentee Parcels	:	11
Oldest Patent Date	:	1/5/1841
Most Recent Patent	:	2/25/1913
Block/Lot Parcels	:	0
Parcels Re-Issued	:	1
Parcels that Overlap	:	0
Cities and Towns	:	1
Cemeteries	:	5

Section 6
- WARBINGTON Jacob B 1841
- WAITS Levi 1846
- RUSSELL David 1841

Section 5
- LEWIS John 1846
- LEWIS John 1859
- HARPER Nancy L 1897
- RUSHING Charles E 1859
- KIMBRELL George 1909
- LEE Felix 1913
- HARPER Nancy L 1897
- LEWIS John 1859
- LEE Felix 1895
- BOLDEN John 1911
- HUTTON Sonny 1902
- HARPER William R 1906

Section 4

Section 7
- MCSHAN Walter J 1895
- WATES Martin 1846
- FREEMAN Gideon A 1860
- JACKSON [126] Green 1906
- FREEMAN Gideon A 1859
- MCSHAN Walter J 1895
- BUSON Samuel A 1859
- FREEMAN Gideon A 1859
- WALKER Moses A 1911
- RUSHING Charles E 1859
- FREEMAN Gideon A 1860
- FREEMAN Gideon A 1860

Section 8

Section 9
- KNIGHT Allen E 1906
- RUSHING Charles E 1859
- LACKEY Samuel L 1898
- RAWSON George W 1899
- GRIFFITH John 1901
- LASLEY James 1897
- LASHLEY James 1860
- GORDON Richmond T 1860
- GRIFFITH John 1901
- LASLEY James 1860
- MCELROY Joel 1890
- GORDON Richmond T 1859
- DAVIS Christopher J 1846
- HARPER Burrell W 1911
- GORDON Charles 1848

Section 18

Section 17
- WHITEHEAD Benjamin F 1892
- WHITEHEAD Benjamin F 1892
- JONES Sandy 1895
- JONES Henrietta 1894
- HENDERSON [106] Blank 1860
- MOTEN Rederick 1890
- MCMULLEN John A 1890
- MCMULLEN John A 1895

Section 16

Section 19
- YOUNG Fred 1905
- KNOX Stephen 1889
- KNOX Stephen 1889
- THOMPSON James 1897
- IRBY Patrick 1895
- HALL Sabra 1895
- HALL Sabra 1895
- WILLIAMS John 1892
- MARSHALL Virinda E 1905
- LARKIN Oliver 1910
- HILL William 1905
- GRANT Andrew J 1906
- JOHNSON Darius G 1891
- JOHNSON Darius G 1891
- MASTERS Jack 1895

Section 20
- HENDERSON Michael P 1846
- HENDERSON Michael P 1846
- WATTS Benjamin F 1859

Section 21
- PRINGLE Abraham 1852
- PRINGLE Samuel T 1848

Section 30

Section 29
- HARPER James M 1898
- HARPER James M 1898
- DEARMAN William R 1896

Section 28
- COKER Levi 1859
- BOWLING [25] William F 1846
- COKER Levi 1859
- PRINGLE Abraham 1852
- PRINGLE Abraham 1852
- COAKER [45] Peyton 1846
- PRINGLE Thomas K 1846

Section 31
- EMBRY Warren 1859
- RICHEY Joseph W 1894
- CRENSHAW Thomas G 1905
- SIMS Primus A 1895
- IRBY John H 1906
- IRBY John H 1906
- SIMS Primus A 1895
- IRBY John H 1906
- MCLEMORE Caleb S 1901

Section 32
- HUNT Henry J 1859
- HUNT Henry J 1859

Section 33
- COKER Peyton 1848
- BRUISTER James W 1848
- BRUISTER James W 1848

Copyright 2006 Boyd IT, Inc. All Rights Reserved

200

Township 6-N Range 17-E (Choctaw) - Map Group 15

Section 3
- RUSHING, Charles E — 1860
- RUSHING, Charles E — 1860
- GORDON, Robert B — 1896
- WELCH, Mina H — 1860
- BURKHALTER, Thomas — 1860
- DURR, Emanuel A — 1852

Section 2
- DURR, Emanuel A — 1852
- PIGFORD, Wright — 1860
- PIGFORD, Wright — 1846
- DURR, Emanuel A — 1852
- GORDON, Richmond T — 1859

Section 1
- BLACKWELL [23], Elemander — 1841
- HARPER [102], Alexander — 1846

Section 10
- RAWSON, Charles — 1846
- TERRELL, Joseph — 1859
- TERRELL, Joseph — 1859
- GORDON, Richmond T — 1859
- GORDON, James — 1859
- GORDON, Richmond T — 1859
- TERRELL, Joseph — 1859

Section 11

Section 12
- BONEY, Kinsey — 1848
- MOORE, Felix E — 1860

Section 15
- GORDON, Charles — 1846
- GORDON, Charles — 1846
- BRIANT [31], Benjamin — 1841
- HOLDEN [112], Ebenezer — 1846

Section 14
- GORDON, Richmond T — 1859
- GORDON, Richmond T — 1859
- WHITE, Thadeus C — 1846
- WHITE, Thadeus C — 1846
- HOLDEN [113], Ebenezer — 1841
- SPINKS, Alfred — 1846

Section 13
- NEWTON, William — 1846
- RODGERS, Bonapart — 1846
- RODGERS, Bonapart — 1846

Section 22
- GORDON, Richmond T — 1859
- PRINGLE, John A — 1859
- PRINGLE, Thomas K — 1846
- PRINGLE, Samuel T — 1848
- PRINGLE, Thomas K — 1846
- MERRITT, Willis — 1859

Section 23
- MERRITT, Willis — 1846
- MERRITT, Willis — 1846
- MERRITT, Willis — 1846

Section 24
- SCARBROUGH, Silas — 1846
- MCGRANEY [146], Philip — 1841
- MCMORY, William S — 1846
- SIMS, George — 1846

Section 27

Section 26
- MCMORY, William S — 1846

Section 25
- COOK [48], Lewis — 1846

Section 34

Section 35
- BRUISTER, James W — 1859
- SPINKS, John C — 1852
- BRUISTER, James W — 1859
- POGUE, James P — 1848

Section 36

Helpful Hints

1. This Map's INDEX can be found on the preceding pages.
2. Refer to Map "C" to see where this Township lies within Lauderdale County, Mississippi.
3. Numbers within square brackets [ ] denote a multi-patentee land parcel (multi-owner). Refer to Appendix "C" for a full list of members in this group.
4. Areas that look to be crowded with Patentees usually indicate multiple sales of the same parcel (Re-issues) or Overlapping parcels. See this Township's Index for an explanation of these and other circumstances that might explain "odd" groupings of Patentees on this map.

Copyright 2006 Boyd IT, Inc. All Rights Reserved

Legend

- ——— Patent Boundary
- ▬▬▬ Section Boundary
- (shaded) No Patents Found (or Outside County)
- 1., 2., 3., ... Lot Numbers (when beside a name)
- [ ] Group Number (see Appendix "C")

Scale: Section = 1 mile X 1 mile (generally, with some exceptions)

201

Family Maps of Lauderdale County, Mississippi

Road Map
T6-N R17-E
Choctaw Meridian
Map Group 15

Cities & Towns
Vimville

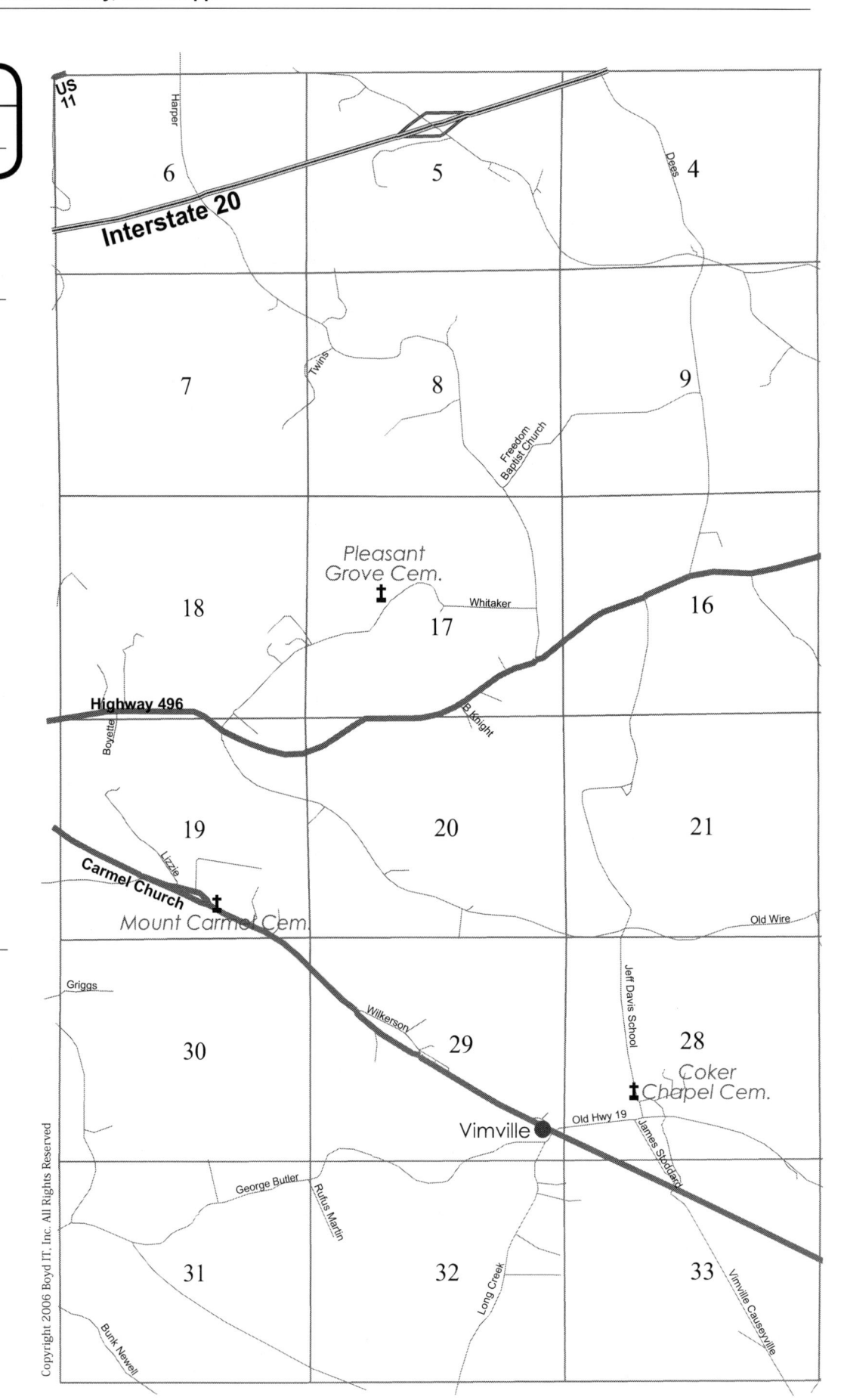

Cemeteries
Coker Chapel Cemetery
Mount Carmel Cemetery
Pigford Cemetery
Pleasant Grove Cemetery
Rawson Cemetery

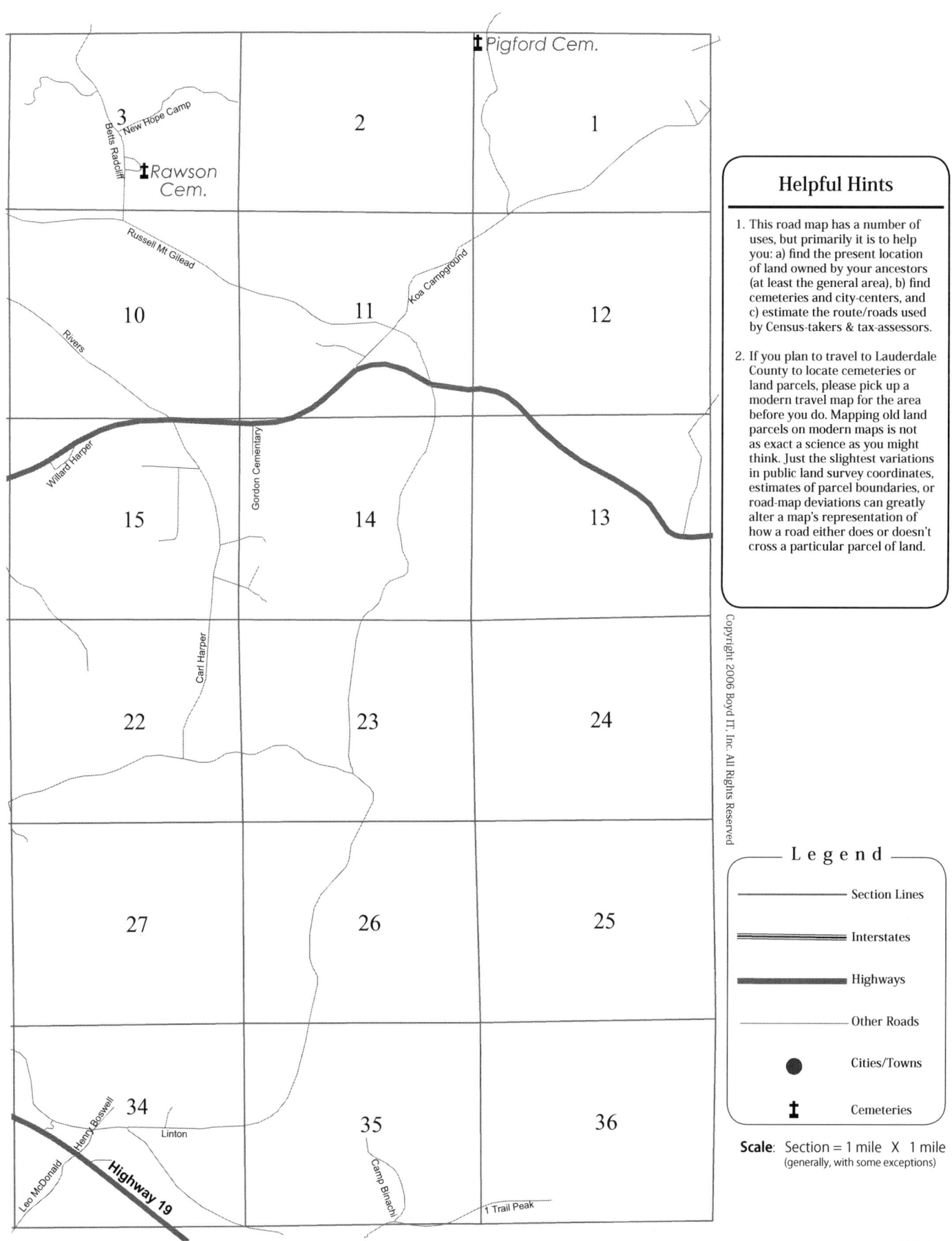

Township 6-N Range 17-E (Choctaw) - Map Group 15

Helpful Hints

1. This road map has a number of uses, but primarily it is to help you: a) find the present location of land owned by your ancestors (at least the general area), b) find cemeteries and city-centers, and c) estimate the route/roads used by Census-takers & tax-assessors.

2. If you plan to travel to Lauderdale County to locate cemeteries or land parcels, please pick up a modern travel map for the area before you do. Mapping old land parcels on modern maps is not as exact a science as you might think. Just the slightest variations in public land survey coordinates, estimates of parcel boundaries, or road-map deviations can greatly alter a map's representation of how a road either does or doesn't cross a particular parcel of land.

Copyright 2006 Boyd IT, Inc All Rights Reserved

Legend

— Section Lines
═ Interstates
━ Highways
— Other Roads
● Cities/Towns
☨ Cemeteries

Scale: Section = 1 mile X 1 mile
(generally, with some exceptions)

203

Family Maps of Lauderdale County, Mississippi

Historical Map

T6-N R17-E
Choctaw Meridian

Map Group 15

Cities & Towns
Vimville

Cemeteries
Coker Chapel Cemetery
Mount Carmel Cemetery
Pigford Cemetery
Pleasant Grove Cemetery
Rawson Cemetery

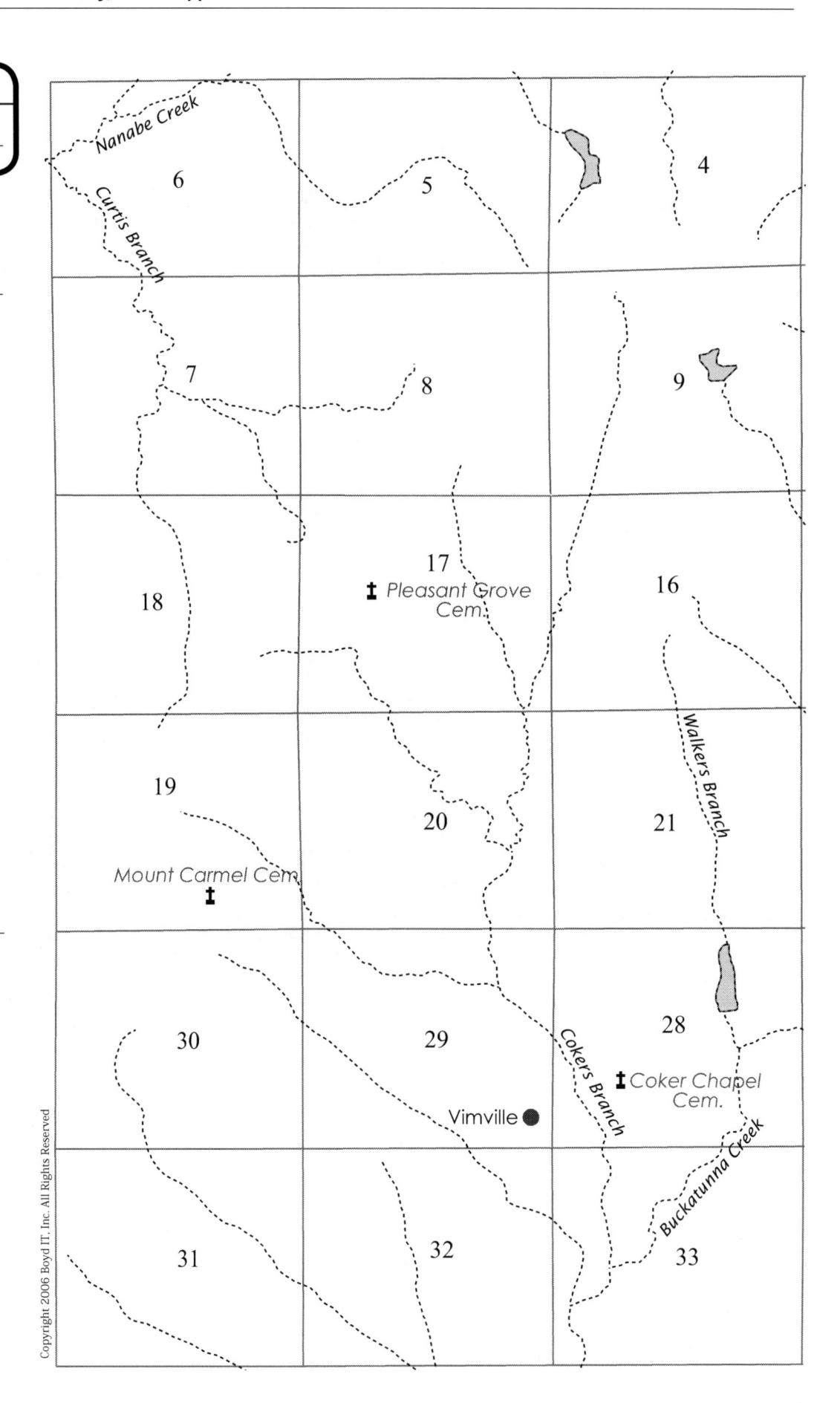

Township 6-N Range 17-E (Choctaw) - Map Group 15

Helpful Hints

1. This Map takes a different look at the same Congressional Township displayed in the preceding two maps. It presents features that can help you better envision the historical development of the area: a) Water-bodies (lakes & ponds), b) Water-courses (rivers, streams, etc.), c) Railroads, d) City/town center-points (where they were oftentimes located when first settled), and e) Cemeteries.

2. Using this "Historical" map in tandem with this Township's Patent Map and Road Map, may lead you to some interesting discoveries. You will often find roads, towns, cemeteries, and waterways are named after nearby landowners: sometimes those names will be the ones you are researching. See how many of these research gems you can find here in Lauderdale County.

Copyright 2006 Boyd IT, Inc. All Rights Reserved

Legend

— Section Lines
—+—+— Railroads
▬ Large Rivers & Bodies of Water
----- Streams/Creeks & Small Rivers
● Cities/Towns
✝ Cemeteries

Scale: Section = 1 mile X 1 mile
(there are some exceptions)

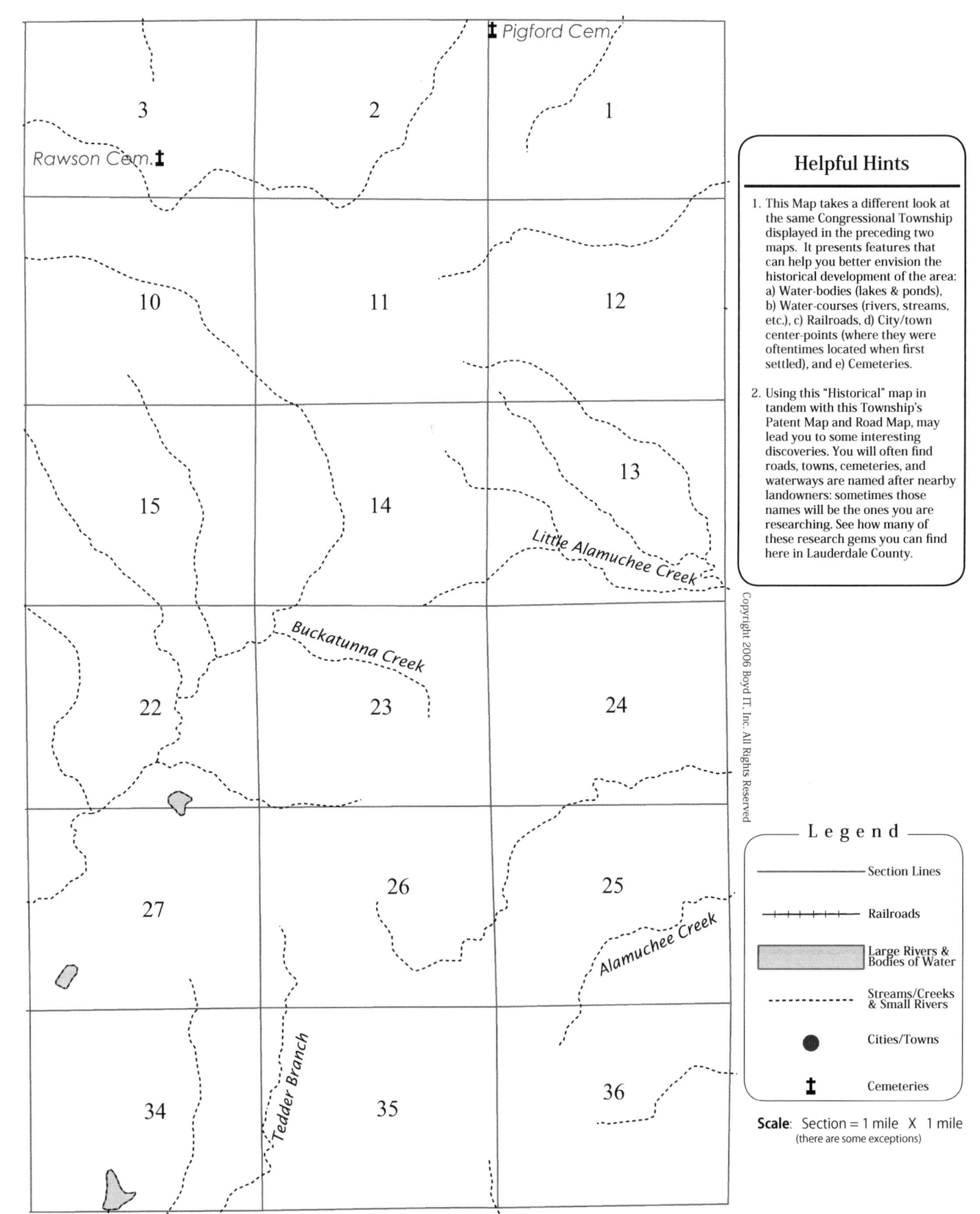

205

Map Group 16: Index to Land Patents
Township 6-North Range 18-East (Choctaw)

After you locate an individual in this Index, take note of the Section and Section Part then proceed to the Land Patent map on the pages immediately following. You should have no difficulty locating the corresponding parcel of land.

The "For More Info" Column will lead you to more information about the underlying Patents. See the *Legend* at right, and the "How to Use this Book" chapter, for more information.

LEGEND
"For More Info . . . " column

- **A** = Authority (Legislative Act, See Appendix "A")
- **B** = Block or Lot (location in Section unknown)
- **C** = Cancelled Patent
- **F** = Fractional Section
- **G** = Group (Multi-Patentee Patent, see Appendix "C")
- **V** = Overlaps another Parcel
- **R** = Re-Issued (Parcel patented more than once)

(A & G items require you to look in the Appendixes referred to above. All other Letter-designations followed by a number require you to locate line-items in this index that possess the ID number found after the letter).

ID	Individual in Patent	Sec.	Sec. Part	Date Issued	Other Counties	For More Info . . .
3057	ANDERSON, James C	20	W½SE	1846-09-01		A1 G10
3076	ANDERSON, John	20	E½SE	1846-09-01		A1 G11
3118	BAILIFF, Robert	21	E½NW	1846-09-01		A1
3119	" "	21	NESW	1846-09-01		A1
3005	BELL, Abraham	18	E½SE	1859-05-02		A1
3043	BOSWELL, Henry	3	SWNW	1846-09-01		A1
3044	" "	3	SWSW	1846-09-01		A1
3170	BOSWELL, Winney	4	E½NE	1859-05-02		A1
3018	BRAGG, Branch K	2	E½NW	1896-05-16		A1 G29
3019	" "	2	W½NE	1896-05-16		A1 G29
3039	BROWER, Franklin P	10	SENW	1859-05-02		A1
3018	" "	2	E½NW	1896-05-16		A1 G29
3019	" "	2	W½NE	1896-05-16		A1 G29
3032	CAMMACK, David N	3	E½SW	1848-09-01		A1
3153	CATLET, William	28	E½SE	1859-05-02		A1
3154	" "	28	NWSE	1859-05-02		A1
3045	CLARK, Henry	31	W½NE	1846-09-01		A1
3104	CORLEY, Nathan	21	W½SW	1846-09-01		A1 G53 R3036
3104	CRAWFORD, Charity	21	W½SW	1846-09-01		A1 G53 R3036
3130	CRENSHAW, Samuel	32	E½NW	1847-04-01		A1
3010	CULPEPPER, Ambrose H	26	11	1860-07-02		A1
3011	" "	26	13	1860-07-02		A1
3012	" "	26	14	1860-07-02		A1
3098	DAUGHTRY, Lawrence	4	NWSE	1860-04-02		A1
3117	DEASON, Rebecca	17	N½SW	1846-09-01		A1 G64
3117	DEASON, Sarah	17	N½SW	1846-09-01		A1 G64
3013	DODD, Arritter	11	W½NW	1841-01-05		A1
3014	" "	11	W½SE	1841-01-05		A1
3096	EASTIS, Joseph M	17	NE	1848-09-01		A1
3068	EAVES, Jesse B	30	NENE	1859-06-01		A1
3069	" "	30	W½NE	1859-06-01		A1
3070	EVES, Jesse B	17	W½SE	1846-09-01		A1
3149	EWING, William A	32	SWSE	1846-09-01		A1
3163	FLETCHER, William R	20	W½NW	1846-09-01		A1 G76
3141	GOODWYN, Thomas H	34	SESW	1859-05-02		A1
3142	" "	34	W½SE	1859-05-02		A1
3024	GRAYSON, Chesterfield	13	N½	1846-09-01		A1 G88 F
3166	HALL, William W	35	12	1846-09-01		A1 F
3167	" "	35	7	1846-09-01		A1 F
3155	HALTON, William	29	W½NW	1841-01-05		A1 G95
3059	HARDEN, James	26	3	1859-11-10		A1
3060	" "	26	5	1859-11-10		A1
3061	" "	26	6	1859-11-10		A1
3062	HARDIN, James	26	1	1846-09-01		A1 F
3063	" "	26	2	1846-09-01		A1 F
3064	" "	26	7	1846-09-01		A1 F

Township 6-N Range 18-E (Choctaw) - Map Group 16

ID	Individual in Patent	Sec.	Sec. Part	Date Issued	Other Counties	For More Info . . .
3065	HARDIN, James (Cont'd)	26	8	1846-09-01		A1 F
3105	HATHORN, Nevin C	6	E½SW	1841-01-05		A1 G104
3129	HATHORN, Samuel B	6	W½SW	1841-01-05		A1 G105
3156	HEARN, William	12		1846-09-01		A1 F
3067	HICKS, James M	11	W½NE	1846-09-01		A1 G109
3077	HICKS, John B	10	E½SE	1846-09-01		A1 G110
3101	HOLLAM, Mary	14	W½SE	1841-01-05		A1 G115
3102	HOLLAND, Mary	10	NE	1841-01-05		A1 G116
3148	HOLLAND, Uriah	11	SW	1842-07-23		A1 G117
3147	"	"	15	W½SE	1842-07-23		A1
3082	HOLLINSON, John	29	E½NE	1841-01-05		A1 G119
3066	HOOKS, James	34	S½NW	1859-06-01		A1
3094	HUMPHREYS, John W	8	E½SE	1860-08-01		A1
3095	"	"	8	SWSE	1860-08-01		A1
3105	HUMPHREYS, William	6	E½SW	1841-01-05		A1 G104
3129	"	"	6	W½SW	1841-01-05		A1 G105
3157	"	"	6	SENW	1848-09-01		A1
3015	HUNT, Augustus G	34	NENE	1860-04-02		A1
3016	"	"	34	W½NE	1860-04-02		A1
3103	HURT, Mary	17	NENW	1919-11-14		A4
3021	JONES, Charles	30	SENW	1852/05/01		A2
3022	"	"	30	SWNW	1852/05/01		A2
3023	"	"	4	SESE	1852/05/01		A2
3128	KIDD, Samuel A	10	SWNW	1859-06-01		A1
3009	LARD, Adam	15	W½NE	1841-01-05		A1 G133
3102	LEWIS, Jane	10	NE	1841-01-05		A1 G116
3148	LEWIS, Stephen	11	SW	1842-07-23		A1 G117
3135	"	"	15	E½SW	1842-07-23		A1 G135
3138	LUCY, Thomas B	32	NWSE	1859-11-10		A1
3139	"	"	32	SESE	1859-11-10		A1
3140	"	"	32	SESW	1859-11-10		A1
3152	LUCY, William B	32	NESW	1846-09-01		A1
3107	MARSH, Peter	14	W½NE	1841-01-05		A1
3108	"	"	14	W½NW	1841-01-05		A1
3101	"	"	14	W½SE	1841-01-05		A1 G115
3009	"	"	15	W½NE	1841-01-05		A1 G133
3114	"	"	15	W½SW	1841-01-05		A1 G138
3109	"	"	22	E½NE	1841-01-05		A1
3135	"	"	15	E½SW	1842-07-23		A1 G135
3113	"	"	10	W½SE	1846-09-01		A1 G140
3024	"	"	13	N½	1846-09-01		A1 G88 F
3110	"	"	22	W½NW	1846-09-01		A1
3115	"	"	23	4	1846-09-01		A1 G137 F
3116	"	"	23	5	1846-09-01		A1 G137 F
3112	"	"	23	6	1846-09-01		A1 F
3111	"	"	23	3	1868-09-01		A1 F
3034	MATHEWS, Edwin	30	SENE	1849-12-01		A1 R3035
3035	"	"	30	SENE	1849-12-01		A1 R3034
3150	MAY, William A	1	1	1897-01-12		A4
3151	"	"	1	2	1897-01-12		A4
3047	MCCOMB, Hugh H	10	E½SW	1846-09-01		A1
3048	"	"	14	SW	1846-09-01		A1
3049	"	"	15	E½SE	1846-09-01		A1
3120	MCCRANEY, Roderick	30	E½SW	1841-05-18		A1 G147
3067	MCCURDY, Archibald P	11	W½NE	1846-09-01		A1 G109
3136	MCGAHA, Tabitha	28	E½NW	1841-01-05		A1 G150
3158	MCKINNON, William	29	E½NW	1841-01-05		A1 G151
3030	MINOR, Cornelius	20	E½NW	1846-09-01		A1
3031	"	"	20	W½NE	1846-09-01		A1
3143	MOODY, Thomas	20	NENE	1859-05-02		A1
3115	NUTT, Jonathan	23	4	1846-09-01		A1 G137 F
3116	"	"	23	5	1846-09-01		A1 G137 F
3037	ODOM, Ephraim	14	E½SE	1841-01-05		A1
3100	ODOM, Malichi	6	NENE	1859-06-01		A1
3042	ONEAL, Gilbert	2	SE	1846-09-01		A1
3159	ONEAL, William	11	E½NW	1841-01-05		A1
3160	"	"	2	E½SW	1841-01-05		A1
3161	"	"	2	SWSW	1846-09-01		A1
3162	"	"	3	SE	1848-09-01		A1 G158
3081	PARISH, John G	31	NESW	1848-09-01		A1
3054	POPE, Jabez	6	SWNW	1848-09-01		A1
3055	"	"	7	NENW	1848-09-01		A1
3033	PORTER, Edmund	11	E½SE	1846-09-01		A1

Family Maps of Lauderdale County, Mississippi

ID	Individual in Patent	Sec.	Sec. Part	Date Issued	Other Counties	For More Info . . .
3053	PRINGLE, Isham K	5	NENE	1852/05/01		A2
3056	RAWLINGS, James B	27	NW	1846-09-01		A1
3075	REED, John A	11	E½NE	1846-09-01		A1
3074	" "	1	3	1849-12-01		A1 F
3106	RHODES, Oliver C	19	W½NE	1841-01-05		A1 G163
3017	RODGERS, Bonaparte	18	W½SW	1841-01-05		A1
3073	RUNNELS, Jesse	29	E½SW	1841-01-05		A1 G164
3099	RUNNELS, Lewis	29	W½NE	1841-01-05		A1 G165
3114	RUSSELL, Isaac	15	W½SW	1841-01-05		A1 G138
3046	SATCHER, Hiram M	4	W½NE	1859-06-01		A1
3132	SCARBROUGH, Silas	28	E½SW	1846-09-01		A1
3020	SHAMBERGER, Brantley	18	NESW	1860-07-02		A1
3006	SHAMBURGER, Absalom	6	SE	1846-09-01		A1
3007	" "	7	NWNE	1846-09-01		A1
3008	" "	8	NWSE	1846-09-01		A1
3057	SHAMBURGER, Elijah	20	W½SE	1846-09-01		A1 G10
3036	" "	21	W½SW	1846-09-01		A1 G169 R3104
3134	SHAMBURGER, Spinks	28	SWSW	1859-11-10		A1
3076	SHANBURGER, Elijah	20	E½SE	1846-09-01		A1 G11
3087	SMITH, John	17	SESW	1846-09-01		A1
3088	" "	17	SWSW	1846-09-01		A1
3086	" "	17	NWNW	1859-06-01		A1
3083	SMITH, John J	14	E½NW	1852/05/01		A2
3084	" "	15	NENW	1852/05/01		A2
3085	" "	15	NWNW	1852/05/01		A2
3133	SMITH, Simkin	17	S½NW	1846-09-01		A1
3144	SMITH, Thomas	21	NE	1846-09-01		A1
3164	SMITH, William	17	E½SE	1846-09-01		A1
3165	" "	20	SENE	1846-09-01		A1
3050	SPINKS, Isaac	30	E½SE	1846-09-01		A1
3051	" "	31	SESW	1848-09-01		A1
3052	" "	31	W½SE	1848-09-01		A1
3092	SPINKS, John	19	E½NE	1846-09-01		A1 G170
3163	" "	20	W½NW	1846-09-01		A1 G76
3089	" "	28	E½NE	1846-09-01		A1
3090	" "	28	W½NE	1846-09-01		A1
3091	" "	28	W½NW	1846-09-01		A1
3078	SPINKS, John C	20	E½SW	1852/05/01		A2
3079	" "	28	NWSW	1852/05/01		A2
3106	SPINKS, Rolley	19	W½NE	1841-01-05		A1 G163
3136	" "	28	E½NW	1841-01-05		A1 G150
3082	" "	29	E½NE	1841-01-05		A1 G119
3158	" "	29	E½NW	1841-01-05		A1 G151
3073	" "	29	E½SW	1841-01-05		A1 G164
3099	" "	29	W½NE	1841-01-05		A1 G165
3155	" "	29	W½NW	1841-01-05		A1 G95
3120	" "	30	E½SW	1841-05-18		A1 G147
3121	" "	21	E½SE	1846-09-01		A1
3122	" "	22	E½NW	1846-09-01		A1
3123	" "	22	NWSE	1846-09-01		A1
3124	" "	22	W½NE	1846-09-01		A1
3125	" "	29	NWSW	1846-09-01		A1
3126	SPINKS, Rolley T	29	NWSE	1848-09-01		A1
3131	SPINKS, Sarah	30	W½SE	1841-01-05		A1 G172
3137	SPINKS, Terry	20	W½SW	1859-06-01		A1
3127	STRANGE, Rolon	10	W½SW	1846-09-01		A1
3097	SUGGS, Joshua	19	W½SW	1846-09-01		A1
3058	SUTTLE, James H	19	E½NW	1846-09-01		A1
3093	TATUM, John	21	W½NW	1846-09-01		A1
3131	TAYLOR, William	30	W½SE	1841-01-05		A1 G172
3036	TERRAL, Celia	21	W½SW	1846-09-01		A1 G169 R3104
3038	THOMSON, Etheldred J	22	SW	1846-09-01		A1
3102	TOOL, Eli	10	NE	1841-01-05		A1 G116
3148	" "	11	SW	1842-07-23		A1 G117
3077	" "	10	E½SE	1846-09-01		A1 G110
3092	TOOMEY, Pinkney H	19	E½NE	1846-09-01		A1 G170
3025	WADKINS, Coleman C	28	SWSE	1846-09-01		A1
3026	" "	33	NWNE	1846-09-01		A1
3113	WALKER, Ezekiel	10	W½SE	1846-09-01		A1 G140
3145	WARD, Thomas	18	SESW	1846-09-01		A1
3146	" "	18	SWSW	1846-09-01		A1
3028	WATKINS, Coleman C	35	11	1846-09-01		A1 F
3029	" "	35	8	1846-09-01		A1 F

Township 6-N Range 18-E (Choctaw) - Map Group 16

ID	Individual in Patent	Sec.	Sec. Part	Date Issued	Other Counties	For More Info . . .
3027	WATKINS, Coleman C (Cont'd)	34	E½SE	1860-04-02		A1
3071	WELCH, Jesse L	4	NWNW	1859-05-02		A1
3072	" "	4	SENW	1859-05-02		A1
3080	WELCH, John C	4	NENW	1849-12-01		A1
3168	WIGGINS, William	14	E½NE	1841-01-05		A1
3169	" "	15	E½NE	1841-01-05		A1
3162	WILKINSON, George W	3	SE	1848-09-01		A1 G158
3041	" "	5	NWNE	1849-12-01		A1
3040	" "	30	W½SW	1859-05-02		A1

Family Maps of Lauderdale County, Mississippi

Patent Map

T6-N R18-E
Choctaw Meridian

Map Group 16

Township Statistics

Parcels Mapped	:	166
Number of Patents	:	138
Number of Individuals	:	113
Patentees Identified	:	107
Number of Surnames	:	88
Multi-Patentee Parcels	:	33
Oldest Patent Date	:	1/5/1841
Most Recent Patent	:	11/14/1919
Block/Lot Parcels	:	21
Parcels Re-Issued	:	2
Parcels that Overlap	:	0
Cities and Towns	:	1
Cemeteries	:	3

Township 6-N Range 18-E (Choctaw) - Map Group 16

Section 1
- Lots-Sec. 1
  - 1 MAY, William A 1897
  - 2 MAY, William A 1897
  - 3 REED, John A 1849

Section 2
- BRAGG [29] Branch K 1896
- BRAGG [29] Branch K 1896
- ONEAL William 1846
- ONEAL William 1841
- ONEAL Gilbert 1846

Section 3
- BOSWELL Henry 1846
- BOSWELL Henry 1846
- CAMMACK David N 1848
- ONEAL [158] William 1848

Section 10
- KIDD Samuel A 1859
- BROWER Franklin P 1859
- STRANGE Rolon 1846
- MCCOMB Hugh H 1846
- MARSH [140] Peter 1846
- HICKS [110] John B 1846

Section 11
- DODD Arritter 1841
- HOLLAND [116] Mary 1841
- ONEAL William 1841
- HICKS [109] James M 1846
- HOLLAND [117] Uriah 1842
- DODD Arritter 1841

Section 12
- HEARN William 1846
- REED John A 1846
- PORTER Edmund 1846

Section 13
- GRAYSON [88] Chesterfield 1846

Section 14
- MARSH Peter 1841
- WIGGINS William 1841
- SMITH John J 1852
- MCCOMB Hugh H 1846
- MARSH Peter 1841
- HOLLAM [115] Mary 1841
- WIGGINS William 1841
- ODOM Ephraim 1841

Section 15
- SMITH John J 1852
- SMITH John J 1852
- LARD [133] Adam 1841
- MARSH [138] Peter 1841
- HOLLAND Uriah 1842
- LEWIS [135] Stephen 1842
- MCCOMB Hugh H 1846

Section 22
- MARSH Peter 1846
- SPINKS Rolley 1846
- SPINKS Rolley 1846
- MARSH Peter 1841
- SPINKS Rolley 1846
- THOMSON Etheldred J 1846

Section 23
- Lots-Sec. 23
  - 3 MARSH, Peter 1868
  - 4 MARSH, Peter [137]1846
  - 5 MARSH, Peter [137]1846
  - 6 MARSH, Peter 1846

Section 26
- Lots-Sec. 26
  - 1 HARDIN, James 1846
  - 2 HARDIN, James 1846
  - 3 HARDEN, James 1859
  - 5 HARDEN, James 1859
  - 6 HARDEN, James 1859
  - 7 HARDIN, James 1846
  - 8 HARDIN, James 1846
  - 11 CULPEPPER, Ambrose H 1860
  - 13 CULPEPPER, Ambrose H 1860
  - 14 CULPEPPER, Ambrose H 1860

Section 27
- RAWLINGS James B 1846

Section 34
- HUNT Augustus G 1860
- HUNT Augustus G 1860
- HOOKS James 1859
- GOODWYN Thomas H 1859
- GOODWYN Thomas H 1859
- WATKINS Coleman C 1860

Section 35
- Lots-Sec. 35
  - 7 HALL, William W 1846
  - 8 WATKINS, Coleman C 1846
  - 11 WATKINS, Coleman C 1846
  - 12 HALL, William W 1846

Helpful Hints

1. This Map's INDEX can be found on the preceding pages.

2. Refer to Map "C" to see where this Township lies within Lauderdale County, Mississippi.

3. Numbers within square brackets [ ] denote a multi-patentee land parcel (multi-owner). Refer to Appendix "C" for a full list of members in this group.

4. Areas that look to be crowded with Patentees usually indicate multiple sales of the same parcel (Re-issues) or Overlapping parcels. See this Township's Index for an explanation of these and other circumstances that might explain "odd" groupings of Patentees on this map.

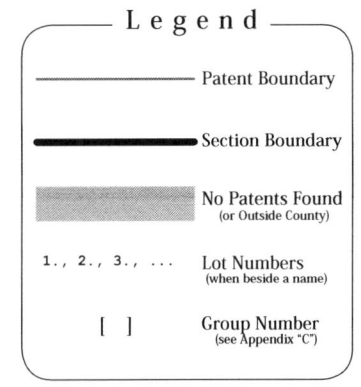

Copyright 2006 Boyd IT, Inc. All Rights Reserved

Legend

— Patent Boundary
— Section Boundary
▓ No Patents Found (or Outside County)
1., 2., 3., ... Lot Numbers (when beside a name)
[ ] Group Number (see Appendix "C")

Scale: Section = 1 mile X 1 mile (generally, with some exceptions)

211

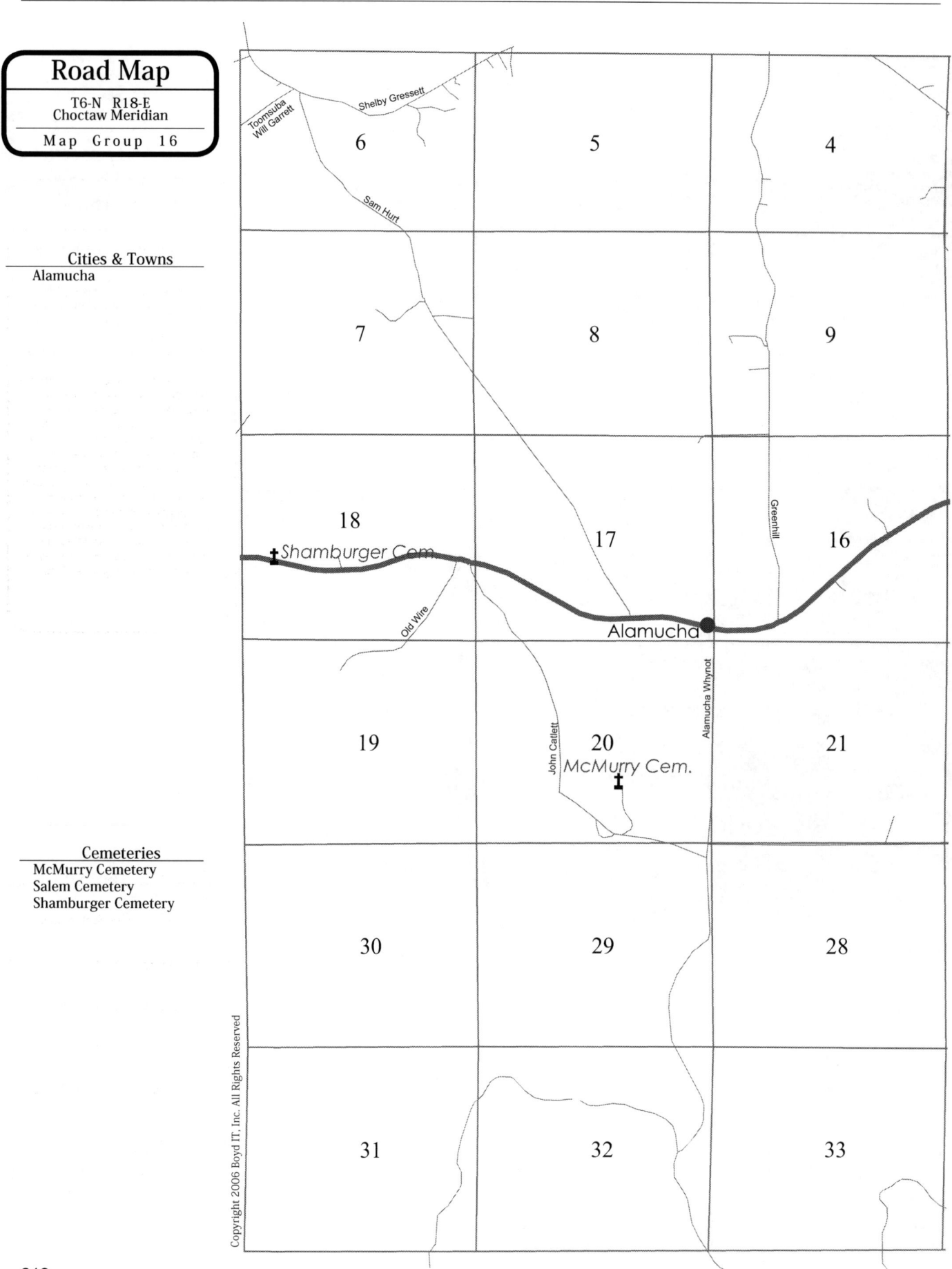

Family Maps of Lauderdale County, Mississippi

Road Map
T6-N R18-E
Choctaw Meridian
Map Group 16

Cities & Towns
Alamucha

Cemeteries
McMurry Cemetery
Salem Cemetery
Shamburger Cemetery

Township 6-N Range 18-E (Choctaw) - Map Group 16

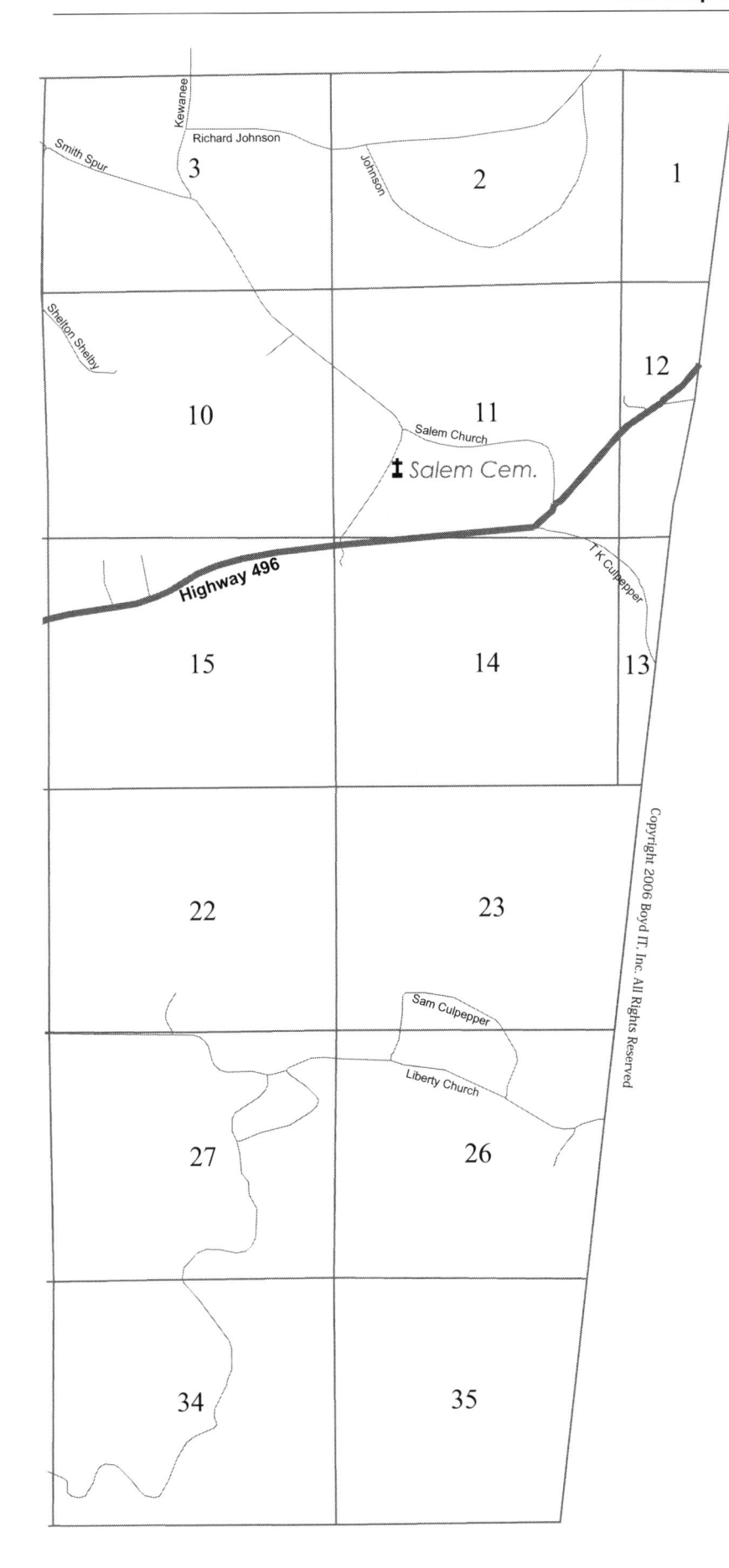

Helpful Hints

1. This road map has a number of uses, but primarily it is to help you: a) find the present location of land owned by your ancestors (at least the general area), b) find cemeteries and city-centers, and c) estimate the route/roads used by Census-takers & tax-assessors.

2. If you plan to travel to Lauderdale County to locate cemeteries or land parcels, please pick up a modern travel map for the area before you do. Mapping old land parcels on modern maps is not as exact a science as you might think. Just the slightest variations in public land survey coordinates, estimates of parcel boundaries, or road-map deviations can greatly alter a map's representation of how a road either does or doesn't cross a particular parcel of land.

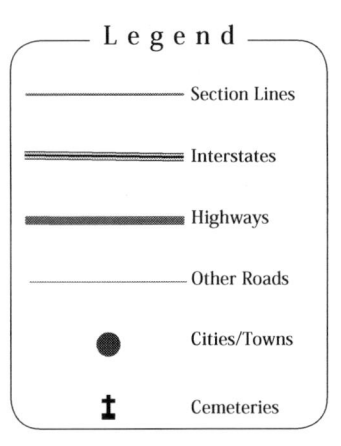

Legend

- Section Lines
- Interstates
- Highways
- Other Roads
- ● Cities/Towns
- ✝ Cemeteries

Scale: Section = 1 mile X 1 mile
(generally, with some exceptions)

213

Family Maps of Lauderdale County, Mississippi

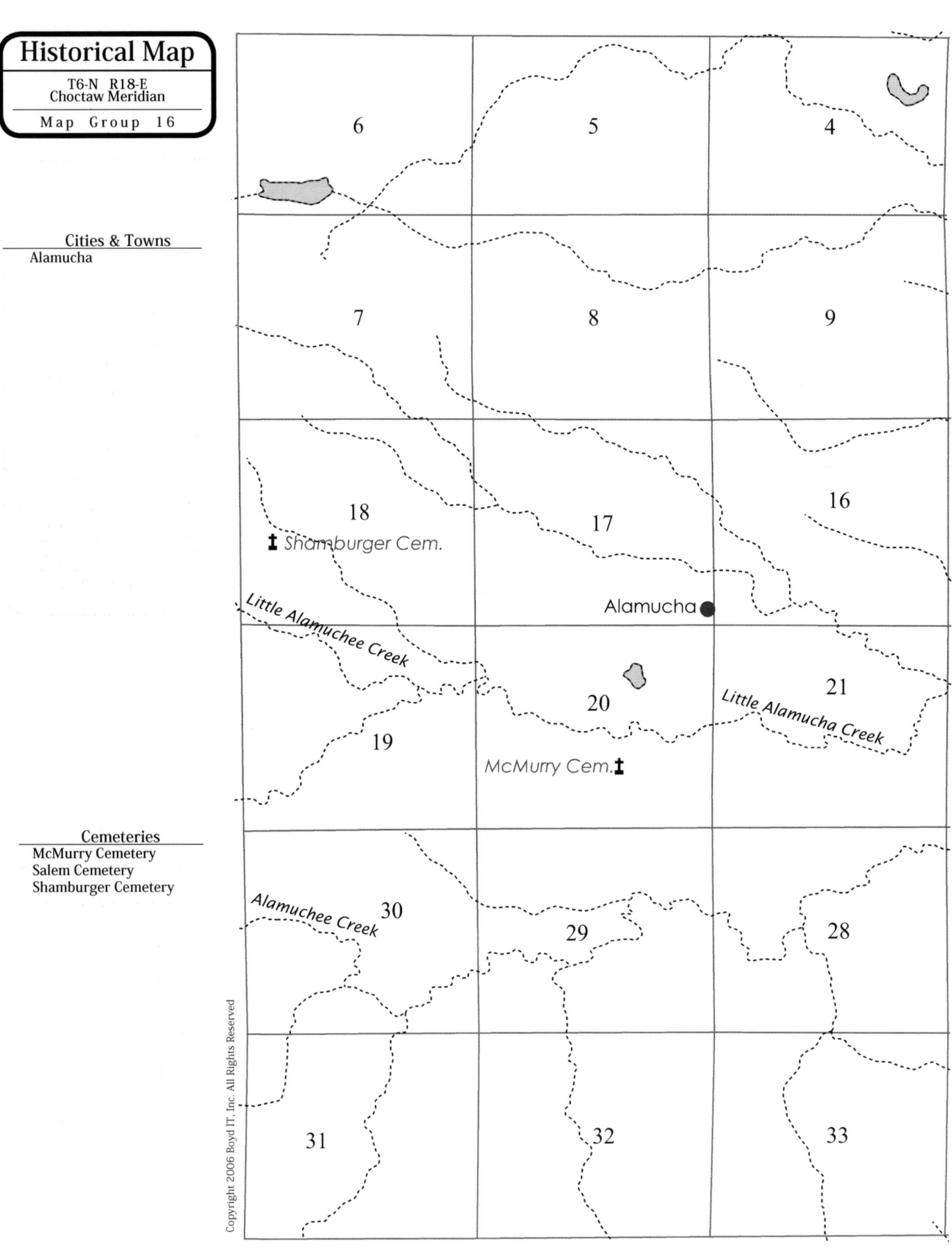

Historical Map
T6-N R18-E
Choctaw Meridian
Map Group 16

Cities & Towns
Alamucha

Cemeteries
McMurry Cemetery
Salem Cemetery
Shamburger Cemetery

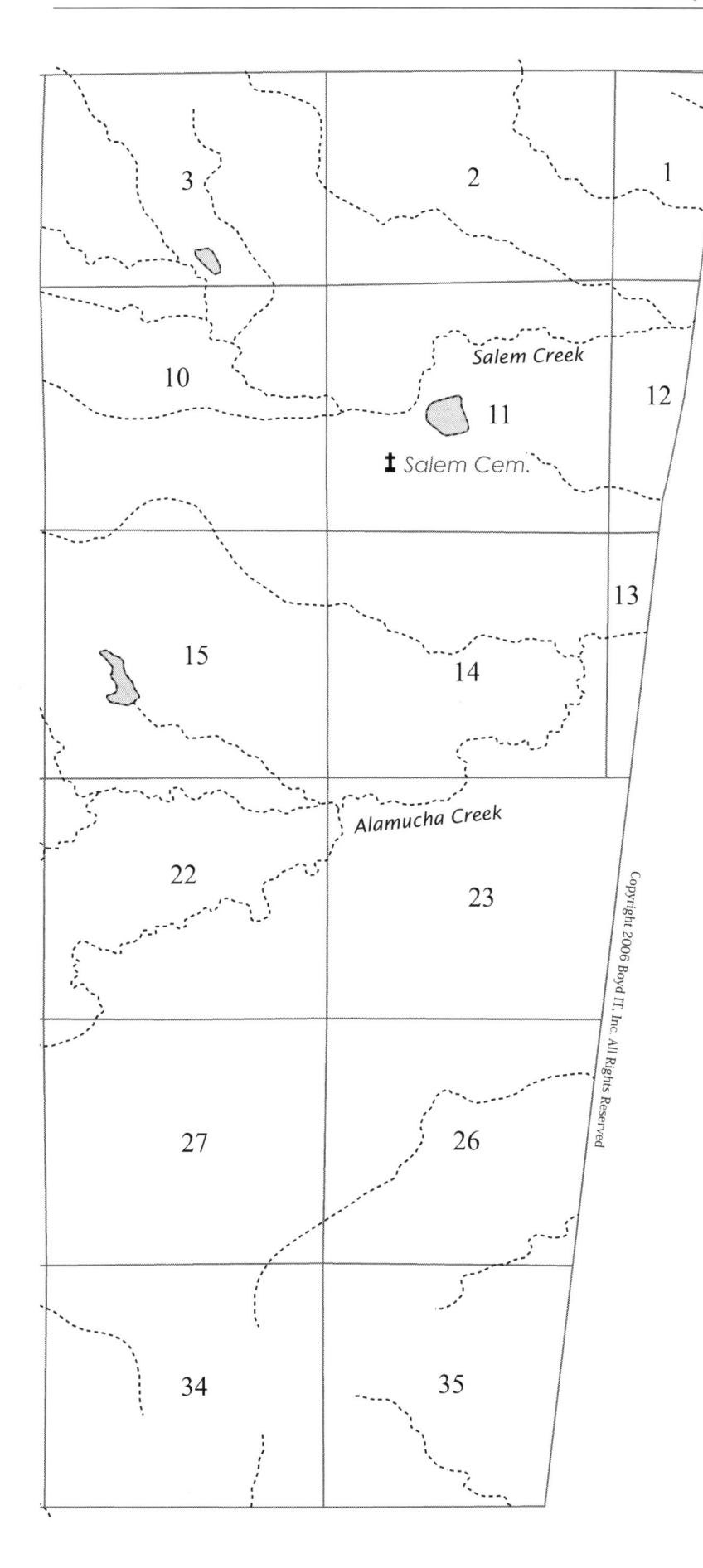

Township 6-N Range 18-E (Choctaw) - Map Group 16

Helpful Hints

1. This Map takes a different look at the same Congressional Township displayed in the preceding two maps. It presents features that can help you better envision the historical development of the area: a) Water-bodies (lakes & ponds), b) Water-courses (rivers, streams, etc.), c) Railroads, d) City/town center-points (where they were oftentimes located when first settled), and e) Cemeteries.

2. Using this "Historical" map in tandem with this Township's Patent Map and Road Map, may lead you to some interesting discoveries. You will often find roads, towns, cemeteries, and waterways are named after nearby landowners: sometimes those names will be the ones you are researching. See how many of these research gems you can find here in Lauderdale County.

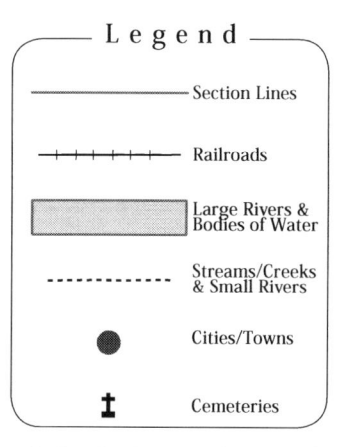

Legend

— Section Lines
+++++ Railroads
▒ Large Rivers & Bodies of Water
----- Streams/Creeks & Small Rivers
● Cities/Towns
✝ Cemeteries

Scale: Section = 1 mile X 1 mile
(there are some exceptions)

215

Map Group 17: Index to Land Patents
Township 5-North Range 14-East (Choctaw)

After you locate an individual in this Index, take note of the Section and Section Part then proceed to the Land Patent map on the pages immediately following. You should have no difficulty locating the corresponding parcel of land.

The "For More Info" Column will lead you to more information about the underlying Patents. See the *Legend* at right, and the "How to Use this Book" chapter, for more information.

LEGEND
"For More Info . . . " column
- **A** = Authority (Legislative Act, See Appendix "A")
- **B** = Block or Lot (location in Section unknown)
- **C** = Cancelled Patent
- **F** = Fractional Section
- **G** = Group (Multi-Patentee Patent, see Appendix "C")
- **V** = Overlaps another Parcel
- **R** = Re-Issued (Parcel patented more than once)

(A & G items require you to look in the Appendixes referred to above. All other Letter-designations followed by a number require you to locate line-items in this index that possess the ID number found after the letter).

ID	Individual in Patent	Sec.	Sec. Part	Date Issued	Other Counties	For More Info . . .
3326	AGLETREE, D	18	NESW	1852/09/10		A2 G70
3327	" "	18	NWSE	1852/09/10		A2 G70
3328	" "	6	SWSE	1852/09/10		A2 G70
3192	ALLEN, Byrd O	15	NENE	1848-09-01		A1
3193	" "	15	NWSW	1859-05-02		A1
3314	ALLEN, Preston Brooks	11	S½SW	1923-09-20		A4
3304	ANDERSON, James M	21	E½NW	1852/09/10		A2 G99
3174	ARRANT, Allen	8	E½SW	1841-01-05		A1
3175	" "	8	S½NW	1841-01-05		A1
3179	ARRINGTON, Arther	34	W½SE	1841-01-05		A1
3241	BARRETT, Henry E	7	NENW	1851-10-01		A1
3309	BENNETT, Nelson	13	SWNW	1892-01-18		A4
3322	BENNETT, Robert	25	SWSW	1897-05-07		A4
3346	BRIANT, William	6	SWNW	1841-01-05		A1
3347	" "	8	NWNW	1841-01-05		A1
3226	BROWN, George	23	NESE	1882-03-30		A4
3250	BROWN, John A	15	NESE	1894-09-28		A4
3251	" "	15	SENE	1894-09-28		A4
3252	" "	15	W½SE	1894-09-28		A4
3352	BROWN, William L	15	SWSW	1882-05-10		A1
3176	CARTER, Allen C	25	SESE	1860-04-02		A1
3180	CHADICK, Asa	10	S½SE	1848-09-01		A1
3239	CHAMBLIS, Henry	17	NENW	1848-09-01		A1
3240	" "	17	SWNE	1848-09-01		A1
3348	CHAMBLISS, William	18	NESE	1850/06/01		A2
3279	CLARK, John W	3	N½SW	1890-08-16		A4
3177	COLE, Amandy	3	S½SW	1890-02-21		A4
3258	CORLEY, John H	6	SENW	1841-01-05		A1
3305	DAUCEY, Milton	1	E½SW	1900-10-04		A4
3199	DEAR, Darling	15	E½SW	1841-01-05		A1
3200	" "	15	SESE	1841-01-05		A1
3201	" "	17	E½SE	1841-01-05		A1
3207	" "	22	E½NW	1841-01-05		A1
3205	" "	20	SWSW	1859-11-10		A1
3202	" "	18	SESE	1860-07-02		A1
3203	" "	20	NESW	1860-07-02		A1
3204	" "	20	SESE	1860-07-02		A1
3206	" "	20	W½NW	1860-07-02		A1
3238	DEAR, Hardy	17	W½SE	1841-01-05		A1
3242	DEAR, Herrin	17	W½NW	1841-01-05		A1
3259	DEAR, John H	30	E½NW	1860-04-10		A1
3260	" "	30	N½NE	1860-04-10		A1
3184	DEBOE, Ben	3	S½SE	1892-04-29		A4
3208	DEER, Darling	17	E½SW	1841-01-05		A1
3209	" "	19	SENE	1848-09-01		A1
3210	" "	19	SESE	1848-09-01		A1

Township 5-N Range 14-E (Choctaw) - Map Group 17

ID	Individual in Patent	Sec.	Sec. Part	Date Issued	Other Counties	For More Info...
3211	DEER, Darling (Cont'd)	20	NWSW	1848-09-01		A1
3212	DEIR, Darling	22	W½NW	1841-01-05		A1
3326	DELK, S	18	NESW	1852/09/10		A2 G70
3327	" "	18	NWSE	1852/09/10		A2 G70
3328	" "	6	SWSE	1852/09/10		A2 G70
3219	DUNN, Elizabeth A	35	SENE	1860-04-02		A1
3255	DUNN, John C	35	NWNW	1860-04-02		A1
3256	DYASS, John	32	W½NE	1841-01-05		A1
3291	DYASS, Joshua	20	SESW	1841-01-05		A1
3292	" "	22	SWSE	1841-01-05		A1
3294	" "	34	W½NE	1841-01-05		A1
3293	" "	27	W½NE	1842-05-06		A1
3271	ELAM, John M	11	N½NW	1895-06-28		A4
3312	ELAM, Preston B	11	NESW	1905-03-30		A4
3313	" "	11	S½NE	1905-03-30		A4
3315	ERNEST, Prince	33	SENW	1898-01-19		A4
3326	EVERETT, Samuel	18	NESW	1852/09/10		A2 G70
3327	" "	18	NWSE	1852/09/10		A2 G70
3328	" "	6	SWSE	1852/09/10		A2 G70
3178	FIKES, Andrew J	25	N½SE	1882-08-03		A4
3230	FIKES, George W	25	NENW	1895-06-19		A4
3231	" "	25	NWNE	1895-06-19		A4
3301	FIKES, Manuel	23	NWSE	1860-10-01		A1
3303	FIKES, Martin	21	W½SE	1859-05-02		A1
3302	" "	21	E½SE	1860-10-01		A1
3311	FORD, Oliver	33	W½NW	1895-01-17		A4
3213	GADDIS, Dicy	1	E½NE	1882-05-20		A4 G78
3213	GADDIS, George	1	E½NE	1882-05-20		A4 G78
3290	GEORGE, Joseph L	6	SESW	1848-09-01		A1
3349	GOUGH, William	28	NESE	1841-01-05		A1
3182	GRAY, Asbury E	6	NWSE	1860-07-02		A1
3183	" "	6	SWNE	1860-07-02		A1
3286	GRICE, Jonathan	5	W½SW	1841-01-05		A1
3287	" "	6	E½SE	1841-01-05		A1
3325	HADEN, Samuel D	13	NWNW	1860-10-01		A1
3350	HALEY, William	7	SWSW	1841-01-05		A1
3304	HANDLEY, Mary F	21	E½NW	1852/09/10		A2 G99
3261	HARRINGTON, John	15	E½NW	1848-09-01		A1
3218	HARRIS, Elisabeth F	25	SESE	1897-02-15		A4
3326	HARRIS, J	18	NESW	1852/09/10		A2 G70
3327	" "	18	NWSE	1852/09/10		A2 G70
3328	" "	6	SWSE	1852/09/10		A2 G70
3353	HARVEY, William O	25	N½SW	1893-08-14		A4
3354	" "	25	S½NW	1893-08-14		A4
3324	HARWELL, Samuel B	5	SESW	1848-09-01		A1
3321	HENDERSON, Richard	1	SWSE	1922-03-25		A4
3181	HERRING, Asa	6	N½NW	1859-05-02		A1
3262	HERRINGTON, John	15	W½NE	1846-09-01		A1
3263	" "	15	W½NW	1846-09-01		A1
3214	HINES, Dover E	1	W½SW	1885-06-20		A4
3215	" "	11	N½NE	1902-01-17		A4 G111
3215	HINES, Narcissa	11	N½NE	1902-01-17		A4 G111
3295	HODGES, Lewis	1	W½NW	1898-12-27		A4
3236	HOLLIMON, Griffin H	36	E½SW	1841-01-05		A1 G118
3237	" "	36	W½SE	1841-01-05		A1 G118
3275	HOWSE, John S	35	SENW	1852/09/10		A2
3276	" "	35	SWNW	1852/09/10		A2
3326	JENNINGS, A	18	NESW	1852/09/10		A2 G70
3327	" "	18	NWSE	1852/09/10		A2 G70
3328	" "	6	SWSE	1852/09/10		A2 G70
3285	JENNINGS, Johnnie Charles	27	S½SW	1917-08-11		A4
3172	JONES, Alfred	23	NENE	1921-11-08		A4
3173	" "	23	S½NE	1921-11-08		A4
3270	JONES, John	1	W½NE	1899-08-30		A4
3334	JONES, Thomas C	28	NWSE	1841-01-05		A1
3335	" "	28	SENE	1841-01-05		A1
3244	KELLEY, Ichabud	6	E½NE	1841-01-05		A1
3332	KIMBAUGH, Shodrick J	13	NWSE	1860-10-01		A1
3331	KIMBROUGH, Shadrack J	13	E½SE	1895-10-09		A4
3196	KING, Clark	13	NENW	1892-06-15		A4
3197	" "	13	NWNE	1892-06-15		A4
3236	KING, James W	36	E½SW	1841-01-05		A1 G118
3237	" "	36	W½SE	1841-01-05		A1 G118

217

Family Maps of Lauderdale County, Mississippi

ID	Individual in Patent	Sec.	Sec. Part	Date Issued	Other Counties	For More Info...
3249	KING, James W (Cont'd)	36	W½SW	1841-01-05		A1
3216	KIRKLAND, Eli W	33	W½SW	1860-10-01		A1
3227	KNIGHT, George	17	NWSW	1852/09/10		A2
3228	" "	18	SENE	1852/09/10		A2
3229	" "	8	SWSE	1852/09/10		A2
3344	LANG, William A	35	S½	1841-01-05		A1
3343	LEE, Wesley	13	S½NE	1888-04-05		A4
3234	LEWIS, Glover	3	NWSE	1905-10-19		A4
3235	" "	3	SWNE	1905-10-19		A4
3247	LEWIS, James	3	NWNE	1920-10-22		A1
3339	LEWIS, Thomas	6	NWNE	1859-05-02		A1
3340	" "	6	SWSW	1859-05-02		A1
3223	LITTLE, Fred A	23	NWNW	1919-06-03		A4
3224	" "	23	NWSW	1919-06-03		A4
3225	" "	23	S½NW	1919-06-03		A4
3185	LLOYD, Benjamin E	27	W½NW	1849-12-01		A1
3186	" "	28	NENW	1849-12-01		A1
3187	" "	28	SWNE	1849-12-01		A1
3195	MAYERHOFF, Charles F	25	NENE	1859-05-02		A1
3243	MCCALL, Hugh H	29	E½SE	1860-04-02		A1
3171	MCDONALD, Alexander	35	SWNE	1859-05-02		A1
3190	MOORE, Berry W	33	E½SE	1895-06-27		A4
3191	" "	33	SENE	1895-06-27		A4
3272	MOORE, John	20	NENE	1850/06/01		A2
3273	" "	20	SWNE	1850/06/01		A2
3310	MOORE, Noah C	33	NENE	1898-05-16		A4
3330	MOORE, Santford O	27	S½SE	1899-11-24		A4
3307	MORRIS, Nathan D	13	SWSE	1841-01-05		A1
3308	" "	13	W½SW	1841-01-05		A1
3274	MUNDELL, John	30	SENE	1859-05-02		A1
3300	MURPHY, Malachi B	8	NENW	1860-04-02		A1
3298	NEW, Luke	3	NWNW	1895-06-19		A4
3299	" "	3	S½NW	1895-06-19		A4
3351	OSBORN, William J	13	NENE	1892-08-01		A1
3355	PARKER, Willie	3	NENW	1912-06-27		A4
3306	PITMAN, Milton	1	E½NW	1898-07-12		A4
3317	PRICE, Reese	27	NESE	1859-06-01		A1
3318	" "	33	SESW	1860-10-01		A1 R3319
3319	" "	33	SESW	1860-10-01		A1 R3318
3320	" "	33	SWSE	1860-10-01		A1
3356	PRICE, Willis	35	NENW	1898-11-11		A4
3316	REECE, Rebecca	33	NENW	1896-01-10		A4 G162
3316	REECE, Squire	33	NENW	1896-01-10		A4 G162
3326	RICH, D E	18	NESW	1852/09/10		A2 G70
3327	" "	18	NWSE	1852/09/10		A2 G70
3328	" "	6	SWSE	1852/09/10		A2 G70
3333	RILEY, Tapley T	9	SENW	1841-01-05		A1
3264	SMITH, John J	17	SWSW	1852/05/01		A2
3265	" "	20	NENW	1852/05/01		A2
3266	" "	20	SWSE	1852/05/01		A2
3267	" "	27	E½NW	1852/05/01		A2
3268	" "	29	NWNE	1852/05/01		A2
3269	" "	31	NWSE	1852/05/01		A2
3248	SPEED, James R	25	SWSE	1888-02-25		A4
3257	SPEED, John E	25	S½NE	1892-06-15		A4
3217	STINSON, Elijah	23	NESW	1860-04-02		A1
3198	STUCKEY, Daniel	27	NWSE	1841-01-05		A1
3277	STUCKEY, John	22	SESW	1841-01-05		A1
3278	" "	27	NESW	1841-01-05		A1
3337	SULLIVAN, Thomas J	29	SWSE	1860-04-02		A1
3245	SUTTLE, Isaac G	13	SESE	1860-10-01		A1
3288	THOMPSON, Joseph B	27	NWSW	1860-07-02		A1
3289	" "	27	SENE	1860-10-01		A1
3323	THOMPSON, Robert	35	N½NE	1895-05-11		A4
3329	THOMPSON, Samuel L	27	NENE	1911-10-09		A1
3336	TILLMAN, Thomas F	11	W½SE	1892-06-15		A4
3220	VAUGHAN, Franklin C	11	NWSW	1859-11-10		A1
3221	" "	11	SWNW	1859-11-10		A1
3222	VAUGHN, Franklin C	11	SENW	1859-11-10		A1
3253	VAUGHN, John A	23	NENW	1878-06-24		A4
3254	" "	23	NWNE	1878-06-24		A4
3194	WALKER, Calton	3	NESE	1860-04-02		A1
3280	WALKER, John	3	E½NE	1902-07-03		A4

Township 5-N Range 14-E (Choctaw) - Map Group 17

ID	Individual in Patent	Sec.	Sec. Part	Date Issued	Other Counties	For More Info . . .
3345	WARD, William A	25	NWNW	1860-04-02		A1
3281	WARREN, John	18	SESW	1859-11-10		A1
3282	" "	18	SWNE	1859-11-10		A1
3283	" "	18	SWSE	1859-11-10		A1
3284	" "	18	W½SW	1859-11-10		A1
3338	WEST, Thomas J	31	SESE	1859-06-01		A1
3296	WHITLOCK, Louisa	13	NESW	1918-04-11		A4
3297	" "	13	SENW	1918-04-11		A4
3246	WILLIAMSON, James C	6	N½SW	1859-05-02		A1
3233	WILSON, George	1	NWSE	1892-04-29		A4
3232	" "	1	E½SE	1897-05-07		A4
3341	WRIGHT, Wadkin	33	NWSE	1919-05-26		A4
3342	" "	33	W½NE	1919-05-26		A4
3188	YEAGER, Berry B	20	NWNE	1859-11-10		A1
3189	" "	20	SENE	1859-11-10		A1

Family Maps of Lauderdale County, Mississippi

Patent Map

T5-N R14-E
Choctaw Meridian

Map Group 17

Township Statistics

Parcels Mapped	:	186
Number of Patents	:	143
Number of Individuals	:	128
Patentees Identified	:	120
Number of Surnames	:	95
Multi-Patentee Parcels	:	9
Oldest Patent Date	:	1/5/1841
Most Recent Patent	:	9/20/1923
Block/Lot Parcels	:	0
Parcels Re-Issued	:	1
Parcels that Overlap	:	0
Cities and Towns	:	0
Cemeteries	:	4

220

Township 5-N Range 14-E (Choctaw) - Map Group 17

Section 3
- NEW Luke 1895
- PARKER Willie 1912
- LEWIS James 1920
- NEW Luke 1895
- LEWIS Glover 1905
- WALKER John 1902
- CLARK John W 1890
- LEWIS Glover 1905
- WALKER Calton 1860
- COLE Amandy 1890
- DEBOE Ben 1892

Section 2
(No Patents Found)

Section 1
- HODGES Lewis 1898
- JONES John 1899
- PITMAN Milton 1898
- GADDIS [78] Dicy 1882
- HINES Dover E 1885
- WILSON George 1892
- WILSON George 1897
- DAUCEY Milton 1900
- HENDERSON Richard 1922

Section 10
(No Patents Found)

Section 11
- ELAM John M 1895
- HINES [111] Dover E 1902
- VAUGHAN Franklin C 1859
- VAUGHN Franklin C 1859
- ELAM Preston B 1905
- VAUGHAN Franklin C 1859
- ELAM Preston B 1905
- TILLMAN Thomas F 1892
- CHADICK Asa 1848
- ALLEN Preston Brooks 1923

Section 12
(No Patents Found)

Section 15
- HERRINGTON John 1846
- HERRINGTON John 1846
- ALLEN Byrd O 1848
- HARRINGTON John 1848
- BROWN John A 1894
- ALLEN Byrd O 1859
- BROWN John A 1894
- BROWN John A 1894
- BROWN William L 1882
- DEAR Darling 1841
- DEAR Darling 1841

Section 14
(No Patents Found)

Section 13
- HADEN Samuel D 1860
- KING Clark 1892
- KING Clark 1892
- OSBORN William J 1892
- BENNETT Nelson 1892
- WHITLOCK Louisa 1918
- LEE Wesley 1888
- MORRIS Nathan D 1841
- WHITLOCK Louisa 1918
- KIMBRAUGH Shodrick J 1860
- KIMBROUGH Shadrack J 1895
- SUTTLE Isaac G 1860
- MORRIS Nathan D 1841

Section 22
- DEIR Darling 1841
- DEAR Darling 1841

Section 23
- LITTLE Fred A 1919
- VAUGHN John A 1878
- VAUGHN John A 1878
- JONES Alfred 1921
- LITTLE Fred A 1919
- JONES Alfred 1921
- LITTLE Fred A 1919
- STINSON Elijah 1860
- FIKES Manuel 1860
- BROWN George 1882

Section 24
(No Patents Found)

Section 27
- STUCKEY John 1841
- DYASS Joshua 1841
- LLOYD Benjamin E 1849
- DYASS Joshua 1842
- THOMPSON Samuel L 1911
- SMITH John J 1852
- THOMPSON Joseph B 1860
- THOMPSON Joseph B 1860
- STUCKEY Daniel 1841
- PRICE Reese 1859
- JENNINGS Johnnie Charles 1917
- MOORE Santford O 1899

Section 26
(No Patents Found)

Section 25
- WARD William A 1860
- FIKES George W 1895
- FIKES George W 1895
- MAYERHOFF Charles F 1859
- HARVEY William O 1893
- SPEED John E 1892
- HARVEY William O 1893
- FIKES Andrew J 1882
- BENNETT Robert 1897
- CARTER Allen C 1860
- SPEED James R 1888
- HARRIS Elisabeth F 1897

Section 34
- DYASS Joshua 1841
- ARRINGTON Arther 1841

Section 35
- DUNN John C 1860
- PRICE Willis 1898
- THOMPSON Robert 1895
- HOWSE John S 1852
- HOWSE John S 1852
- MCDONALD Alexander 1859
- DUNN Elizabeth A 1860
- LANG William A 1841

Section 36
- KING James W 1841
- HOLLIMON [118] Griffin H 1841
- HOLLIMON [118] Griffin H 1841

Helpful Hints

1. This Map's INDEX can be found on the preceding pages.
2. Refer to Map "C" to see where this Township lies within Lauderdale County, Mississippi.
3. Numbers within square brackets [ ] denote a multi-patentee land parcel (multi-owner). Refer to Appendix "C" for a full list of members in this group.
4. Areas that look to be crowded with Patentees usually indicate multiple sales of the same parcel (Re-issues) or Overlapping parcels. See this Township's Index for an explanation of these and other circumstances that might explain "odd" groupings of Patentees on this map.

Copyright 2006 Boyd IT, Inc. All Rights Reserved

Legend

- ——— Patent Boundary
- ——— Section Boundary
- (shaded) No Patents Found (or Outside County)
- 1., 2., 3., ... Lot Numbers (when beside a name)
- [ ] Group Number (see Appendix "C")

Scale: Section = 1 mile X 1 mile (generally, with some exceptions)

Family Maps of Lauderdale County, Mississippi

Road Map
T5-N R14-E
Choctaw Meridian
Map Group 17

Cities & Towns
None

Cemeteries
Concord Cemetery
Goodwater Cemetery
Osborne Cemetery
Saint Marks Cemetery

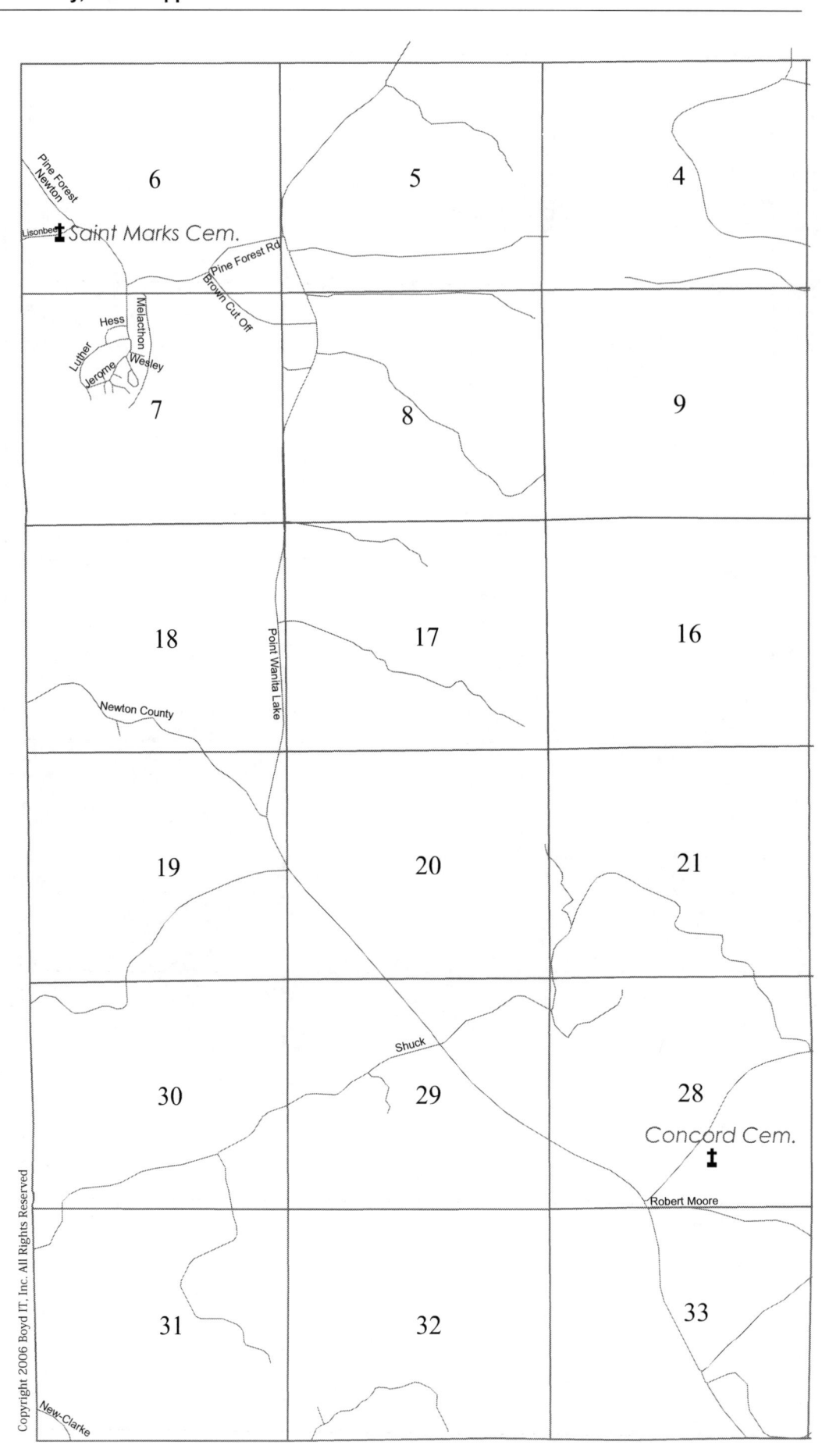

222

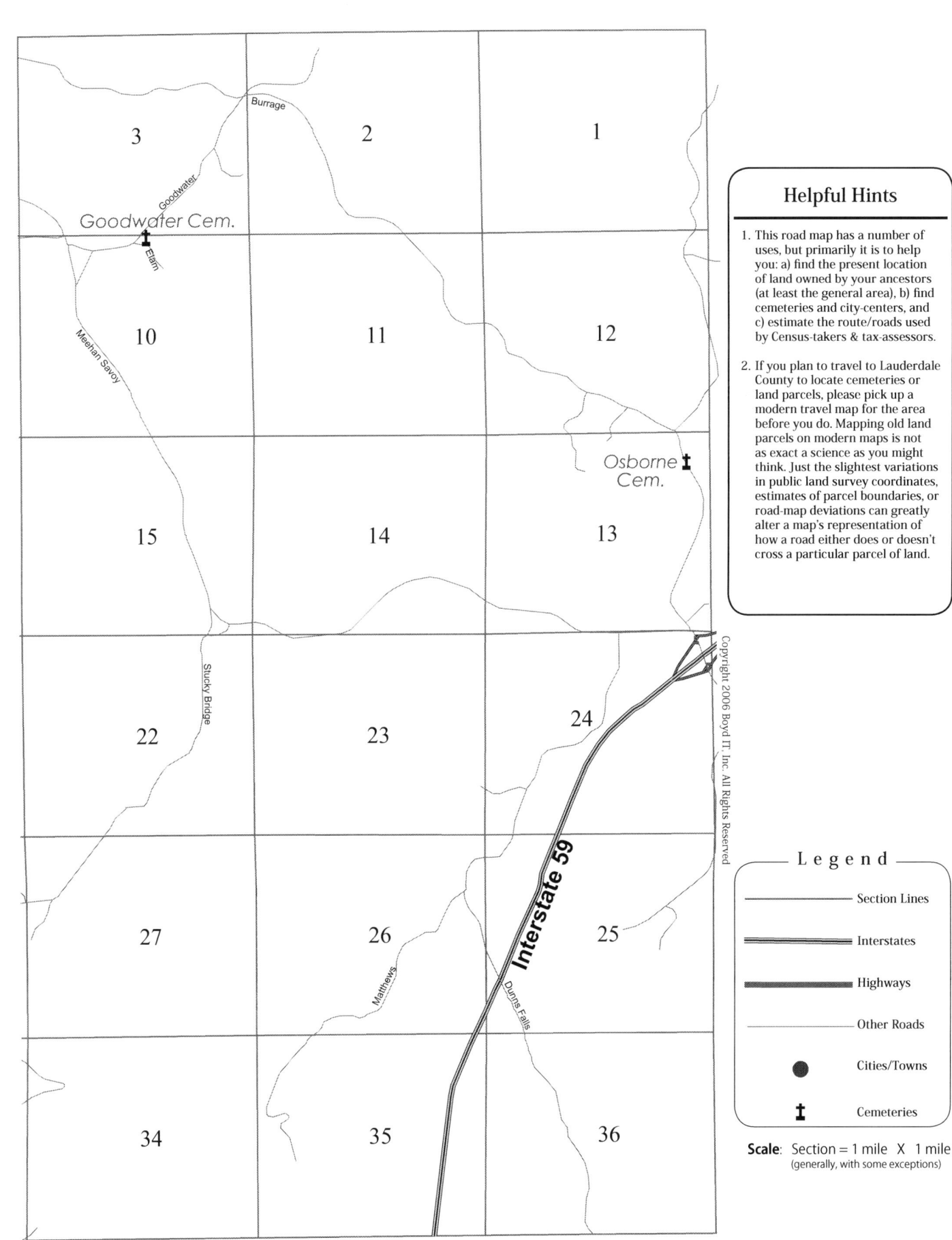

Township 5-N Range 14-E (Choctaw) - Map Group 17

Helpful Hints

1. This road map has a number of uses, but primarily it is to help you: a) find the present location of land owned by your ancestors (at least the general area), b) find cemeteries and city-centers, and c) estimate the route/roads used by Census-takers & tax-assessors.

2. If you plan to travel to Lauderdale County to locate cemeteries or land parcels, please pick up a modern travel map for the area before you do. Mapping old land parcels on modern maps is not as exact a science as you might think. Just the slightest variations in public land survey coordinates, estimates of parcel boundaries, or road-map deviations can greatly alter a map's representation of how a road either does or doesn't cross a particular parcel of land.

Legend
- Section Lines
- Interstates
- Highways
- Other Roads
- ● Cities/Towns
- ✝ Cemeteries

Scale: Section = 1 mile X 1 mile
(generally, with some exceptions)

Family Maps of Lauderdale County, Mississippi

Historical Map
T5-N R14-E
Choctaw Meridian
Map Group 17

Cities & Towns
None

Cemeteries
Concord Cemetery
Goodwater Cemetery
Osborne Cemetery
Saint Marks Cemetery

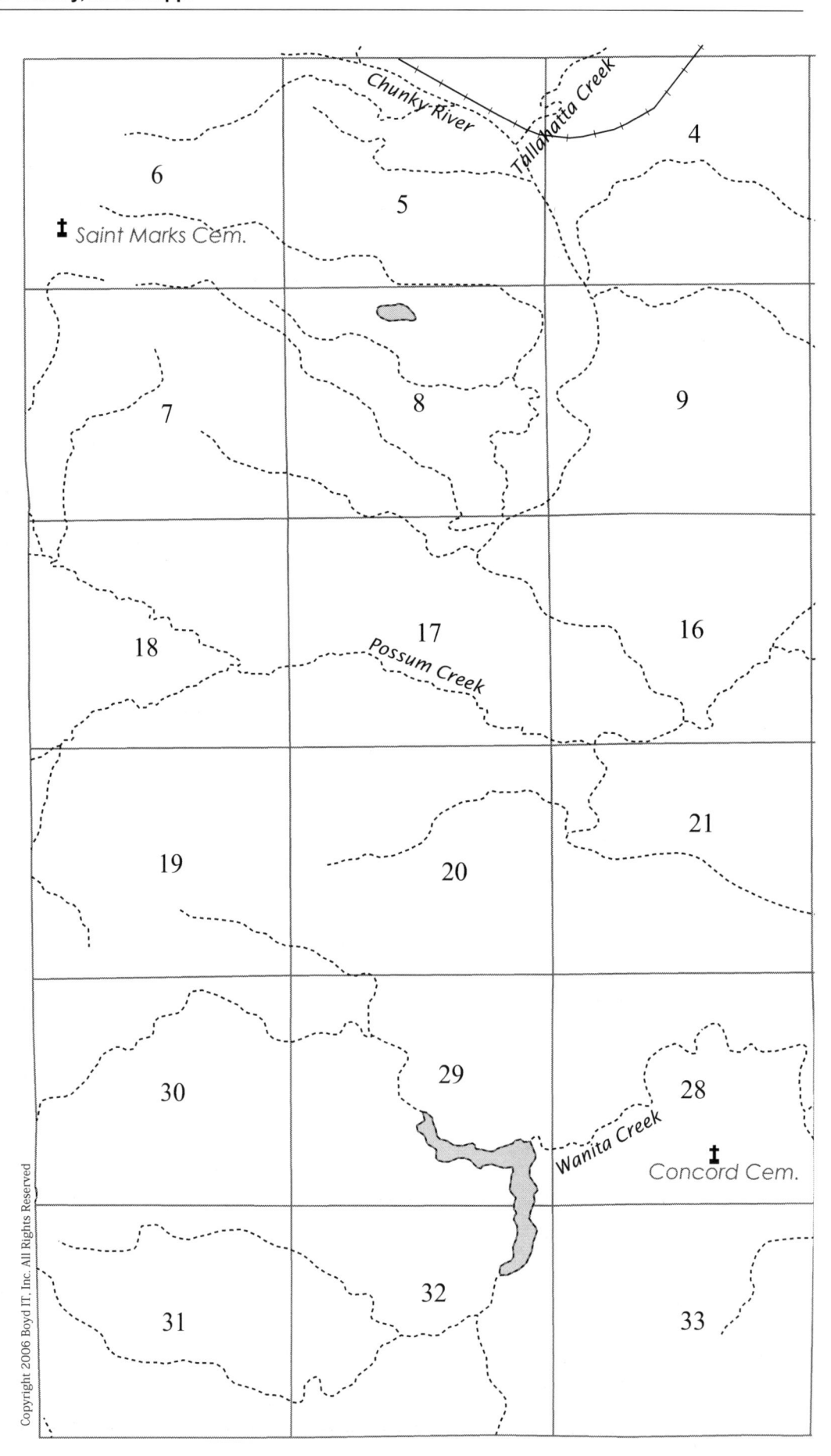

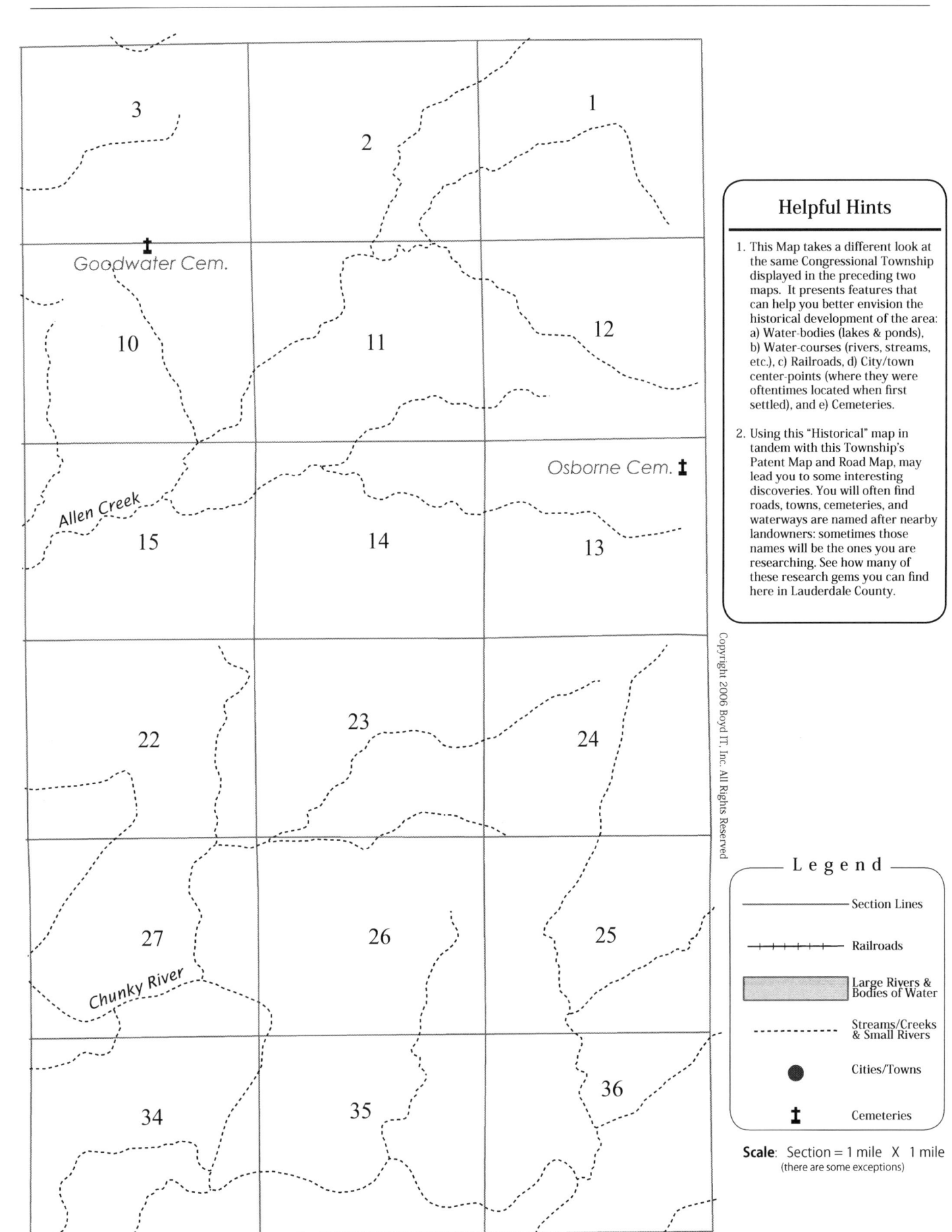

Township 5-N Range 14-E (Choctaw) - Map Group 17

Helpful Hints

1. This Map takes a different look at the same Congressional Township displayed in the preceding two maps. It presents features that can help you better envision the historical development of the area: a) Water-bodies (lakes & ponds), b) Water-courses (rivers, streams, etc.), c) Railroads, d) City/town center-points (where they were oftentimes located when first settled), and e) Cemeteries.

2. Using this "Historical" map in tandem with this Township's Patent Map and Road Map, may lead you to some interesting discoveries. You will often find roads, towns, cemeteries, and waterways are named after nearby landowners: sometimes those names will be the ones you are researching. See how many of these research gems you can find here in Lauderdale County.

Goodwater Cem.
Osborne Cem.
Allen Creek
Chunky River

Copyright 2006 Boyd IT, Inc. All Rights Reserved

Legend

— Section Lines
+++ Railroads
▓ Large Rivers & Bodies of Water
---- Streams/Creeks & Small Rivers
● Cities/Towns
✝ Cemeteries

Scale: Section = 1 mile X 1 mile
(there are some exceptions)

225

Family Maps of Lauderdale County, Mississippi

Map Group 18: Index to Land Patents
Township 5-North Range 15-East (Choctaw)

After you locate an individual in this Index, take note of the Section and Section Part then proceed to the Land Patent map on the pages immediately following. You should have no difficulty locating the corresponding parcel of land.

The "For More Info" Column will lead you to more information about the underlying Patents. See the *Legend* at right, and the "How to Use this Book" chapter, for more information.

```
LEGEND
           "For More Info . . . " column
A = Authority (Legislative Act, See Appendix "A")
B = Block or Lot (location in Section unknown)
C = Cancelled Patent
F = Fractional Section
G = Group (Multi-Patentee Patent, see Appendix "C")
V = Overlaps another Parcel
R = Re-Issued (Parcel patented more than once)

(A & G items require you to look in the Appendixes referred
to above. All other Letter-designations followed by a number
require you to locate line-items in this index that possess
the ID number found after the letter).
```

ID	Individual in Patent	Sec.	Sec. Part	Date Issued	Other Counties	For More Info . . .
3357	BALLARD, Alexander	4	NWSE	1849-12-01		A1
3358	" "	4	SESE	1849-12-01		A1
3392	BANES, George W	13	N½SE	1911-01-12		A1
3393	" "	13	SWNE	1911-01-12		A1
3492	BAXTON, Martin C	4	NWSW	1850/06/01		A2
3453	BENNETT, Jerry	33	NWSW	1904-07-02		A4
3544	BERRY, William J	20	NENW	1848-09-01		A1
3526	BOZEMAN, Samuel J	1	W½NE	1882-03-30		A4
3470	BROWN, John H	7	NENE	1859-05-02		A1
3493	BROWN, Monroe	11	N½SE	1906-06-04		A4
3360	BRUNSON, Amos	13	S½SE	1897-06-07		A4
3463	BULLARD, John A	33	E½SW	1892-04-29		A4
3464	BULLARD, John C	29	E½SE	1911-10-09		A4
3454	BYNUM, Jesse	3	NESE	1841-01-05		A1
3455	" "	3	SWSE	1841-01-05		A1
3449	CARAWAY, James W	25	NWSE	1910-06-23		A1 V3491
3491	CARAWAY, Malona J	25	SE	1906-06-16		A4 G38 V3449
3491	CARAWAY, William	25	SE	1906-06-16		A4 G38 V3449
3536	CARSON, William	3	NWSW	1841-01-05		A1
3537	" "	9	SWSE	1841-01-05		A1
3361	CARTER, Asa R	21	E½SW	1859-05-02		A1 G40
3362	" "	21	W½SE	1859-05-02		A1 G40
3504	CARTER, Rachael	11	SESW	1909-11-26		A4 G41 C R3506
3465	CARYE, John	8	E½NE	1841-01-05		A1
3466	" "	8	NESE	1841-01-05		A1
3479	COATES, John W	1	NWNW	1860-04-02		A1
3505	COATS, Rachael	11	S½SE	1901-06-25		A4
3504	" "	11	SESW	1909-11-26		A4 G41 C R3506
3506	" "	11	SESW	1910-03-10		A4 R3504
3539	COATS, William	11	S½NE	1892-06-30		A4
3538	" "	11	NENE	1895-06-27		A4
3495	COLE, Nelson	19	W½SW	1892-07-11		A4
3482	COOPER, Joseph	31	E½SW	1841-01-05		A1 C R3549
3481	" "	30	E½SW	1852-10-09		A1
3445	CORLEY, James R	23	NWNE	1859-06-01		A1
3446	CUBLEY, James R	35	NESW	1892-06-30		A4
3447	" "	35	SWNW	1892-06-30		A4
3448	" "	35	W½SW	1892-06-30		A4
3478	DANIEL, John T	33	SWSW	1892-06-30		A1
3361	DEASE, Oliver C	21	E½SW	1859-05-02		A1 G40
3362	" "	21	W½SE	1859-05-02		A1 G40
3371	DICKERSON, Charles	35	SESW	1901-03-23		A4
3456	DUBOSE, Jesse	19	NESW	1894-04-10		A4
3457	" "	19	NWSE	1894-04-10		A4
3458	" "	19	SENW	1894-04-10		A4
3459	" "	19	SWNE	1894-04-10		A4

Township 5-N Range 15-E (Choctaw) - Map Group 18

ID	Individual in Patent	Sec.	Sec. Part	Date Issued	Other Counties	For More Info . . .
3540	DUBOSE, William	17	W½SW	1898-06-01		A4
3545	DUBOSE, William M	19	NENW	1898-12-27		A4
3546	" "	19	NWNE	1898-12-27		A4
3547	" "	19	W½NW	1898-12-27		A4
3405	ELLIS, Henry	1	SWNW	1897-11-01		A4
3406	" "	1	W½SW	1897-11-01		A4
3487	FAIRCHILD, Lofton E	21	SWNW	1896-04-03		A4
3488	" "	21	W½SW	1896-04-03		A4
3427	FAIRCHILDS, Jack	23	SWNW	1859-05-02		A1
3428	" "	25	NESW	1859-06-01		A1
3391	FAY, George	1	E½NW	1890-02-21		A4
3430	FLINN, Jackson	7	E½SE	1891-05-20		A4
3527	FLINN, Thomas	7	SESW	1896-09-29		A4
3528	" "	7	SWSE	1896-09-29		A4
3377	GADDIS, Cornelius M	5	NENE	1859-05-02		A1
3378	" "	5	SENE	1860-08-01		A1
3460	GADDIS, Joe	19	E½SE	1893-09-01		A4
3461	" "	19	SESW	1893-09-01		A4
3462	" "	19	SWSE	1893-09-01		A4
3468	GORDON, John	13	E½NW	1901-12-04		A4
3469	" "	13	NWNE	1901-12-04		A4
3365	GRAHAM, Benjamin	2	N½SW	1841-01-05		A1
3370	GRAHAM, Caroline	31	E½NE	1892-03-17		A4 G83
3370	GRAHAM, Edward	31	E½NE	1892-03-17		A4 G83
3397	GRANT, Greene W	5	NW	1841-01-05		A1 G85
3398	" "	5	W½NE	1841-01-05		A1 G85
3440	GREEN, James E	29	NW	1898-12-01		A4
3443	GREEN, James L	29	E½SW	1902-01-17		A4
3444	" "	29	W½SE	1902-01-17		A4
3511	GREEN, Samantha N	31	SENW	1895-01-17		A4
3512	" "	31	SWNE	1895-01-17		A4
3388	HALL, Emsy	11	NENW	1897-11-22		A4
3389	" "	11	NWNE	1897-11-22		A4
3394	HALL, Gilbert	7	W½NW	1896-04-28		A4
3426	HAND, Hiram	21	N½NW	1898-03-15		A4
3441	HAND, James	17	E½NE	1898-09-28		A4
3399	HARDY, Greenville	17	W½SE	1896-07-11		A4 C R3401
3400	HARVEY, Greenville	17	E½SE	1901-06-25		A4
3401	" "	17	W½SE	1902-07-29		A4 R3399
3472	HAWKINS, John J	25	N½NW	1908-10-26		A4
3473	" "	25	SENW	1908-10-26		A4
3397	HERNDON, Thomas H	5	NW	1841-01-05		A1 G85
3398	" "	5	W½NE	1841-01-05		A1 G85
3471	HERREN, John	26	W½NW	1841-01-05		A1
3483	HOLLY, Laura	11	NESW	1901-08-12		A4 G120
3484	" "	11	SENW	1901-08-12		A4 G120
3380	HOPKINS, David	17	NW	1893-12-21		A4
3402	HOPKINS, Gus	5	N½SE	1901-12-17		A4
3403	" "	5	N½SW	1901-12-17		A4
3437	HOPKINS, James A	25	SWNW	1859-05-02		A1
3407	JEMISON, Henry	15	W½SE	1841-01-05		A1 G127
3408	" "	2	E½NE	1841-01-05		A1 G127
3409	" "	2	S½NW	1841-01-05		A1 G127
3410	" "	21	E½SE	1841-01-05		A1 G127
3411	" "	22	E½NW	1841-01-05		A1 G127
3412	" "	22	SW	1841-01-05		A1 G127
3413	" "	22	W½NE	1841-01-05		A1 G127
3414	" "	23	E½NW	1841-01-05		A1 G127
3415	" "	23	W½SW	1841-01-05		A1 G127
3416	" "	27	NWNE	1841-01-05		A1 G127
3417	" "	27	SENE	1841-01-05		A1 G127
3418	" "	28	SE	1841-01-05		A1 G127
3419	" "	28	W½NE	1841-01-05		A1 G127
3420	" "	4	E½NW	1841-01-05		A1 G127
3421	" "	4	W½NE	1841-01-05		A1 G127
3422	" "	9	W½SW	1841-01-05		A1 G127
3359	JOHNSON, Allen	7	SENE	1906-06-21		A4
3500	KING, Peyton	21	SENW	1859-05-02		A1
3501	" "	21	W½NE	1859-05-02		A1
3499	" "	15	NESE	1860-04-02		A1
3375	LANEY, Colbert M	27	NENE	1841-01-05		A1
3376	" "	27	SWNE	1841-01-05		A1
3486	MANDEVILLE, Lloyd	31	E½SE	1884-12-30		A4

Family Maps of Lauderdale County, Mississippi

ID	Individual in Patent	Sec.	Sec. Part	Date Issued	Other Counties	For More Info . . .
3379	MCCARY, Crafton D	33	E½SE	1897-02-15		A4
3497	MCDONALD, Pendleton	9	NWSE	1861-07-01		A1
3368	MCLAUGHLIN, Burwell	7	NENW	1907-04-17		A4
3369	"	7	NWNE	1907-04-17		A4
3373	MCLAUGHLIN, Charles	5	S½SW	1899-09-30		A4
3475	MCLAUGHLIN, John	17	NWNE	1898-03-15		A4
3509	MITCHELL, Robert G	35	NENE	1909-01-11		A4
3510	MOORE, Robert S	5	S½SE	1896-07-11		A4
3476	NEWMAN, John	19	SENE	1879-12-15		A4
3431	NICHOLAS, Jacob	15	NWSW	1841-01-05		A1
3432	" "	21	E½NE	1841-01-05		A1
3433	" "	28	E½SW	1841-01-05		A1
3434	" "	33	W½SE	1841-01-05		A1
3435	" "	4	E½SW	1841-01-05		A1
3436	" "	4	SWSE	1841-01-05		A1
3438	NICHOLAS, James A	4	NWNW	1841-01-05		A1
3477	NICHOLAS, John	22	W½NW	1841-01-05		A1
3442	ODEN, James J	29	W½SW	1888-02-25		A4
3439	PARKER, James D	27	SW	1841-01-05		A1
3480	PRIVETT, John W	35	SWNE	1906-06-26		A4
3423	REESE, Henry W	11	SWNW	1859-05-02		A1
3424	" "	11	W½SW	1859-05-02		A1
3425	" "	9	E½NE	1859-05-02		A1
3383	REW, Edward I	9	SWNW	1859-05-02		A1
3372	RUSHING, Charles E	3	SESE	1859-05-02		A1
3529	SCARBROUGH, Thomas W	25	S½SW	1899-06-28		A4
3395	SCOTT, Gillam G	4	NESE	1841-01-05		A1
3396	"	9	SESE	1841-01-05		A1
3382	SCRUGGS, Edward H	35	SESE	1904-10-27		A4 R3532
3507	SHURDEN, Robert A	31	NENW	1894-04-10		A4
3508	" "	31	NWNE	1894-04-10		A4
3485	SMITH, Levi	2	NWSE	1841-01-05		A1
3548	SPEED, William P	7	SWSW	1884-12-30		A4
3367	STEINWINDER, Burrell B	25	NE	1901-08-12		A4
3390	STEINWINDER, George A	23	NWSE	1914-01-14		A1
3374	SWILLY, Charles	13	W½NW	1902-03-07		A4
3513	TABB, Samuel G	10		1841-01-05		A1
3514	" "	15	E½SW	1841-01-05		A1
3515	" "	15	NE	1841-01-05		A1
3516	" "	15	NW	1841-01-05		A1
3517	" "	15	SWSW	1841-01-05		A1
3518	" "	3	E½SW	1841-01-05		A1
3519	" "	3	N½	1841-01-05		A1
3520	" "	3	NWSE	1841-01-05		A1
3521	" "	3	SWSW	1841-01-05		A1
3522	" "	33	NE	1841-01-05		A1
3523	" "	4	E½NE	1841-01-05		A1
3524	" "	4	SWNW	1841-01-05		A1
3525	" "	9	E½SE	1841-01-05		A1
3530	TAYLOR, Wade B	13	E½NE	1899-04-17		A4
3467	THOMAS, John G	27	SE	1905-10-19		A4
3364	THOMPSON, Benjamin F	1	W½SE	1899-08-03		A4
3363	" "	1	E½SE	1906-04-14		A4
3498	TRUST, Peter	1	E½NE	1892-07-20		A4
3489	TURNER, Mack	31	W½NW	1899-06-28		A4
3490	" "	31	W½SW	1899-06-28		A4
3502	VAUGHN, Pinkney	27	NW	1841-01-05		A1
3503	" "	28	E½NE	1841-01-05		A1
3483	WALKER, Laura	11	NESW	1901-08-12		A4 G120
3484	" "	11	SENW	1901-08-12		A4 G120
3531	WALKER, Wiley M	35	SENE	1891-06-30		A4
3532	" "	35	SESE	1891-06-30		A4 R3382
3533	" "	35	W½SE	1891-06-30		A4
3407	WALLACE, James Y	15	W½SE	1841-01-05		A1 G127
3408	" "	2	E½NE	1841-01-05		A1 G127
3409	" "	2	S½NW	1841-01-05		A1 G127
3410	" "	21	E½SE	1841-01-05		A1 G127
3411	" "	22	E½NW	1841-01-05		A1 G127
3412	" "	22	SW	1841-01-05		A1 G127
3413	" "	22	W½NE	1841-01-05		A1 G127
3414	" "	23	E½NW	1841-01-05		A1 G127
3415	" "	23	W½SW	1841-01-05		A1 G127
3416	" "	27	NWNE	1841-01-05		A1 G127

Township 5-N Range 15-E (Choctaw) - Map Group 18

ID	Individual in Patent	Sec.	Sec. Part	Date Issued	Other Counties	For More Info . . .
3417	WALLACE, James Y (Cont'd)	27	SENE	1841-01-05		A1 G127
3418	" "	28	SE	1841-01-05		A1 G127
3419	" "	28	W½NE	1841-01-05		A1 G127
3420	" "	4	E½NW	1841-01-05		A1 G127
3421	" "	4	W½NE	1841-01-05		A1 G127
3422	" "	9	W½SW	1841-01-05		A1 G127
3550	WASHINGTON, Willis	31	W½SE	1889-05-23		A4
3549	" "	31	E½SW	1892-04-16		A4 R3482
3404	WEBB, Harvey	1	E½SW	1892-04-29		A4 G180
3404	WEBB, Mary E	1	E½SW	1892-04-29		A4 G180
3384	WELLBORN, Elijah J	7	NESW	1901-12-30		A4
3385	" "	7	NWSE	1901-12-30		A4
3386	" "	7	SENW	1901-12-30		A4
3387	" "	7	SWNE	1901-12-30		A4
3541	WEST, William H	35	E½NW	1898-12-27		A4
3542	" "	35	NWNE	1898-12-27		A4
3543	" "	35	NWNW	1898-12-27		A4
3450	WHITEHEAD, James	2	NENW	1841-01-05		A1
3451	" "	2	NWNW	1841-01-05		A1
3452	" "	2	W½NE	1841-01-05		A1
3381	WHITLOCK, Dennis	7	NWSW	1920-10-09		A4
3474	WHITLOCK, John L	23	NWNW	1900-11-28		A4
3534	WHITLOCK, William A	23	E½SW	1906-06-26		A4
3535	" "	23	S½SE	1906-06-26		A4
3366	WILLIAMS, Billy	17	E½SW	1901-03-23		A4
3494	WILLIAMS, Moses	11	NWNW	1882-03-30		A4
3429	WOODWARD, Jack	17	SWNE	1904-12-31		A4
3496	YOUNG, Paul	13	SW	1904-10-27		A4

Family Maps of Lauderdale County, Mississippi

Patent Map

T5-N R15-E
Choctaw Meridian

Map Group 18

Township Statistics

Parcels Mapped	:	194
Number of Patents	:	138
Number of Individuals	:	115
Patentees Identified	:	108
Number of Surnames	:	87
Multi-Patentee Parcels	:	26
Oldest Patent Date	:	1/5/1841
Most Recent Patent	:	10/9/1920
Block/Lot Parcels	:	0
Parcels Re - Issued	:	4
Parcels that Overlap	:	2
Cities and Towns	:	3
Cemeteries	:	3

Copyright 2006 Boyd IT, Inc. All Rights Reserved

230

Township 5-N Range 15-E (Choctaw) - Map Group 18

Family Maps of Lauderdale County, Mississippi

Road Map
T5-N R15-E
Choctaw Meridian
Map Group 18

Cities & Towns
Arunde
Savoy
Sterling

Cemeteries
Graham Cemetery
Oak Grove Cemetery
Sageville Cemetery

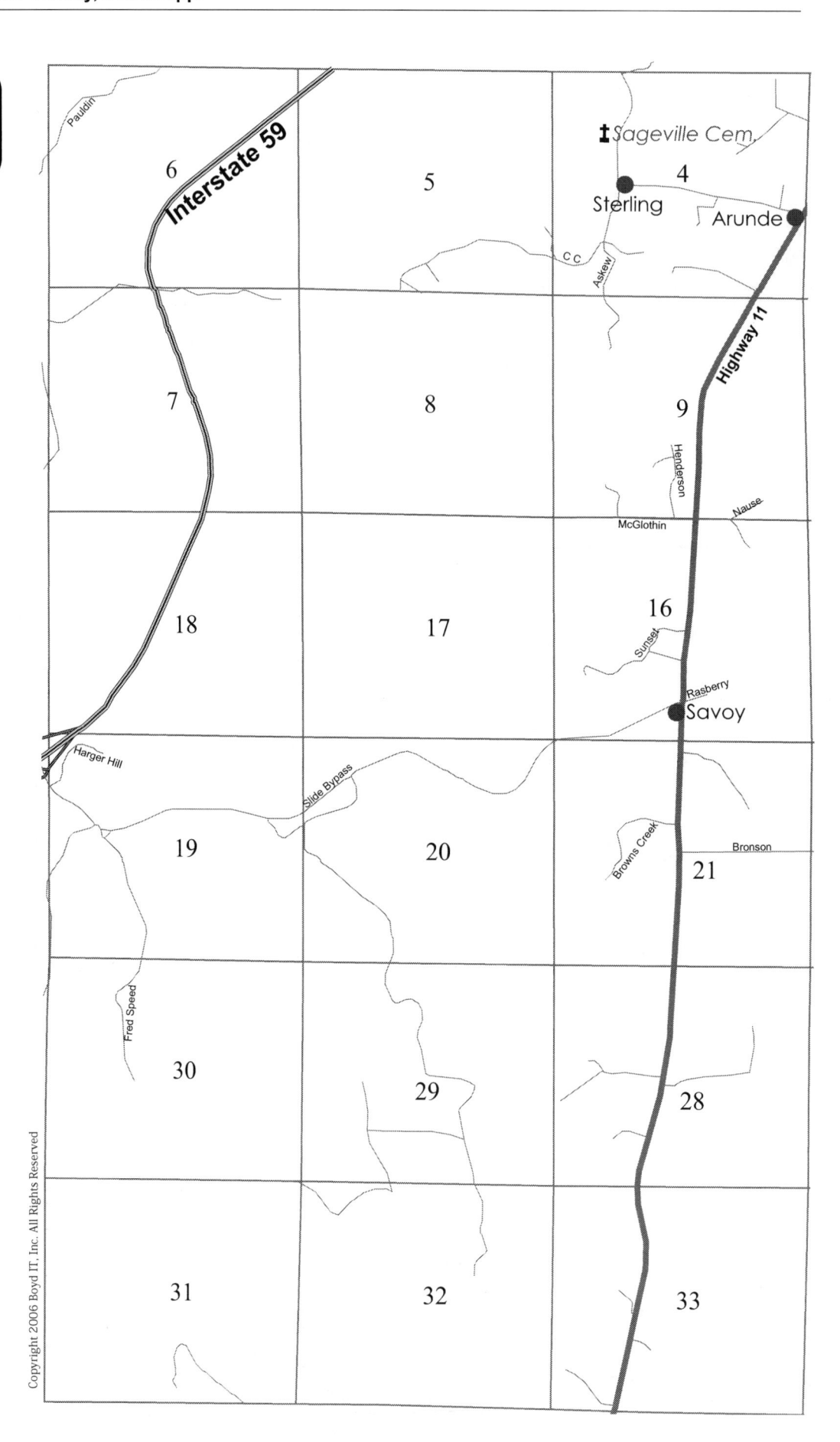

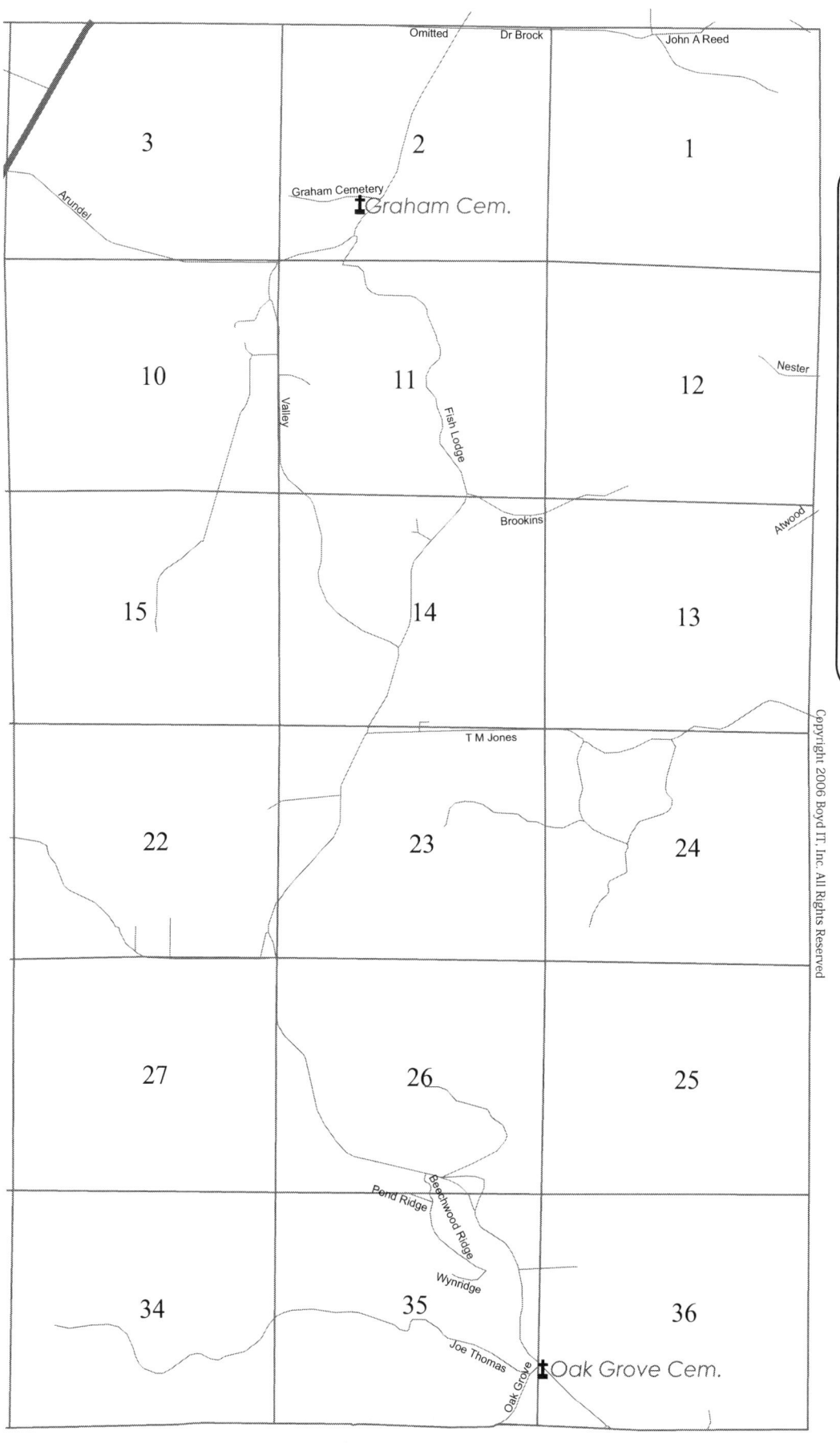

Township 5-N Range 15-E (Choctaw) - Map Group 18

Helpful Hints

1. This road map has a number of uses, but primarily it is to help you: a) find the present location of land owned by your ancestors (at least the general area), b) find cemeteries and city-centers, and c) estimate the route/roads used by Census-takers & tax-assessors.

2. If you plan to travel to Lauderdale County to locate cemeteries or land parcels, please pick up a modern travel map for the area before you do. Mapping old land parcels on modern maps is not as exact a science as you might think. Just the slightest variations in public land survey coordinates, estimates of parcel boundaries, or road-map deviations can greatly alter a map's representation of how a road either does or doesn't cross a particular parcel of land.

Legend

- Section Lines
- Interstates
- Highways
- Other Roads
- ● Cities/Towns
- ☩ Cemeteries

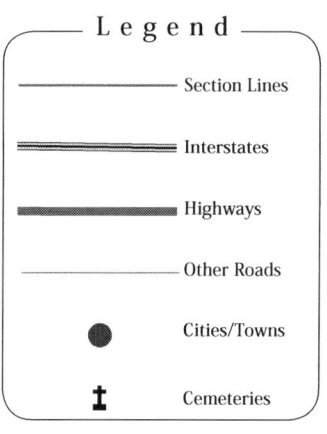

Scale: Section = 1 mile X 1 mile
(generally, with some exceptions)

Copyright 2006 Boyd IT, Inc. All Rights Reserved

Family Maps of Lauderdale County, Mississippi

Historical Map

T5-N R15-E
Choctaw Meridian

Map Group 18

Cities & Towns
Arunde
Savoy
Sterling

Cemeteries
Graham Cemetery
Oak Grove Cemetery
Sageville Cemetery

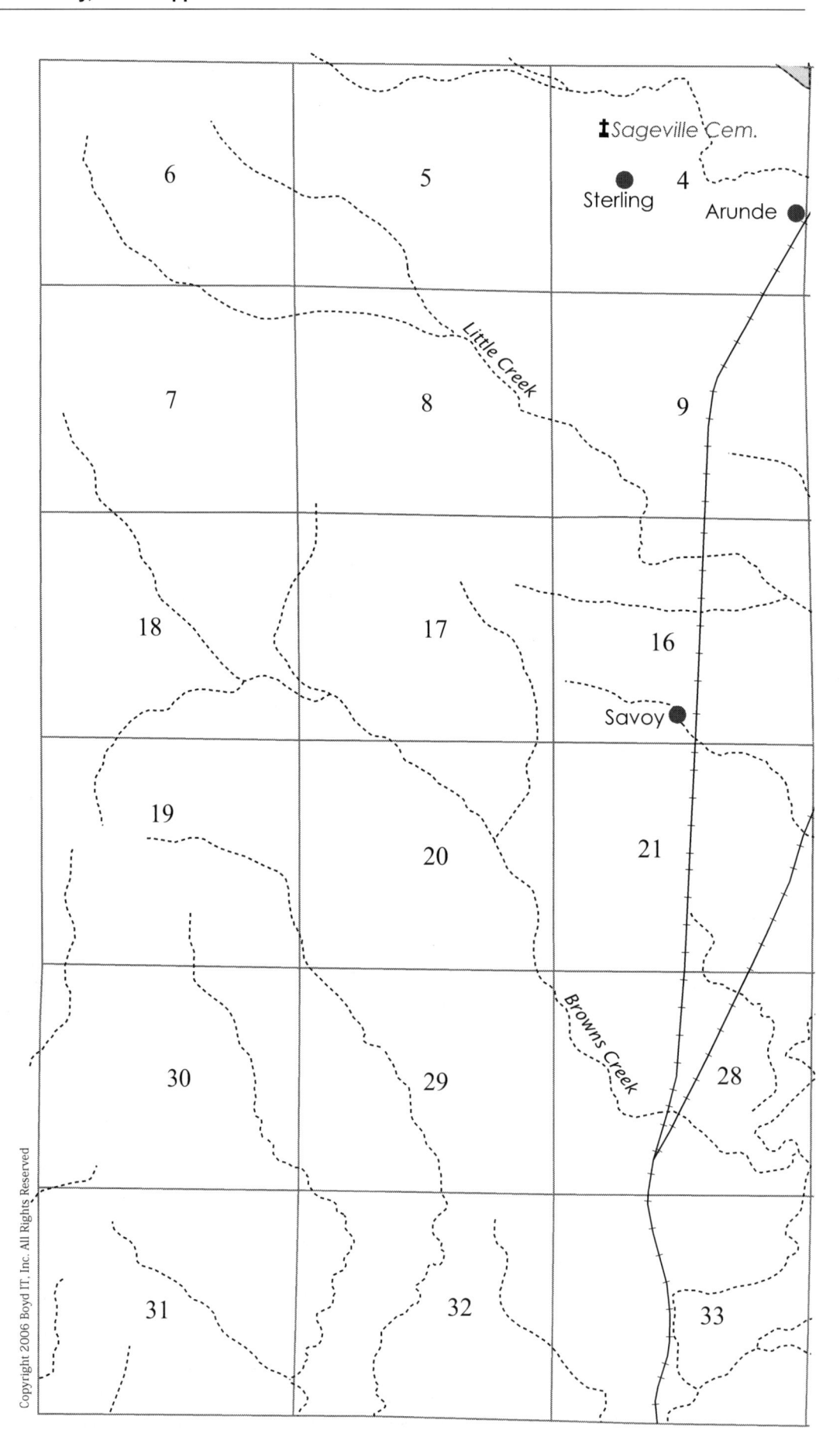

Township 5-N Range 15-E (Choctaw) - Map Group 18

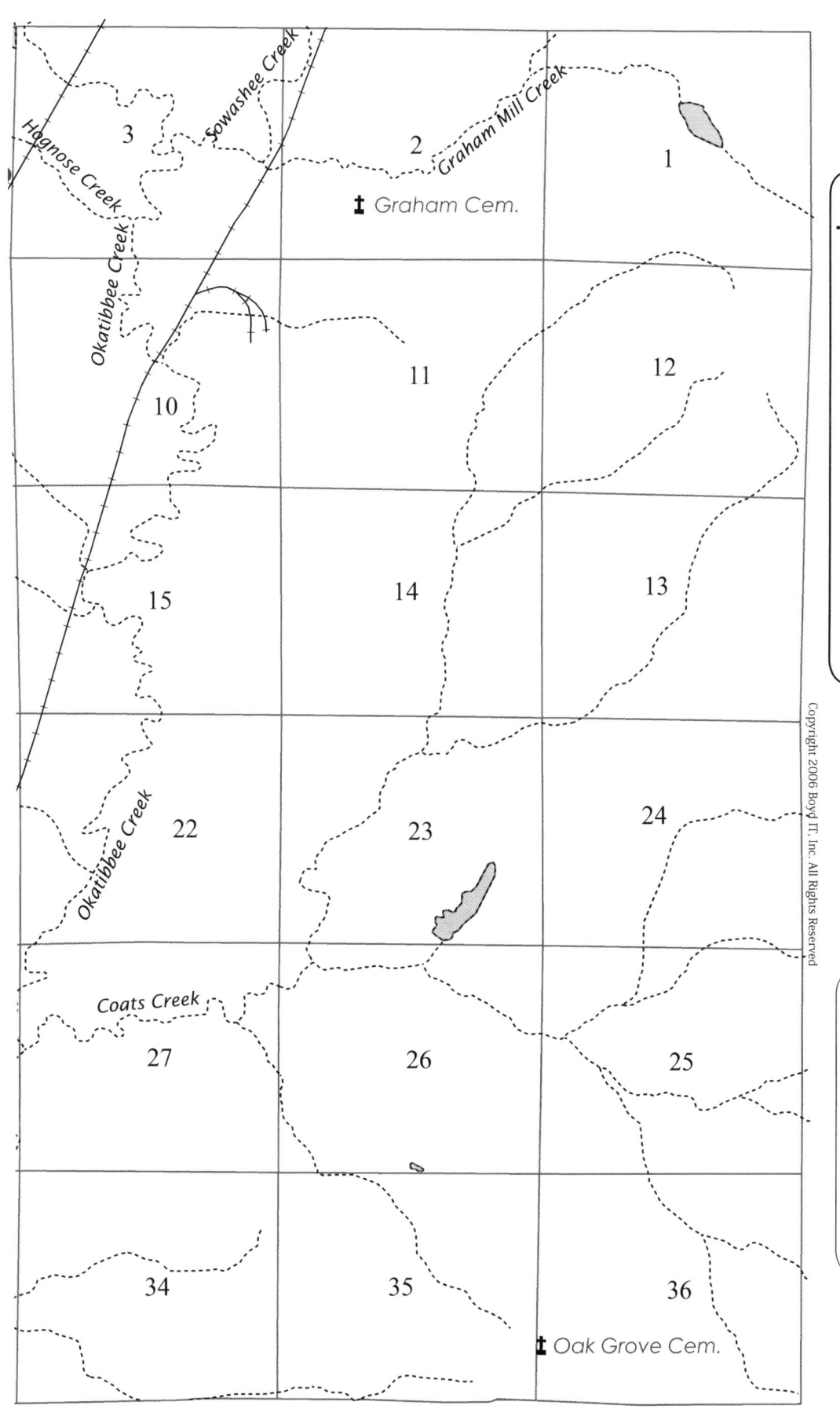

Helpful Hints

1. This Map takes a different look at the same Congressional Township displayed in the preceding two maps. It presents features that can help you better envision the historical development of the area: a) Water-bodies (lakes & ponds), b) Water-courses (rivers, streams, etc.), c) Railroads, d) City/town center-points (where they were oftentimes located when first settled), and e) Cemeteries.

2. Using this "Historical" map in tandem with this Township's Patent Map and Road Map, may lead you to some interesting discoveries. You will often find roads, towns, cemeteries, and waterways are named after nearby landowners: sometimes those names will be the ones you are researching. See how many of these research gems you can find here in Lauderdale County.

Legend

- Section Lines
- Railroads
- Large Rivers & Bodies of Water
- Streams/Creeks & Small Rivers
- Cities/Towns
- Cemeteries

Scale: Section = 1 mile X 1 mile
(there are some exceptions)

Map Group 19: Index to Land Patents
Township 5-North Range 16-East (Choctaw)

After you locate an individual in this Index, take note of the Section and Section Part then proceed to the Land Patent map on the pages immediately following. You should have no difficulty locating the corresponding parcel of land.

The "For More Info" Column will lead you to more information about the underlying Patents. See the *Legend* at right, and the "How to Use this Book" chapter, for more information.

```
LEGEND
       "For More Info . . . " column
A = Authority (Legislative Act, See Appendix "A")
B = Block or Lot (location in Section unknown)
C = Cancelled Patent
F = Fractional Section
G = Group  (Multi-Patentee Patent, see Appendix "C")
V = Overlaps another Parcel
R = Re-Issued (Parcel patented more than once)

(A & G items require you to look in the Appendixes referred
to above. All other Letter-designations followed by a number
require you to locate line-items in this index that possess
the ID number found after the letter).
```

ID	Individual in Patent	Sec.	Sec. Part	Date Issued	Other Counties	For More Info . . .
3621	ALEXANDER, John	15	W½NW	1889-01-12		A4
3552	ALLSOBROOKS, Albert G	5	W½SW	1893-07-31		A4
3553	ANDERSON, Alonzo A	29	SW	1901-12-04		A4
3664	ANDERSON, Samuel J	29	S½NW	1899-08-03		A4
3622	BARRON, John	29	E½NE	1891-05-20		A4
3554	BLACKWELL, Amanda P	29	W½NE	1905-03-30		A4 G22
3632	BLACKWELL, Joshua	17	N½NE	1897-09-09		A4
3582	BRUNSON, Emma M	31	NWNW	1897-11-22		A4
3585	BUNYARD, George G	1	N½NW	1890-06-25		A4
3586	" "	1	NWSW	1896-04-03		A4
3587	" "	1	SWNW	1896-04-03		A4
3616	BUNYARD, James W	1	W½NE	1890-06-25		A4
3556	BYNUM, Bessie Viola	7	NENE	1917-08-11		A4
3615	CORLEY, James R	31	SWNE	1859-06-01		A1
3675	COVINGTON, Thomas J	11	W½NW	1882-05-10		A1
3623	CRAWFORD, John H	19	W½SW	1892-05-16		A4
3571	CREEL, Collins W	9	NESW	1899-06-28		A4
3572	" "	9	SENW	1899-06-28		A4
3658	CROSBY, Roland B	9	SENE	1841-01-05		A1
3659	CROSBY, Rowland B	10	NWNW	1841-01-05		A1
3661	" "	10	SWNW	1841-01-05		A1 G55
3660	" "	9	NENE	1841-01-05		A1
3662	ETHRRIDGE, Samuel	35	NWSW	1849-12-01		A1 R3663
3663	ETTURIDGE, Samuel	35	NWSW	1849-12-01		A1 R3662
3603	FAIRCHILDS, Jack	31	NWSE	1859-06-01		A1
3605	" "	31	SWSW	1860-10-01		A1 G72
3604	" "	33	SWNW	1860-10-01		A1
3666	FISHER, Southy	11	E½NW	1841-01-05		A1
3668	" "	14	W½NW	1841-01-05		A1
3669	" "	14	W½SW	1841-01-05		A1
3670	" "	23	E½NW	1841-01-05		A1
3667	" "	14	E½NW	1859-06-01		A1
3674	GARDNER, Thomas C	33	NWNW	1860-10-01		A1
3606	GAY, Jack	3	W½SW	1892-06-15		A4
3600	GILLINER, Hugh	35	SWNW	1841-01-05		A1
3601	" "	35	SWSW	1841-01-05		A1
3598	GRAHAM, Hilliard L	5	E½NE	1895-01-17		A4
3599	" "	5	E½SE	1895-01-17		A4
3555	GRAY, Arthur	9	NWNE	1860-10-01		A1
3595	GRAY, Henry T	31	NWSW	1908-07-23		A4
3575	HAWKINS, David A	19	NWSE	1897-11-22		A4
3563	HELVESTON, Charles E	17	SESW	1896-04-28		A4
3564	" "	17	SWNE	1896-04-28		A4
3565	" "	17	W½SE	1896-04-28		A4
3573	HILL, Cornelius A	9	S½SW	1895-06-27		A4
3583	HOOKS, Evan	17	SESE	1859-11-10		A1

Township 5-N Range 16-E (Choctaw) - Map Group 19

ID	Individual in Patent	Sec.	Sec. Part	Date Issued	Other Counties	For More Info...
3691	HOOKS, William M	17	SWSW	1860-10-01		A1
3661	HOUZE, William J	10	SWNW	1841-01-05		A1 G55
3620	HYDE, Jesse	4	W½SE	1841-01-05		A1
3551	IRBY, Aaron	1	NENE	1895-02-21		A4
3648	JAMES, Phillip	31	NWNE	1841-01-05		A1
3609	JONES, James F	29	SE	1906-06-21		A4
3676	JONES, Thomas M	21	E½SW	1896-10-10		A4
3677	" "	21	NWSE	1896-10-10		A4
3678	" "	21	NWSW	1896-10-10		A4
3688	JONES, William	21	SWSW	1906-06-16		A4
3602	KELLY, Isaiah	7	NWSW	1901-12-04		A4
3614	LASLEY, James	33	NWSW	1860-10-01		A1
3652	LEMORE, Richard M	26	W½SW	1850/06/01		A2
3579	LINTON, Edwin P	33	NENW	1904-09-28		A4
3580	" "	33	NWSE	1904-09-28		A4
3581	" "	33	W½NE	1904-09-28		A4
3635	MARTIN, Marious	3	W½NW	1841-01-05		A1
3636	" "	7	E½SE	1841-01-05		A1
3637	" "	7	W½SE	1841-01-05		A1
3640	MARTIN, Murdock M	15	SESW	1841-01-05		A1
3641	MARTIN, Norman	15	NESW	1841-01-05		A1
3642	" "	15	SENW	1841-01-05		A1
3643	" "	15	W½NE	1841-01-05		A1
3644	" "	15	W½SE	1841-01-05		A1
3645	" "	15	W½SW	1841-01-05		A1
3646	" "	22	W½NE	1841-01-05		A1
3679	MARTIN, Virginious	10	W½NE	1841-01-05		A1
3680	" "	3	W½SE	1841-01-05		A1
3681	" "	4	E½NE	1841-01-05		A1
3568	MCDONALD, Charles J	21	E½SE	1895-02-21		A4
3647	MCDONALD, Pendleton	22	NESW	1860-07-02		A1
3692	MCEACHIN, William M	5	SWNW	1901-06-25		A4
3684	MCGAW, William H	17	NESE	1897-09-09		A4
3685	" "	17	SENE	1897-09-09		A4
3588	MCINNIS, George W	3	NESW	1860-04-02		A1
3589	" "	3	SESW	1860-10-01		A1
3624	MCINNIS, John	10	E½NW	1841-01-05		A1
3625	" "	10	NESW	1841-01-05		A1
3626	" "	10	SESE	1841-01-05		A1
3627	" "	10	SWSE	1841-01-05		A1
3653	MCLEMORE, Richard	23	SWSW	1850/06/01		A2
3654	" "	26	W½NW	1850/06/01		A2
3554	MITCHAM, Amanda P	29	W½NE	1905-03-30		A4 G22
3557	MOLPUS, Buchanan F	1	NESW	1895-06-27		A4
3558	" "	1	NWSE	1895-06-27		A4
3559	" "	1	SENW	1895-06-27		A4
3617	MOLPUS, James W	31	SWSE	1901-06-25		A4
3593	PICKARD, Henry A	9	N½NW	1890-03-28		A4
3689	PICKARD, William L	9	NWSW	0012-00-00		A4
3690	" "	9	SWNW	0012-00-00		A4
3596	PRESTON, Henry T	5	NWNW	1897-02-17		A1
3610	RILEY, James F	31	NENW	1898-02-24		A4
3672	ROBBINS, Thomas A	19	S½NW	1898-08-27		A4
3673	" "	19	SWNE	1898-08-27		A4
3566	RUSHING, Charles E	5	SESW	1860-04-02		A1
3567	RUSSING, Charles E	21	SENE	1859-06-01		A1
3590	SCARBOROUGH, George W	33	SWSE	1906-08-10		A4
3665	SCARBOROUGH, Silas	35	SENE	1852/05/01		A2
3695	SCARBOROUGH, William T	21	NW	1896-10-10		A4
3584	SCARBROUGH, Frederick L	29	NWNW	1913-02-14		A4
3574	SIMS, Daniel H	1	SWSW	1891-06-30		A4
3696	SKIPPER, Zachariah L	9	NWSE	1898-03-15		A4
3697	" "	9	S½SE	1898-03-15		A4
3698	" "	9	SWNE	1898-03-15		A4
3618	SMITH, Jeptha L	21	NENE	1892-05-26		A4
3619	" "	21	W½NE	1892-05-26		A4
3569	STAFFORD, Charles	5	E½NW	1891-05-20		A4
3570	" "	5	W½NE	1891-05-20		A4
3633	STAFFORD, Julian M	5	NESW	1899-06-28		A4
3631	STEINWINDER, Joseph R	19	N½NW	1895-06-27		A4
3562	STINSON, Burwell J	17	N½NW	1895-06-28		A4
3629	STINSON, John	7	SENE	1860-04-02		A1
3628	" "	7	NESW	1894-01-20		A4

237

Family Maps of Lauderdale County, Mississippi

ID	Individual in Patent	Sec.	Sec. Part	Date Issued	Other Counties	For More Info . . .
3630	STINSON, John (Cont'd)	7	SWNE	1894-01-20		A4
3638	STONE, Miles	3	W½NE	1860-04-02		A1
3639	STONE, Moses	3	E½NE	1859-06-01		A1
3592	TAYLOR, Hanon	7	W½NW	1891-05-20		A4
3612	TAYLOR, James H	17	N½SW	1894-02-01		A4
3613	" "	17	S½NW	1894-02-01		A4
3634	TAYLOR, Marion	7	SWSW	1892-07-20		A4
3649	TAYLOR, Priestley	19	N½NE	1895-02-21		A4
3650	" "	19	NESE	1895-02-21		A4
3651	" "	19	SENE	1895-02-21		A4
3605	THOMPSON, George W	31	SWSW	1860-10-01		A1 G72
3611	TOURELL, James F	5	W½SE	1895-02-21		A4
3682	WALKER, William E	31	E½NE	1895-02-21		A4
3683	" "	31	E½SE	1895-02-21		A4
3597	WEATHERFORD, Henry	29	NENW	1860-04-02		A1
3594	WEATHERS, Henry J	7	SESW	1895-02-07		A1
3560	WHITE, Burrell E	19	S½SE	1893-08-23		A4
3561	" "	19	SESW	1893-08-23		A4
3671	WHITE, Steven C	19	NESW	1860-10-01		A1
3694	WHITE, William S	11	W½NE	1891-05-20		A4
3693	" "	11	NENE	1895-05-11		A4
3591	WHITEHEAD, Granberry B	7	SENW	1895-06-27		A4
3607	WHITEHEAD, James E	7	NENW	1894-12-17		A4
3608	" "	7	NWNE	1894-12-17		A4
3576	WILLIAMS, David	23	NWSW	1848-09-01		A1
3577	" "	23	SWNW	1848-09-01		A1
3578	WINGATE, Edward T	11	W½SW	1841-01-05		A1
3656	WINGATE, Robert H	2	NWSE	1841-01-05		A1
3657	" "	2	SESE	1841-01-05		A1
3655	WOODS, Richard	3	E½NW	1882-03-30		A4
3686	YARBROUGH, William J	31	E½SW	1894-12-17		A4
3687	" "	31	S½NW	1894-12-17		A4

Family Maps of Lauderdale County, Mississippi

Patent Map

T5-N R16-E
Choctaw Meridian

Map Group 19

Township Statistics

Parcels Mapped	:	148
Number of Patents	:	113
Number of Individuals	:	97
Patentees Identified	:	96
Number of Surnames	:	70
Multi-Patentee Parcels	:	3
Oldest Patent Date	:	1/5/1841
Most Recent Patent	:	8/11/1917
Block/Lot Parcels	:	0
Parcels Re-Issued	:	1
Parcels that Overlap	:	0
Cities and Towns	:	5
Cemeteries	:	5

Section 6
(unassigned parcels)

Section 5
- PRESTON, Henry T — 1897
- MCEACHIN, William M — 1901
- STAFFORD, Charles — 1891
- STAFFORD, Charles — 1891
- ALLSOBROOKS, Albert G — 1893
- STAFFORD, Julian M — 1899
- TOURELL, James F — 1895
- RUSHING, Charles E — 1860
- GRAHAM, Hilliard L — 1895
- GRAHAM, Hilliard L — 1895

Section 4
- MARTIN, Virginious — 1841
- HYDE, Jesse — 1841

Section 7
- TAYLOR, Hanon — 1891
- WHITEHEAD, James E — 1894
- WHITEHEAD, James E — 1894
- BYNUM, Bessie Viola — 1917
- WHITEHEAD, Granberry B — 1895
- STINSON, John — 1894
- STINSON, John — 1860
- KELLY, Isaiah — 1901
- STINSON, John — 1894
- MARTIN, Marious — 1841
- TAYLOR, Marion — 1892
- WEATHERS, Henry J — 1895
- MARTIN, Marious — 1841

Section 8
(unassigned parcels)

Section 9
- PICKARD, Henry A — 1890
- GRAY, Arthur — 1860
- CROSBY, Rowland B — 1841
- PICKARD, William L — 0012
- CREEL, Collins W — 1899
- SKIPPER, Zachariah L — 1898
- CROSBY, Roland B — 1841
- PICKARD, William L — 0012
- CREEL, Collins W — 1899
- SKIPPER, Zachariah L — 1898
- HILL, Cornelius A — 1895
- SKIPPER, Zachariah L — 1898

Section 18
(unassigned parcels)

Section 17
- STINSON, Burwell J — 1895
- BLACKWELL, Joshua — 1897
- TAYLOR, James H — 1894
- HELVESTON, Charles E — 1896
- MCGAW, William H — 1897
- TAYLOR, James H — 1894
- HELVESTON, Charles E — 1896
- MCGAW, William H — 1897
- HOOKS, William M — 1860
- HELVESTON, Charles E — 1896
- HOOKS, Evan — 1859

Section 16
(unassigned parcels)

Section 19
- STEINWINDER, Joseph R — 1895
- TAYLOR, Priestley — 1895
- ROBBINS, Thomas A — 1898
- ROBBINS, Thomas A — 1898
- TAYLOR, Priestley — 1895
- CRAWFORD, John H — 1892
- WHITE, Steven C — 1860
- HAWKINS, David A — 1897
- TAYLOR, Priestley — 1895
- WHITE, Burrell E — 1893
- WHITE, Burrell E — 1893

Section 20
(unassigned parcels)

Section 21
- SMITH, Jeptha L — 1892
- SMITH, Jeptha L — 1892
- SCARBOROUGH, William T — 1896
- RUSSING, Charles E — 1859
- JONES, Thomas M — 1896
- JONES, Thomas M — 1896
- MCDONALD, Charles J — 1895
- JONES, William — 1906
- JONES, Thomas M — 1896

Section 30
(unassigned parcels)

Section 29
- SCARBROUGH, Frederick L — 1913
- WEATHERFORD, Henry — 1860
- BLACKWELL, Amanda P — 1905
- [22]
- ANDERSON, Samuel J — 1899
- BARRON, John — 1891
- ANDERSON, Alonzo A — 1901
- JONES, James F — 1906

Section 28
(unassigned parcels)

Section 31
- BRUNSON, Emma M — 1897
- RILEY, James F — 1898
- JAMES, Phillip — 1841
- YARBROUGH, William J — 1894
- CORLEY, James R — 1859
- WALKER, William E — 1895
- GRAY, Henry T — 1908
- FAIRCHILDS, Jack — 1859
- FAIRCHILDS [72], Jack — 1860
- YARBROUGH, William J — 1894
- MOLPUS, James W — 1901
- WALKER, William E — 1895

Section 32
(unassigned parcels)

Section 33
- GARDNER, Thomas C — 1860
- LINTON, Edwin P — 1904
- LINTON, Edwin P — 1904
- FAIRCHILDS, Jack — 1860
- LASLEY, James — 1860
- LINTON, Edwin P — 1904
- SCARBOROUGH, George W — 1906

Copyright 2006 Boyd IT, Inc. All Rights Reserved

240

Township 5-N Range 16-E (Choctaw) - Map Group 19

Section 3
- MARTIN Marious 1841
- STONE Miles 1860
- WOODS Richard 1882
- STONE Moses 1859
- GAY Jack 1892
- MCINNIS George W 1860
- MARTIN Virginious 1841
- MCINNIS George W 1860

Section 2

Section 1
- BUNYARD George G 1890
- BUNYARD James W 1890
- IRBY Aaron 1895
- BUNYARD George G 1896
- MOLPUS Buchanan F 1895
- BUNYARD George G 1896
- MOLPUS Buchanan F 1895
- MOLPUS Buchanan F 1895
- SIMS Daniel H 1891

Section 10
- CROSBY Rowland B 1841
- MARTIN Virginious 1841
- CROSBY [55] Rowland B 1841
- MCINNIS John 1841
- MCINNIS John 1841
- MCINNIS John 1841
- MCINNIS John 1841

Section 11
- COVINGTON Thomas J 1882
- WHITE William S 1891
- WHITE William S 1895
- FISHER Southy 1841
- WINGATE Edward T 1841
- WINGATE Robert H 1841
- WINGATE Robert H 1841

Section 12

Section 15
- ALEXANDER John 1889
- MARTIN Norman 1841
- MARTIN Norman 1841
- MARTIN Norman 1841
- MARTIN Norman 1841
- MARTIN Norman 1841
- MARTIN Murdock M 1841

Section 14
- FISHER Southy 1841
- FISHER Southy 1859
- FISHER Southy 1841

Section 13

Section 22
- MARTIN Norman 1841
- MCDONALD Pendleton 1860

Section 23
- WILLIAMS David 1848
- FISHER Southy 1841
- WILLIAMS David 1848
- MCLEMORE Richard 1850

Section 24

Section 27

Section 26
- MCLEMORE Richard 1850
- LEMORE Richard M 1850

Section 25

Section 34

Section 35
- GILLINER Hugh 1841
- SCARBOROUGH Silas 1852
- ETHRRIDGE Samuel 1849
- ETTURIDGE Samuel 1849
- GILLINER Hugh 1841

Section 36

Helpful Hints

1. This Map's INDEX can be found on the preceding pages.
2. Refer to Map "C" to see where this Township lies within Lauderdale County, Mississippi.
3. Numbers within square brackets [ ] denote a multi-patentee land parcel (multi-owner). Refer to Appendix "C" for a full list of members in this group.
4. Areas that look to be crowded with Patentees usually indicate multiple sales of the same parcel (Re-issues) or Overlapping parcels. See this Township's Index for an explanation of these and other circumstances that might explain "odd" groupings of Patentees on this map.

Copyright 2006 Boyd IT, Inc. All Rights Reserved

Legend

— Patent Boundary

— Section Boundary

▨ No Patents Found (or Outside County)

1., 2., 3., ... Lot Numbers (when beside a name)

[ ] Group Number (see Appendix "C")

Scale: Section = 1 mile X 1 mile (generally, with some exceptions)

Family Maps of Lauderdale County, Mississippi

Road Map
T5-N R16-E
Choctaw Meridian
Map Group 19

Cities & Towns
Enzor
Pleasant Hill
Stinson
Wolf Springs
Zero

Cemeteries
Long Creek Cemetery
Mount Horeb Cemetery
New Hope Cemetery
Pine Grove Cemetery
Stinson Cemetery

Township 5-N Range 16-E (Choctaw) - Map Group 19

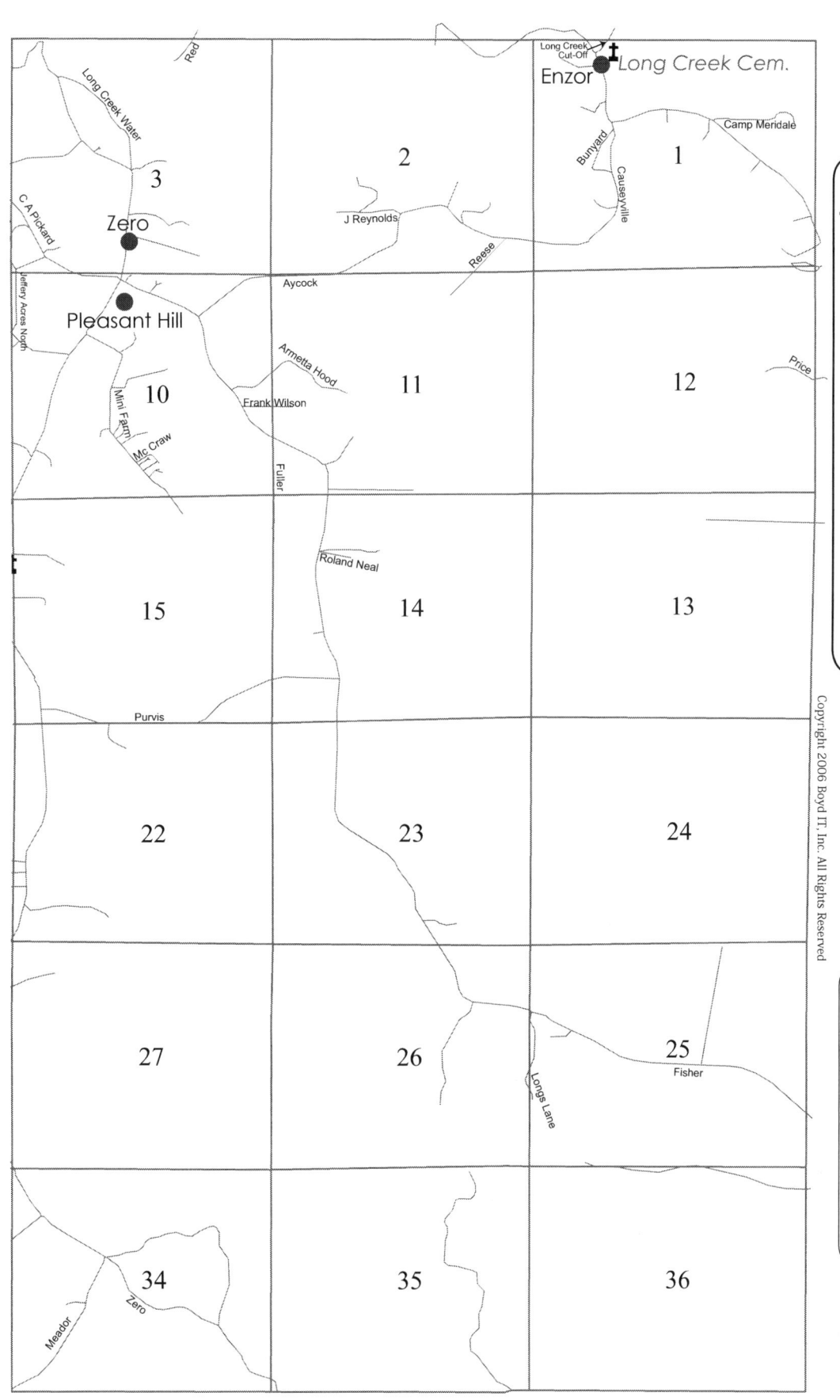

Helpful Hints

1. This road map has a number of uses, but primarily it is to help you: a) find the present location of land owned by your ancestors (at least the general area), b) find cemeteries and city-centers, and c) estimate the route/roads used by Census-takers & tax-assessors.

2. If you plan to travel to Lauderdale County to locate cemeteries or land parcels, please pick up a modern travel map for the area before you do. Mapping old land parcels on modern maps is not as exact a science as you might think. Just the slightest variations in public land survey coordinates, estimates of parcel boundaries, or road-map deviations can greatly alter a map's representation of how a road either does or doesn't cross a particular parcel of land.

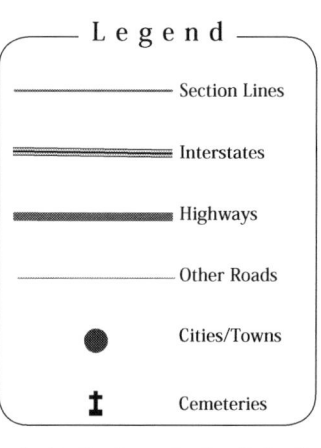

Copyright 2006 Boyd IT, Inc. All Rights Reserved

Legend

———	Section Lines
═══	Interstates
▬▬▬	Highways
———	Other Roads
●	Cities/Towns
✝	Cemeteries

Scale: Section = 1 mile X 1 mile
(generally, with some exceptions)

243

Family Maps of Lauderdale County, Mississippi

Historical Map
T5-N R16-E
Choctaw Meridian

Map Group 19

Cities & Towns
Enzor
Pleasant Hill
Stinson
Wolf Springs
Zero

Cemeteries
Long Creek Cemetery
Mount Horeb Cemetery
New Hope Cemetery
Pine Grove Cemetery
Stinson Cemetery

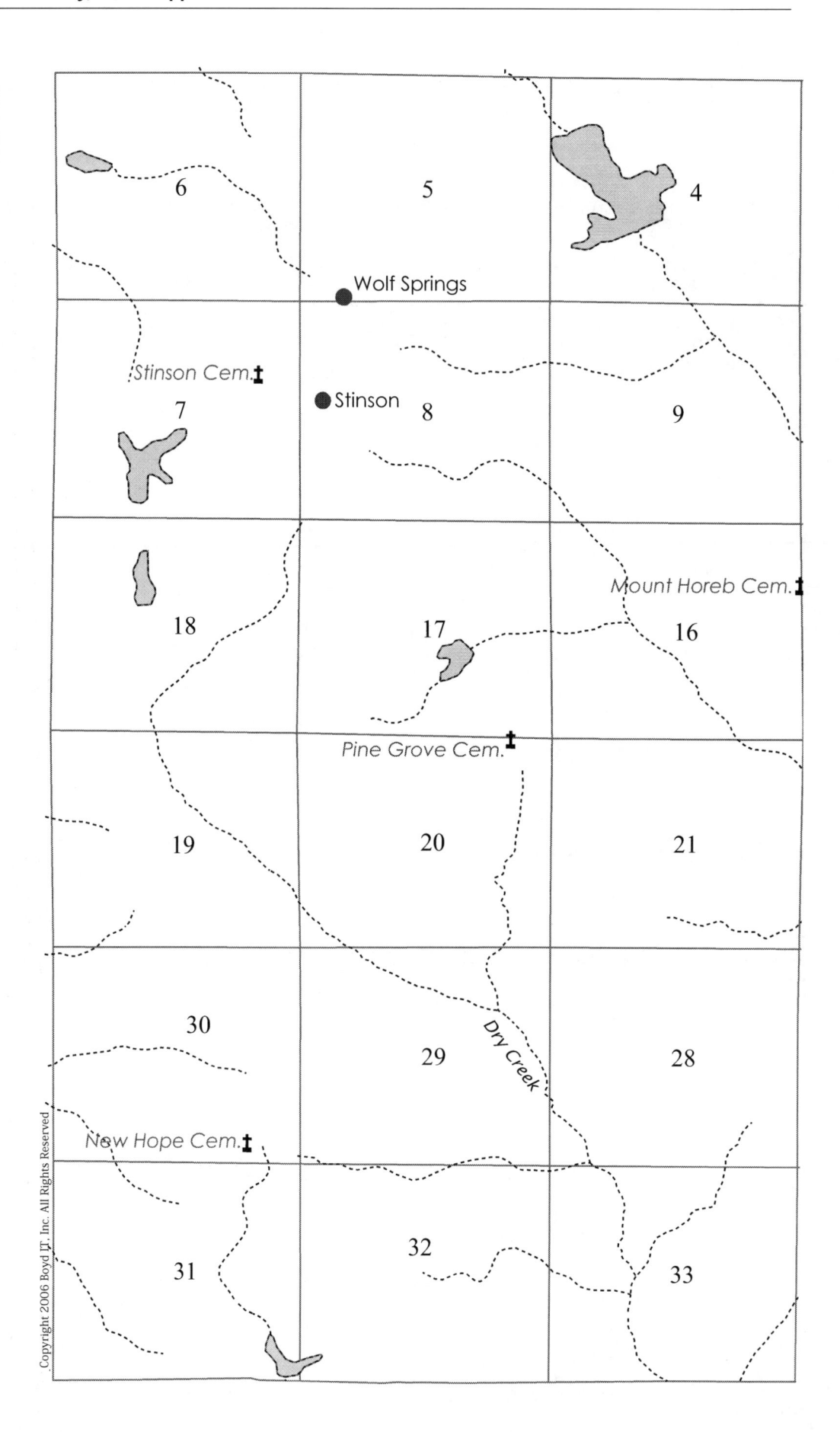

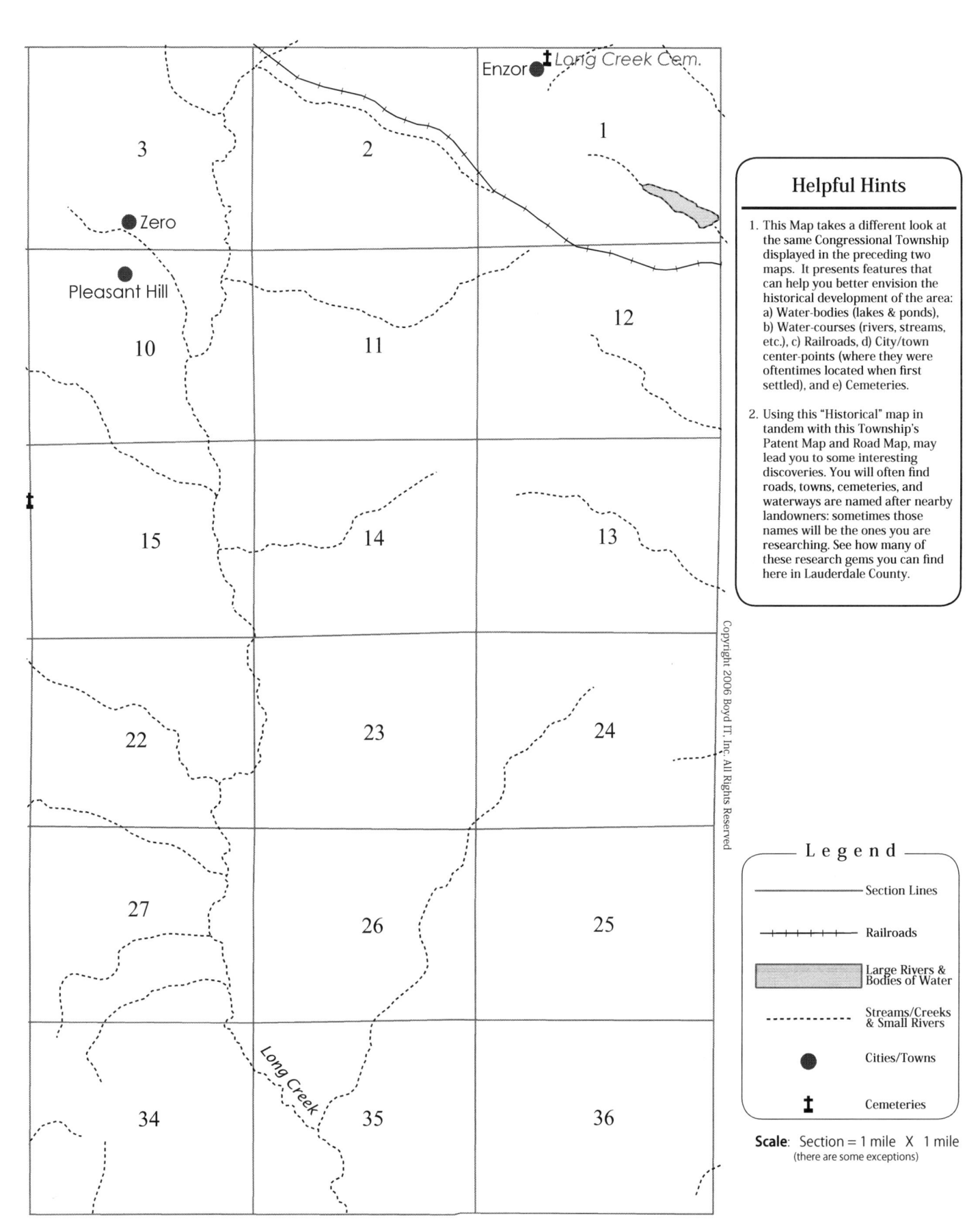

Township 5-N Range 16-E (Choctaw) - Map Group 19

Helpful Hints

1. This Map takes a different look at the same Congressional Township displayed in the preceding two maps. It presents features that can help you better envision the historical development of the area: a) Water-bodies (lakes & ponds), b) Water-courses (rivers, streams, etc.), c) Railroads, d) City/town center-points (where they were oftentimes located when first settled), and e) Cemeteries.

2. Using this "Historical" map in tandem with this Township's Patent Map and Road Map, may lead you to some interesting discoveries. You will often find roads, towns, cemeteries, and waterways are named after nearby landowners: sometimes those names will be the ones you are researching. See how many of these research gems you can find here in Lauderdale County.

Legend

- Section Lines
- +++++ Railroads
- ▓ Large Rivers & Bodies of Water
- ----- Streams/Creeks & Small Rivers
- ● Cities/Towns
- ✝ Cemeteries

Scale: Section = 1 mile X 1 mile
(there are some exceptions)

Features shown on map:
- Enzor
- Long Creek Cem.
- Zero
- Pleasant Hill
- Long Creek

Sections: 1, 2, 3, 10, 11, 12, 13, 14, 15, 22, 23, 24, 25, 26, 27, 34, 35, 36

Copyright 2006 Boyd IT, Inc. All Rights Reserved

245

Map Group 20: Index to Land Patents
Township 5-North Range 17-East (Choctaw)

After you locate an individual in this Index, take note of the Section and Section Part then proceed to the Land Patent map on the pages immediately following. You should have no difficulty locating the corresponding parcel of land.

The "For More Info" Column will lead you to more information about the underlying Patents. See the *Legend* at right, and the "How to Use this Book" chapter, for more information.

```
                    LEGEND
          "For More Info . . . " column
A = Authority (Legislative Act, See Appendix "A")
B = Block or Lot (location in Section unknown)
C = Cancelled Patent
F = Fractional Section
G = Group  (Multi-Patentee Patent, see Appendix "C")
V = Overlaps another Parcel
R = Re-Issued (Parcel patented more than once)

(A & G items require you to look in the Appendixes referred
to above. All other Letter-designations followed by a number
require you to locate line-items in this index that possess
the ID number found after the letter).
```

ID	Individual in Patent	Sec.	Sec. Part	Date Issued	Other Counties	For More Info . . .
3740	ANDERSON, William J	19	NWNE	1849-12-01		A1 R3741
3741	" "	19	NWNE	1849-12-01		A1 R3740
3708	BANYARD, Isaac	28	SESE	1849-12-01		A1 R3709
3709	" "	28	SESE	1849-12-01		A1 R3708
3742	BANYARD, William M	33	SWNW	1849-12-01		A1 R3743
3743	" "	33	SWNW	1849-12-01		A1 R3742
3729	BOUTWELL, Sarah	21	NWSW	1848-09-01		A1
3730	" "	21	S½NW	1848-09-01		A1
3704	BRUISTER, Christopher C	22	W½NE	1849-12-01		A1
3746	BUNYARD, William M	33	SWNE	1848-09-01		A1
3744	" "	32	N½NE	1860-04-10		A1
3745	" "	32	SENE	1860-04-10		A1
3735	BUTLER, William	28	SW	1850/06/01		A2
3736	" "	28	W½SE	1850/06/01		A2
3747	CARR, William S	29	S½SE	1849-12-01		A1 R3748
3748	" "	29	S½SE	1849-12-01		A1 R3747
3726	ERBY, John S	33	NWNE	1849-12-01		A1
3728	GILMORE, Mary	10	W½NW	1846-09-01		A1
3737	GODWIN, William C	26	SWSE	1860-04-02		A1
3702	HAYES, Benjamin W	32	SWNE	1859-05-02		A1
3716	HAYS, James W	32	NWSE	1860-04-02		A1
3717	" "	32	NWSW	1860-04-02		A1
3718	HAYS, John	31	SENE	1848-09-01		A1
3719	" "	32	SWNW	1848-09-01		A1
3720	HAYSE, John	32	NESW	1849-12-01		A1
3721	" "	32	SENW	1849-12-01		A1
3703	JONES, Charles	1	NWSE	1852/05/01		A2
3734	KENNERLY, Thomas	12	NW	1859-05-02		A1
3700	LILLEY, Benjamin F	32	SESE	1859-06-01		A1
3714	LILLEY, James	32	E½SE	1859-05-02		A1
3715	LILLY, James	32	SWSE	1846-09-01		A1
3707	MARTIN, Daniel	34	E½NE	1841-01-05		A1 G142
3706	" "	26	SWSW	1859-05-02		A1
3699	MCDOUGAL, Archibald	10	E½NW	1848-09-01		A1
3724	MCINNIS, John	4	W½NW	1841-01-05		A1
3723	" "	4	SENW	1846-09-01		A1
3725	" "	4	W½SW	1846-09-01		A1
3701	MCKELVIN, Benjamin	33	W½SE	1850/06/01		A2
3722	MCKLEVAIN, John L	33	NENE	1850/06/01		A2
3707	MITCHELL, Robert	34	E½NE	1841-01-05		A1 G142
3727	POGUE, Martin J	20	E½SE	1848-09-01		A1
3710	PRINGLE, Isham K	1	E½SW	1848-09-01		A1
3711	" "	1	NWSW	1848-09-01		A1
3712	" "	1	SENW	1852/05/01		A2
3713	" "	1	SWNW	1852/05/01		A2
3733	RAINER, Thomas G	22	SESW	1859-11-10		A1

Township 5-N Range 17-E (Choctaw) - Map Group 20

ID	Individual in Patent	Sec.	Sec. Part	Date Issued	Other Counties	For More Info . . .
3731	WHEAT, Solomon	15	E½SW	1841-01-05		A1
3732	" "	15	W½SE	1841-01-05		A1
3738	WHITE, William H	2	NENW	1860-04-02		A1
3739	" "	2	NWNE	1860-04-02		A1
3705	WILLIS, Cincinattus	1	SESE	1849-12-01		A1

Family Maps of Lauderdale County, Mississippi

Patent Map

T5-N R17-E
Choctaw Meridian
Map Group 20

Township Statistics

Parcels Mapped	:	50
Number of Patents	:	42
Number of Individuals	:	32
Patentees Identified	:	32
Number of Surnames	:	29
Multi-Patentee Parcels	:	1
Oldest Patent Date	:	1/5/1841
Most Recent Patent	:	4/10/1860
Block/Lot Parcels	:	0
Parcels Re - Issued	:	4
Parcels that Overlap	:	0
Cities and Towns	:	2
Cemeteries	:	1

Township 5-N Range 17-E (Choctaw) - Map Group 20

Section	Patentees
2	WHITE William H 1860; WHITE William H 1860
1	PRINGLE Isham K 1852; PRINGLE Isham K 1852; PRINGLE Isham K 1848; JONES Charles 1852; PRINGLE Isham K 1848; WILLIS Cincinattus 1849
10	GILMORE Mary 1846; MCDOUGAL Archibald 1848
12	KENNERLY Thomas 1859
15	WHEAT Solomon 1841; WHEAT Solomon 1841
22	BRUISTER Christopher C 1849; RAINER Thomas G 1859
26	MARTIN Daniel 1859; GODWIN William C 1860
34	MARTIN [142] Daniel 1841

Helpful Hints

1. This Map's INDEX can be found on the preceding pages.

2. Refer to Map "C" to see where this Township lies within Lauderdale County, Mississippi.

3. Numbers within square brackets [ ] denote a multi-patentee land parcel (multi-owner). Refer to Appendix "C" for a full list of members in this group.

4. Areas that look to be crowded with Patentees usually indicate multiple sales of the same parcel (Re-issues) or Overlapping parcels. See this Township's Index for an explanation of these and other circumstances that might explain "odd" groupings of Patentees on this map.

Copyright 2006 Boyd IT, Inc. All Rights Reserved

Legend

― Patent Boundary
━ Section Boundary
▓ No Patents Found (or Outside County)
1., 2., 3., ... Lot Numbers (when beside a name)
[ ] Group Number (see Appendix "C")

Scale: Section = 1 mile X 1 mile (generally, with some exceptions)

249

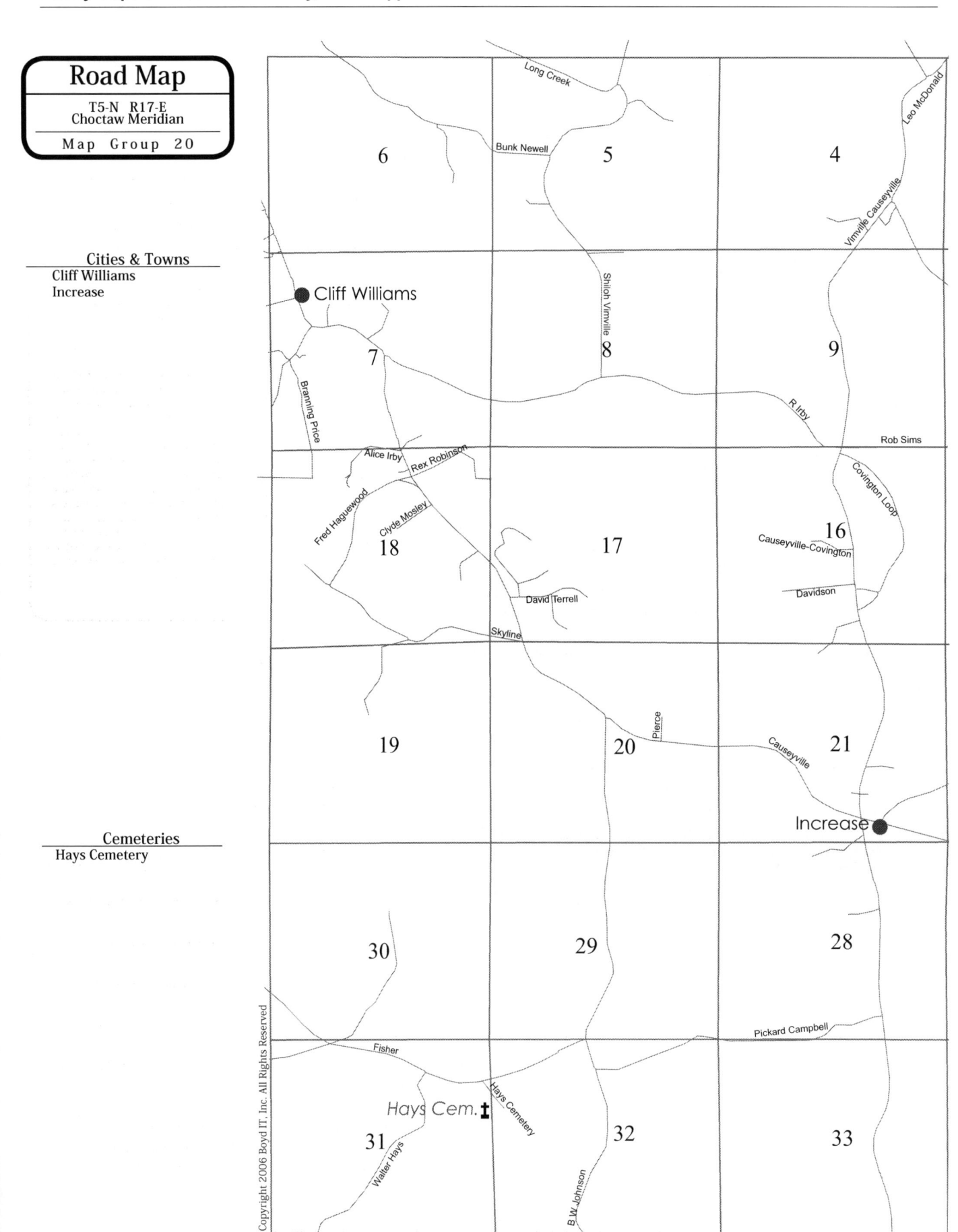

Family Maps of Lauderdale County, Mississippi

Road Map
T5-N R17-E
Choctaw Meridian
Map Group 20

Cities & Towns
Cliff Williams
Increase

Cemeteries
Hays Cemetery

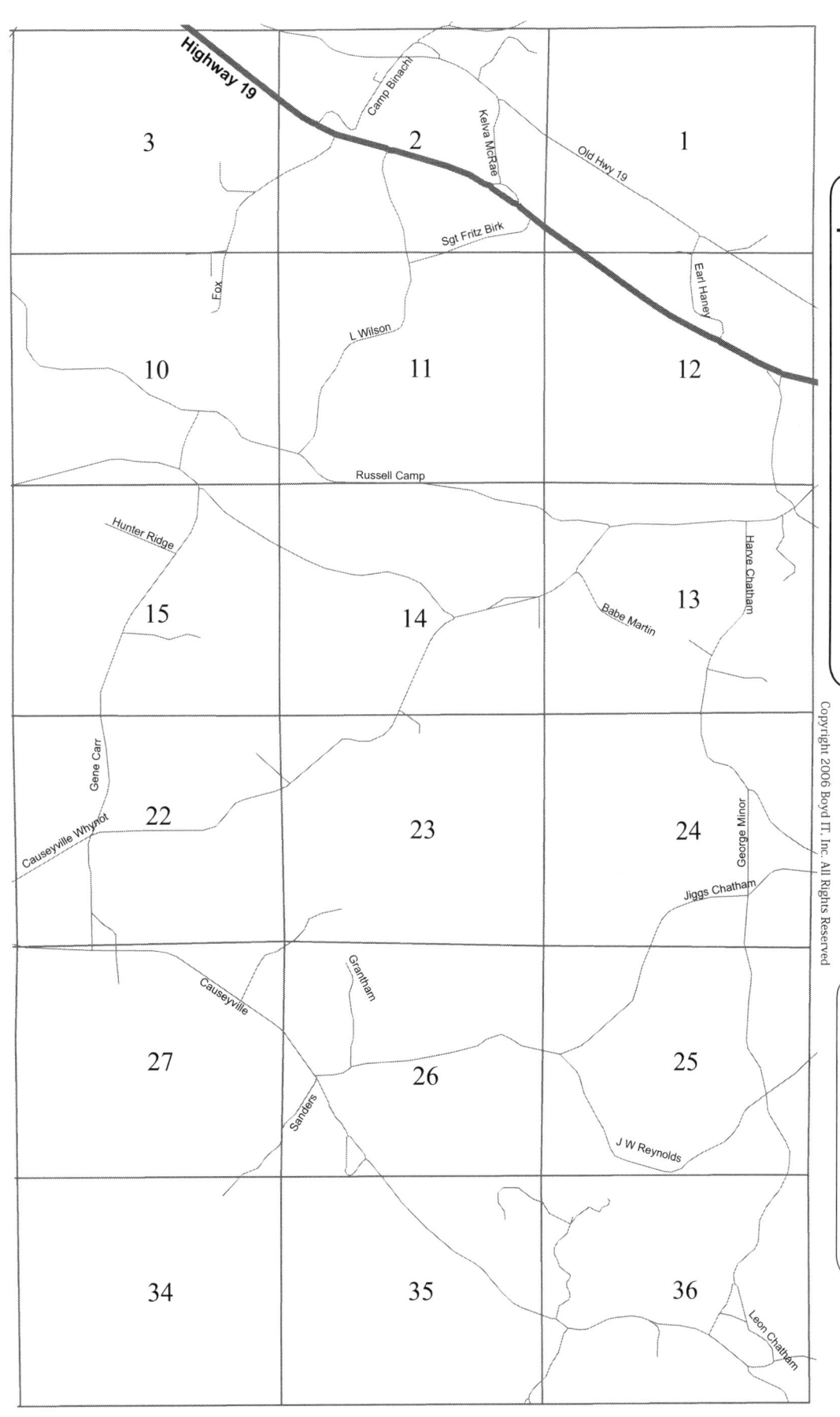

Township 5-N Range 17-E (Choctaw) - Map Group 20

Helpful Hints

1. This road map has a number of uses, but primarily it is to help you: a) find the present location of land owned by your ancestors (at least the general area), b) find cemeteries and city-centers, and c) estimate the route/roads used by Census-takers & tax-assessors.

2. If you plan to travel to Lauderdale County to locate cemeteries or land parcels, please pick up a modern travel map for the area before you do. Mapping old land parcels on modern maps is not as exact a science as you might think. Just the slightest variations in public land survey coordinates, estimates of parcel boundaries, or road-map deviations can greatly alter a map's representation of how a road either does or doesn't cross a particular parcel of land.

Legend

- Section Lines
- Interstates
- Highways
- Other Roads
- ● Cities/Towns
- ☦ Cemeteries

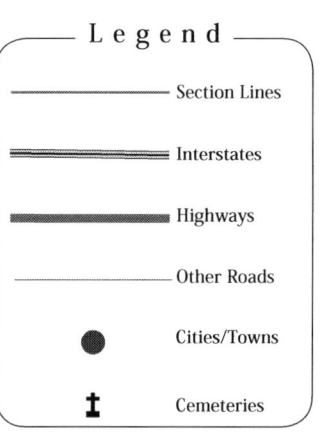

Scale: Section = 1 mile X 1 mile
(generally, with some exceptions)

251

Family Maps of Lauderdale County, Mississippi

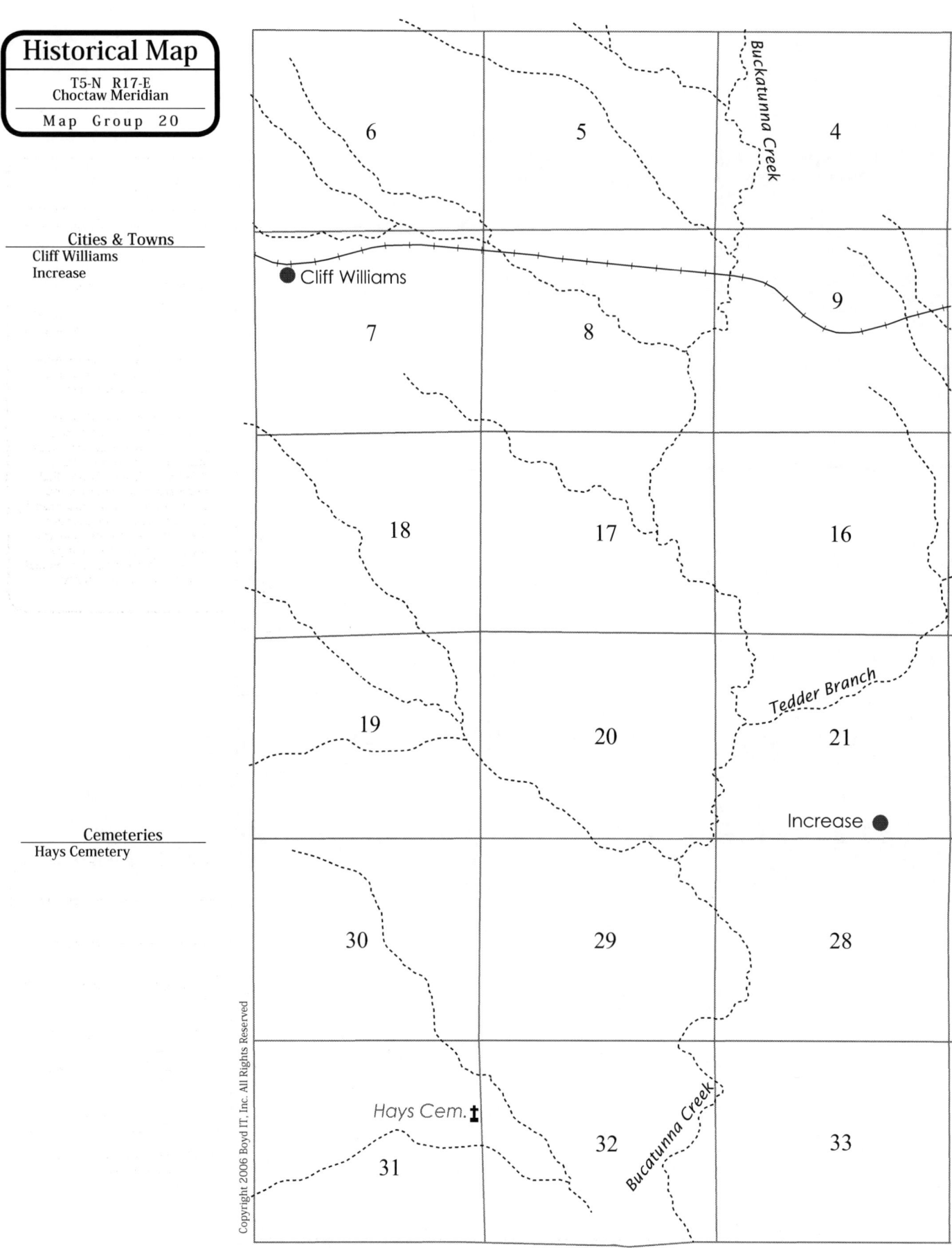

Historical Map
T5-N R17-E
Choctaw Meridian
Map Group 20

Cities & Towns
Cliff Williams
Increase

Cemeteries
Hays Cemetery

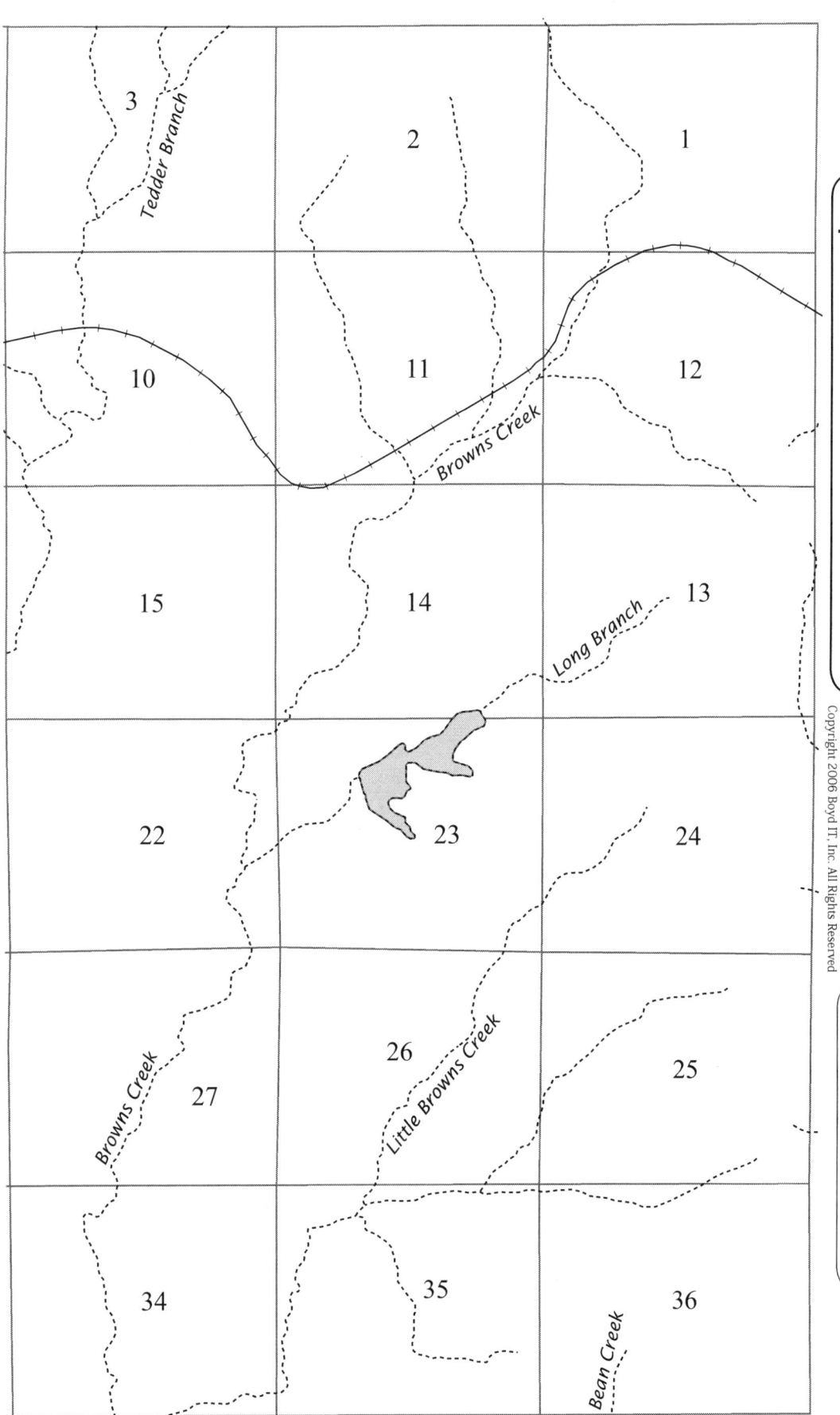

Township 5-N Range 17-E (Choctaw) - Map Group 20

Helpful Hints

1. This Map takes a different look at the same Congressional Township displayed in the preceding two maps. It presents features that can help you better envision the historical development of the area: a) Water-bodies (lakes & ponds), b) Water-courses (rivers, streams, etc.), c) Railroads, d) City/town center-points (where they were oftentimes located when first settled), and e) Cemeteries.

2. Using this "Historical" map in tandem with this Township's Patent Map and Road Map, may lead you to some interesting discoveries. You will often find roads, towns, cemeteries, and waterways are named after nearby landowners: sometimes those names will be the ones you are researching. See how many of these research gems you can find here in Lauderdale County.

Copyright 2006 Boyd IT, Inc. All Rights Reserved

Legend

— Section Lines
+++++ Railroads
▓ Large Rivers & Bodies of Water
- - - - Streams/Creeks & Small Rivers
● Cities/Towns
☦ Cemeteries

Scale: Section = 1 mile X 1 mile
(there are some exceptions)

253

Map Group 21: Index to Land Patents
Township 5-North Range 18-East (Choctaw)

After you locate an individual in this Index, take note of the Section and Section Part then proceed to the Land Patent map on the pages immediately following. You should have no difficulty locating the corresponding parcel of land.

The "For More Info" Column will lead you to more information about the underlying Patents. See the *Legend* at right, and the "How to Use this Book" chapter, for more information.

```
LEGEND
        "For More Info . . . " column
A = Authority (Legislative Act, See Appendix "A")
B = Block or Lot (location in Section unknown)
C = Cancelled Patent
F = Fractional Section
G = Group (Multi-Patentee Patent, see Appendix "C")
V = Overlaps another Parcel
R = Re-Issued (Parcel patented more than once)

(A & G items require you to look in the Appendixes referred
to above. All other Letter-designations followed by a number
require you to locate line-items in this index that possess
the ID number found after the letter).
```

ID	Individual in Patent	Sec.	Sec. Part	Date Issued	Other Counties	For More Info . . .
3822	BAILEY, Jones W	22	SESW	1859-05-02		A1
3823	" "	22	SWSE	1859-05-02		A1
3758	BLANKS, Charles F	33	NENE	1897-12-10		A4
3767	BLANKS, Dempsey	34	SWNE	1846-09-01		A1
3768	" "	34	W½SE	1846-09-01		A1
3770	BLANKS, Ervin R	34	NWNW	1859-05-02		A1
3784	BLANKS, James L	34	NESW	1859-05-02		A1
3785	" "	34	SWNW	1859-05-02		A1
3786	" "	35	1	1859-05-02		A1
3783	" "	33	SENE	1861-02-01		A1
3791	BLANKS, James W	34	NWNE	1859-05-02		A1
3793	BLANKS, Jefferson	34	SESW	1860-07-02		A1
3828	BLANKS, Joseph J	33	SWNE	1889-11-29		A1
3781	BREWSTER, James	4	SESW	1859-05-02		A1
3782	" "	8	NENE	1859-05-02		A1
3807	BROCK, John	15	NWSE	1853/01/24		A2 G39
3808	" "	22	NWNE	1853/01/24		A2 G39
3809	" "	26	2	1853/01/24		A2 G39 C
3810	" "	29	SENE	1853/01/24		A2 G39
3794	" "	23	5	1859-05-02		A1
3795	" "	23	8	1859-05-02		A1
3796	" "	26	1	1859-05-02		A1
3798	" "	27	SENE	1859-05-02		A1
3797	" "	27	NENE	1860-10-01		A1
3772	BRONSON, Harvey P	29	N½SE	1849-12-01		A1 R3773
3773	" "	29	N½SE	1849-12-01		A1 R3772
3759	BROWN, Collin B	15	SENE	1882-10-10		A1
3842	BROWN, Robert C	15	NWNE	1882-08-03		A1
3765	BRYAN, David	14	7	1859-05-02		A1
3766	" "	14	8	1859-05-02		A1
3787	BUCKLEW, James M	15	SESW	1859-05-02		A1
3807	CARR, John P	15	NWSE	1853/01/24		A2 G39
3808	" "	22	NWNE	1853/01/24		A2 G39
3809	" "	26	2	1853/01/24		A2 G39 C
3810	" "	29	SENE	1853/01/24		A2 G39
3847	CLARK, Samuel	5	NWNE	1846-09-01		A1
3807	CLIET, John	15	NWSE	1853/01/24		A2 G39
3808	" "	22	NWNE	1853/01/24		A2 G39
3809	" "	26	2	1853/01/24		A2 G39 C
3810	" "	29	SENE	1853/01/24		A2 G39
3800	DAY, John	27	NWNE	1859-06-01		A1
3799	" "	27	NESE	1860-04-02		A1
3801	" "	27	SWNE	1860-04-02		A1
3802	" "	27	SWSE	1860-04-02		A1
3850	DUNBAR, Thomas	33	NWNE	1859-05-02		A1
3775	GARDNER, Isaiah	21	SESE	1859-05-02		A1

Township 5-N Range 18-E (Choctaw) - Map Group 21

ID	Individual in Patent	Sec.	Sec. Part	Date Issued	Other Counties	For More Info . . .
3777	GARDNER, Isaiah (Cont'd)	27	W½NW	1859-05-02		A1
3776	"	27	SENW	1860-10-01		A1
3835	GOODMAN, Loyied G	33	SESW	1904-11-15		A4
3848	GOODWYN, Samuel P	10	S½SE	1859-05-02		A1 V3752
3803	GUNN, John	2	W½SW	1841-01-05		A1 G93 F
3836	GUNN, Margaret	2	E½SW	1841-01-05		A1 G94 F
3792	HAGOOD, James W	20	SWSE	1859-05-02		A1
3756	HARRELL, Bogan C	22	NESW	1859-11-10		A1
3757	"	22	NWSE	1859-11-10		A1
3846	HARRISON, Robert	28	E½NW	1841-01-05		A1 G103
3845	"	28	W½SW	1849-12-01		A1
3843	"	21	SWSE	1859-05-02		A1
3844	"	28	NWNE	1859-05-02		A1
3749	HUNTER, Ann	11	6	1859-05-02		A1
3750	"	14	2	1859-05-02		A1
3752	HUNTER, Anthony D	10	SESE	1846-09-01		A1 V3848
3754	"	11	7	1846-09-01		A1 F
3755	"	15	NENE	1850/06/01		A2
3751	"	10	NESE	1859-05-02		A1
3753	"	10	SWNE	1859-05-02		A1
3769	HUNTER, Edward	11	4	1860-04-02		A1
3830	HUNTER, Libbens	2	5	1859-11-10		A1
3831	"	2	6	1859-11-10		A1
3832	"	2	7	1859-11-10		A1
3833	HUNTER, Libbeus	11	2	1846-09-01		A1 F
3834	HUNTER, Libius	2	4	1850/06/01		A2 C
3852	HUNTER, William T	11	5	1918-04-11		A4
3790	JOHNSON, James S	11	1	1846-09-01		A1 F
3840	JONES, Redding	3	NWSE	1846-09-01		A1
3836	LEE, Robert	2	E½SW	1841-01-05		A1 G94 F
3788	MCRAE, James	34	E½SE	1841-01-05		A1
3789	"	35	2	1846-09-01		A1 F
3806	MCRAE, John	3	E½SE	1841-01-05		A1
3816	POWELL, John	27	SESE	1860-04-02		A1
3779	PRINGLE, Isham K	6	SWNW	1846-09-01		A1
3778	"	6	NWNW	1848-09-01		A1
3763	REYNOLDS, Daniel	33	NESW	1860-10-01		A1
3774	ROBERSON, Henry	21	NESE	1859-06-01		A1
3771	ROBERTSON, Harry	21	NWSE	1859-06-01		A1
3804	ROBINSON, John H	22	NENW	1849-12-01		A1 R3821
3821	ROBINSON, John X	22	NENW	1849-12-01		A1 R3804
3837	ROBISON, Ransom H	4	E½NE	1859-11-10		A1
3838	ROBISON, Ranson X	4	N½SE	1860-04-02		A1
3839	"	4	NWNE	1860-04-02		A1
3846	SAPP, Martha	28	E½NW	1841-01-05		A1 G103
3849	SCARBOROUGH, Silas	6	SESE	1852/05/01		A2
3805	SHANNON, John M	22	SENW	1859-06-01		A1
3813	SMITH, John P	15	NESE	1849-12-01		A1
3807	"	15	NWSE	1853/01/24		A2 G39
3808	"	22	NWNE	1853/01/24		A2 G39
3809	"	26	2	1853/01/24		A2 G39 C
3810	"	29	SENE	1853/01/24		A2 G39
3811	"	10	W½SW	1859-05-02		A1
3814	"	15	SWNE	1859-06-01		A1
3812	"	15	N½SW	1860-04-02		A1
3815	"	15	SWSE	1860-08-01		A1
3803	THOMPSON, David B	2	W½SW	1841-01-05		A1 G93 F
3764	"	5	E½SE	1846-09-01		A1
3841	THOMPSON, Reuben P	15	SWSW	1860-04-02		A1
3824	UPCHURCH, Jordon	14	6	1849-12-01		A1 F R3825
3825	"	14	6	1849-12-01		A1 F R3824
3826	"	23	1	1849-12-01		A1 F R3827
3827	"	23	1	1849-12-01		A1 F R3826
3761	WALLACE, Daniel C	22	SWNE	1859-05-02		A1
3760	"	15	SESE	1860-04-02		A1
3762	"	23	6	1860-04-02		A1
3851	WALLACE, William H	33	SE	1888-02-25		A4
3780	WELCH, Jacob P	10	SENW	1859-06-01		A1
3817	WILKINSON, John W	11	8	1859-05-02		A1
3818	"	14	1	1859-05-02		A1
3819	"	14	4	1859-05-02		A1
3820	"	33	W½SW	1860-10-01		A1
3829	WILLIAMS, Joseph	33	E½NW	1860-10-01		A1

Family Maps of Lauderdale County, Mississippi

Patent Map

T5-N R18-E
Choctaw Meridian

Map Group 21

Township Statistics

Parcels Mapped	:	104
Number of Patents	:	85
Number of Individuals	:	61
Patentees Identified	:	59
Number of Surnames	:	42
Multi-Patentee Parcels	:	7
Oldest Patent Date	:	1/5/1841
Most Recent Patent	:	4/11/1918
Block/Lot Parcels	:	27
Parcels Re-Issued	:	4
Parcels that Overlap	:	2
Cities and Towns	:	0
Cemeteries	:	1

Section 6
- PRINGLE, Isham K — 1848
- PRINGLE, Isham K — 1846
- SCARBOROUGH, Silas — 1852

Section 5
- CLARK, Samuel — 1846
- THOMPSON, David B — 1846
- BREWSTER, James — 1859

Section 4
- ROBISON, Ranson X — 1860
- ROBISON, Ransom H — 1859
- ROBISON, Ranson X — 1860
- BREWSTER, James — 1859

Section 7

Section 8

Section 9

Section 18

Section 17

Section 16

Section 19

Section 20
- HAGOOD, James W — 1859

Section 21
- ROBERTSON, Harry — 1859
- ROBERSON, Henry — 1859
- HARRISON, Robert — 1859
- GARDNER, Isaiah — 1859

Section 30

Section 29
- CARR [39], John P — 1853
- BRONSON, Harvey P — 1849

Section 28
- HARRISON, Robert — 1859
- HARRISON [103], Robert — 1841
- HARRISON, Robert — 1849

Section 31

Section 32

Section 33
- DUNBAR, Thomas — 1859
- BLANKS, Charles F — 1897
- WILLIAMS, Joseph — 1860
- BLANKS, Joseph J — 1889
- BLANKS, James L — 1861
- WILKINSON, John W — 1860
- REYNOLDS, Daniel — 1860
- WALLACE, William H — 1888
- GOODMAN, Loyied G — 1904

Copyright 2006 Boyd IT, Inc. All Rights Reserved

Township 5-N Range 18-E (Choctaw) - Map Group 21

Section 2
Lots-Sec. 2
- 4 HUNTER, Libius 1850
- 5 HUNTER, Libbens 1859
- 6 HUNTER, Libbens 1859
- 7 HUNTER, Libbens 1859

Section 3
- JONES, Redding 1846
- MCRAE, John 1841
- GUNN [93], John 1841
- GUNN [94], Margaret 1841

Section 10
- WELCH, Jacob P 1859
- HUNTER, Anthony D 1859
- SMITH, John P 1859
- HUNTER, Anthony D 1859
- GOODWYN, Samuel P 1859
- HUNTER, Anthony D 1846

Section 11
Lots-Sec. 11
1. JOHNSON, James S 1846
2. HUNTER, Libbeus 1846
4. HUNTER, Edward 1860
5. HUNTER, William T 1918
6. HUNTER, Ann 1859
7. HUNTER, Anthony D 1846
8. WILKINSON, John W 1859

Section 14
Lots-Sec. 14
1. WILKINSON, John W 1859
2. HUNTER, Ann 1859
4. WILKINSON, John W 1859
6. UPCHURCH, Jordon 1849
6. UPCHURCH, Jordon 1849
7. BRYAN, David 1859
8. BRYAN, David 1859

Section 15
- BROWN, Robert C 1882
- HUNTER, Anthony D 1850
- SMITH, John P 1859
- BROWN, Collin B 1882
- SMITH, John P 1860
- CARR [39], John P 1853
- SMITH, John P 1849
- THOMPSON, Reuben P 1860
- BUCKLEW, James M 1859
- SMITH, John P 1860
- WALLACE, Daniel C 1860

Section 22
- ROBINSON, John H 1849
- ROBINSON, John X 1849
- CARR [39], John P 1853
- SHANNON, John M 1859
- WALLACE, Daniel C 1859
- HARRELL, Bogan C 1859
- HARRELL, Bogan C 1859
- BAILEY, Jones W 1859
- BAILEY, Jones W 1859

Section 23
Lots-Sec. 23
1. UPCHURCH, Jordon 1849
2. UPCHURCH, Jordon 1849
5. BROCK, John 1859
6. WALLACE, Daniel C 1860
8. BROCK, John 1859

Section 26
Lots-Sec. 26
1. BROCK, John 1859
2. CARR, John P [39] 1853

- DAY, John 1859
- BROCK, John 1860
- GARDNER, Isaiah 1860
- DAY, John 1860
- BROCK, John 1859
- DAY, John 1860
- DAY, John 1860
- POWELL, John 1860

Section 27
- GARDNER, Isaiah 1859

Section 34
- BLANKS, Ervin R 1859
- BLANKS, James W 1859
- BLANKS, James L 1859
- BLANKS, Dempsey 1846
- BLANKS, James L 1859
- BLANKS, Dempsey 1846
- BLANKS, Jefferson 1860

Section 35
Lots-Sec. 35
1. BLANKS, James L 1859
2. MCRAE, James 1846

- MCRAE, James 1841

Helpful Hints

1. This Map's INDEX can be found on the preceding pages.

2. Refer to Map "C" to see where this Township lies within Lauderdale County, Mississippi.

3. Numbers within square brackets [ ] denote a multi-patentee land parcel (multi-owner). Refer to Appendix "C" for a full list of members in this group.

4. Areas that look to be crowded with Patentees usually indicate multiple sales of the same parcel (Re-issues) or Overlapping parcels. See this Township's Index for an explanation of these and other circumstances that might explain "odd" groupings of Patentees on this map.

Copyright 2006 Boyd IT, Inc. All Rights Reserved

Legend

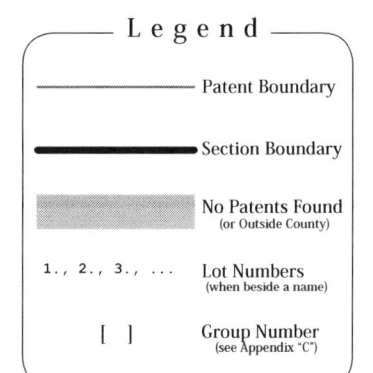

- ──────── Patent Boundary
- ▬▬▬▬▬▬ Section Boundary
- ▓▓▓▓▓▓ No Patents Found (or Outside County)
- 1., 2., 3., ... Lot Numbers (when beside a name)
- [ ] Group Number (see Appendix "C")

Scale: Section = 1 mile X 1 mile
(generally, with some exceptions)

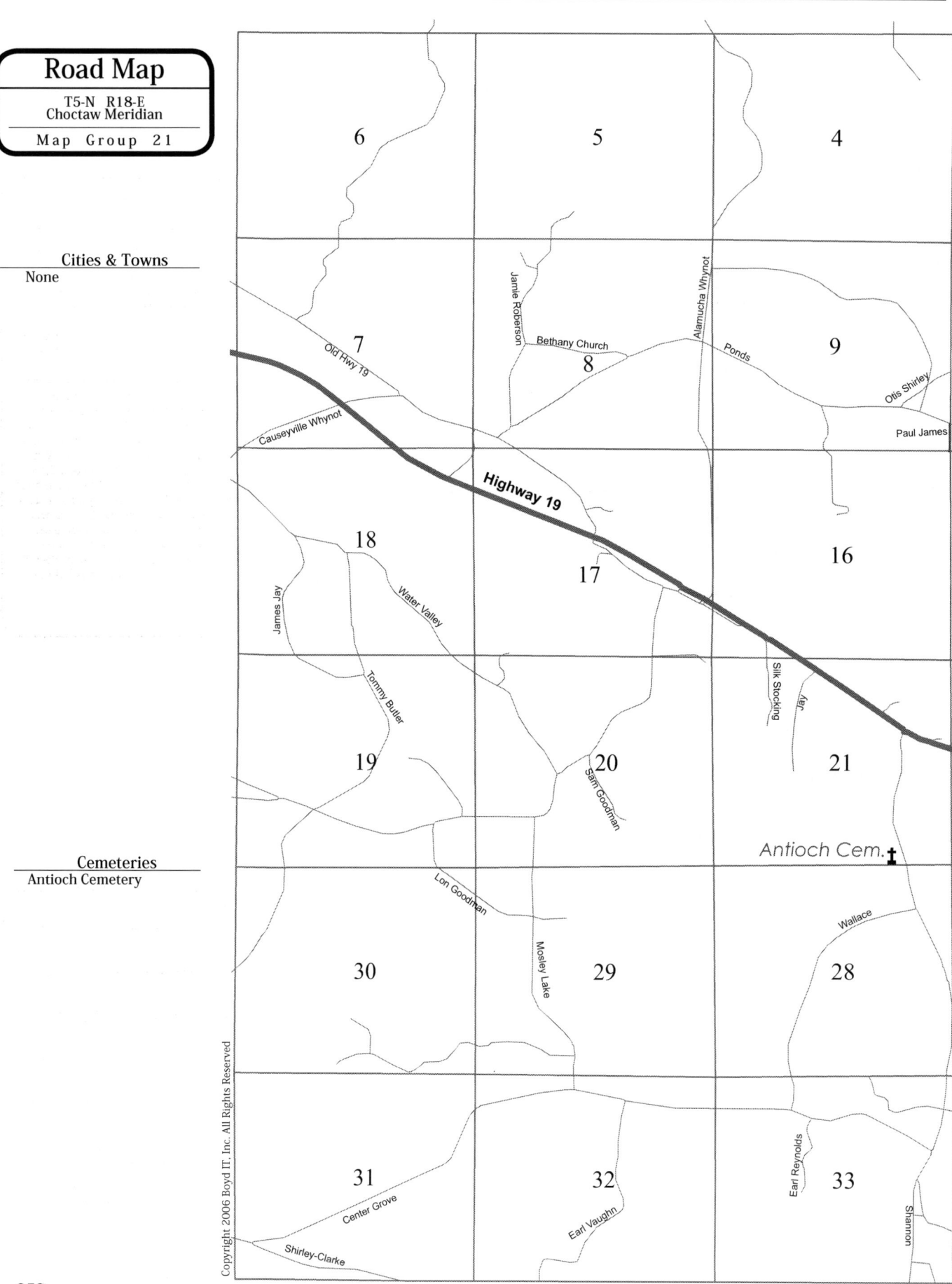

Family Maps of Lauderdale County, Mississippi

Road Map
T5-N R18-E Choctaw Meridian
Map Group 21

Cities & Towns
None

Cemeteries
Antioch Cemetery

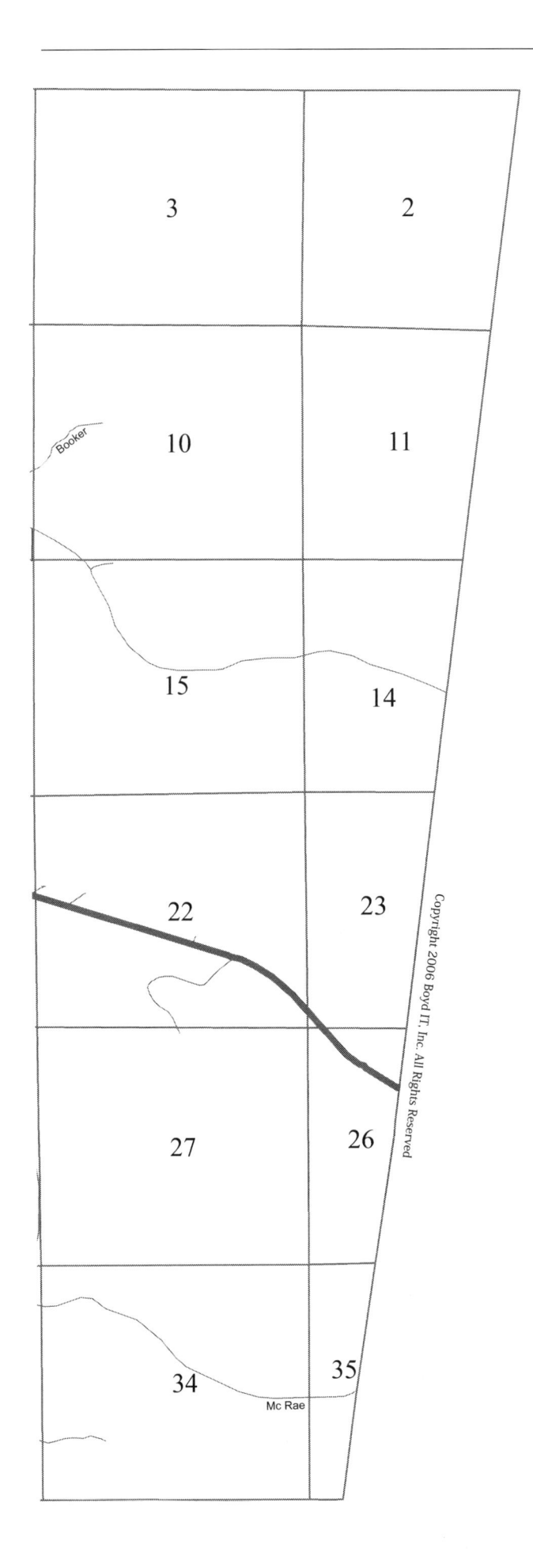

Township 5-N Range 18-E (Choctaw) - Map Group 21

Helpful Hints

1. This road map has a number of uses, but primarily it is to help you: a) find the present location of land owned by your ancestors (at least the general area), b) find cemeteries and city-centers, and c) estimate the route/roads used by Census-takers & tax-assessors.

2. If you plan to travel to Lauderdale County to locate cemeteries or land parcels, please pick up a modern travel map for the area before you do. Mapping old land parcels on modern maps is not as exact a science as you might think. Just the slightest variations in public land survey coordinates, estimates of parcel boundaries, or road-map deviations can greatly alter a map's representation of how a road either does or doesn't cross a particular parcel of land.

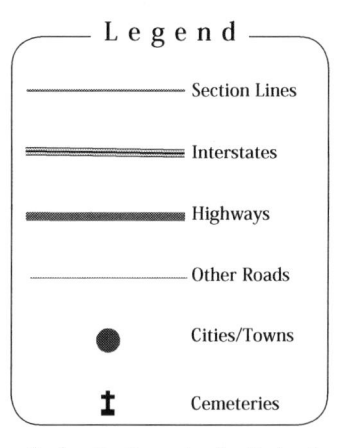

Copyright 2006 Boyd IT, Inc. All Rights Reserved

Legend

— Section Lines
═ Interstates
▬ Highways
— Other Roads
● Cities/Towns
✝ Cemeteries

Scale: Section = 1 mile X 1 mile
(generally, with some exceptions)

259

Family Maps of Lauderdale County, Mississippi

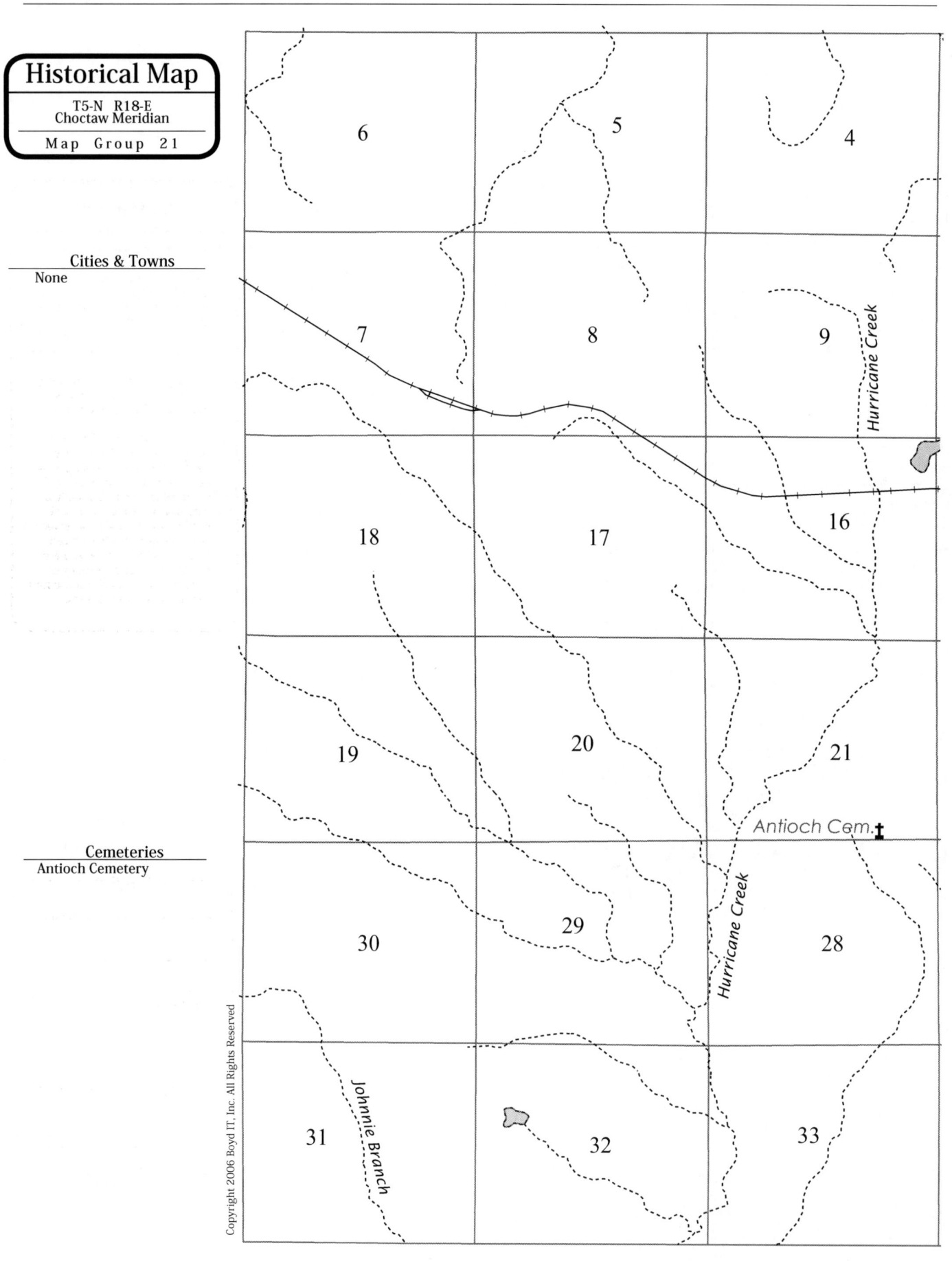

Historical Map
T5-N R18-E
Choctaw Meridian
Map Group 21

Cities & Towns
None

Cemeteries
Antioch Cemetery

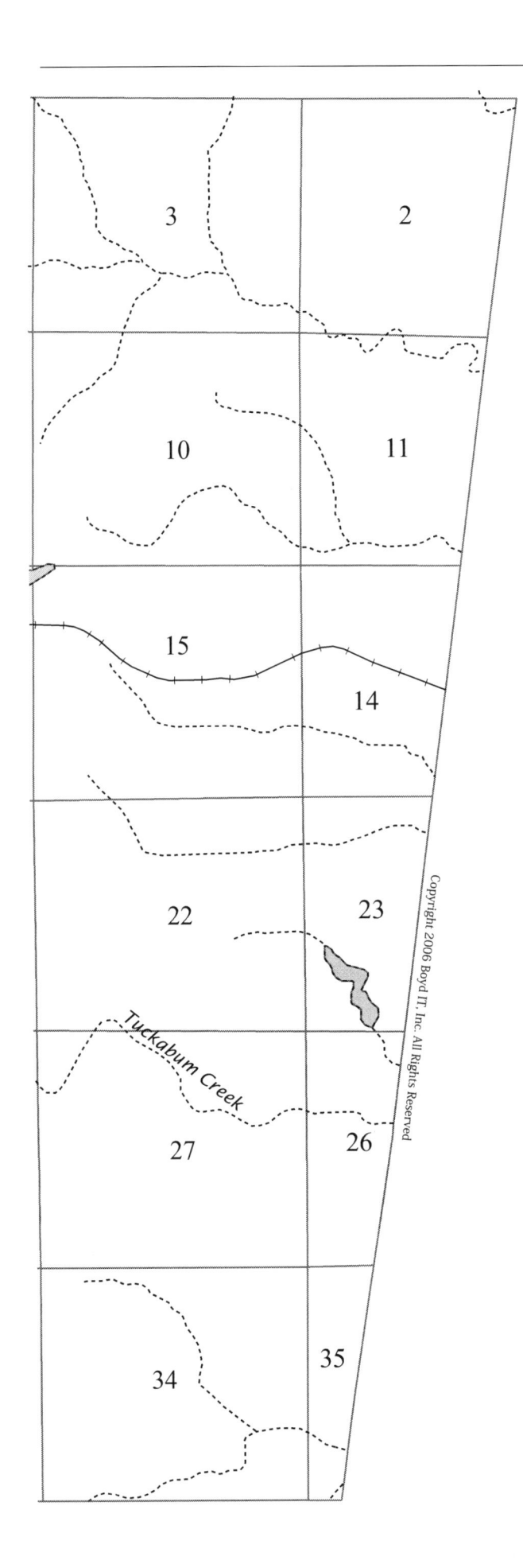

Township 5-N Range 18-E (Choctaw) - Map Group 21

Helpful Hints

1. This Map takes a different look at the same Congressional Township displayed in the preceding two maps. It presents features that can help you better envision the historical development of the area: a) Water-bodies (lakes & ponds), b) Water-courses (rivers, streams, etc.), c) Railroads, d) City/town center-points (where they were oftentimes located when first settled), and e) Cemeteries.

2. Using this "Historical" map in tandem with this Township's Patent Map and Road Map, may lead you to some interesting discoveries. You will often find roads, towns, cemeteries, and waterways are named after nearby landowners: sometimes those names will be the ones you are researching. See how many of these research gems you can find here in Lauderdale County.

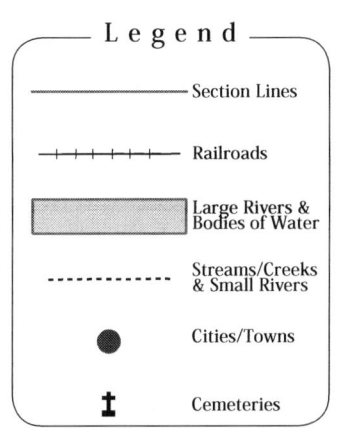

Legend

— Section Lines
+++++ Railroads
▬ Large Rivers & Bodies of Water
----- Streams/Creeks & Small Rivers
● Cities/Towns
☦ Cemeteries

Scale: Section = 1 mile X 1 mile
(there are some exceptions)

Tuckabum Creek

Appendices

Appendix A - Acts of Congress Authorizing the Patents Contained in this Book

The following Acts of Congress are referred to throughout the Indexes in this book. The text of the Federal Statutes referred to below can usually be found on the web. For more information on such laws, check out the publishers's web-site at *www.arphax.com*, go to the "Research" page, and click on the "Land-Law" link.

Ref. No.	Date and Act of Congress	Number of Parcels of Land
1	April 24, 1820: Sale-Cash Entry (3 Stat. 566)	3020
2	CHOCTAW SCRIP	123
3	March 3, 1855: ScripWarrant Act of 1855 (10 Stat. 701)	2
4	May 20, 1862: Homestead EntryOriginal (12 Stat. 392)	707

Appendix B - Section Parts (Aliquot Parts)

The following represent the various abbreviations we have found thus far in describing the parts of a Public Land Section. Some of these are very obscure and rarely used, but we wanted to list them for just that reason. A full section is 1 square mile or 640 acres.

Section Part	Description	Acres
<none>	Full Acre (if no Section Part is listed, presumed a full Section)	640
<1-??>	A number represents a Lot Number and can be of various sizes	?
E½	East Half-Section	320
E½E½	East Half of East Half-Section	160
E½E½SE	East Half of East Half of Southeast Quarter-Section	40
E½N½	East Half of North Half-Section	160
E½NE	East Half of Northeast Quarter-Section	80
E½NENE	East Half of Northeast Quarter of Northeast Quarter-Section	20
E½NENW	East Half of Northeast Quarter of Northwest Quarter-Section	20
E½NESE	East Half of Northeast Quarter of Southeast Quarter-Section	20
E½NESW	East Half of Northeast Quarter of Southwest Quarter-Section	20
E½NW	East Half of Northwest Quarter-Section	80
E½NWNE	East Half of Northwest Quarter of Northeast Quarter-Section	20
E½NWNW	East Half of Northwest Quarter of Northwest Quarter-Section	20
E½NWSE	East Half of Northwest Quarter of Southeast Quarter-Section	20
E½NWSW	East Half of Northwest Quarter of Southwest Quarter-Section	20
E½S½	East Half of South Half-Section	160
E½SE	East Half of Southeast Quarter-Section	80
E½SENE	East Half of Southeast Quarter of Northeast Quarter-Section	20
E½SENW	East Half of Southeast Quarter of Northwest Quarter-Section	20
E½SESE	East Half of Southeast Quarter of Southeast Quarter-Section	20
E½SESW	East Half of Southeast Quarter of Southwest Quarter-Section	20
E½SW	East Half of Southwest Quarter-Section	80
E½SWNE	East Half of Southwest Quarter of Northeast Quarter-Section	20
E½SWNW	East Half of Southwest Quarter of Northwest Quarter-Section	20
E½SWSE	East Half of Southwest Quarter of Southeast Quarter-Section	20
E½SWSW	East Half of Southwest Quarter of Southwest Quarter-Section	20
E½W½	East Half of West Half-Section	160
N½	North Half-Section	320
N½E½NE	North Half of East Half of Northeast Quarter-Section	40
N½E½NW	North Half of East Half of Northwest Quarter-Section	40
N½E½SE	North Half of East Half of Southeast Quarter-Section	40
N½E½SW	North Half of East Half of Southwest Quarter-Section	40
N½N½	North Half of North Half-Section	160
N½NE	North Half of Northeast Quarter-Section	80
N½NENE	North Half of Northeast Quarter of Northeast Quarter-Section	20
N½NENW	North Half of Northeast Quarter of Northwest Quarter-Section	20
N½NESE	North Half of Northeast Quarter of Southeast Quarter-Section	20
N½NESW	North Half of Northeast Quarter of Southwest Quarter-Section	20
N½NW	North Half of Northwest Quarter-Section	80
N½NWNE	North Half of Northwest Quarter of Northeast Quarter-Section	20
N½NWNW	North Half of Northwest Quarter of Northwest Quarter-Section	20
N½NWSE	North Half of Northwest Quarter of Southeast Quarter-Section	20
N½NWSW	North Half of Northwest Quarter of Southwest Quarter-Section	20
N½S½	North Half of South Half-Section	160
N½SE	North Half of Southeast Quarter-Section	80
N½SENE	North Half of Southeast Quarter of Northeast Quarter-Section	20
N½SENW	North Half of Southeast Quarter of Northwest Quarter-Section	20
N½SESE	North Half of Southeast Quarter of Southeast Quarter-Section	20

Family Maps of Lauderdale County, Mississippi

Section Part	Description	Acres
N½SESW	North Half of Southeast Quarter of Southwest Quarter-Section	20
N½SESW	North Half of Southeast Quarter of Southwest Quarter-Section	20
N½SW	North Half of Southwest Quarter-Section	80
N½SWNE	North Half of Southwest Quarter of Northeast Quarter-Section	20
N½SWNW	North Half of Southwest Quarter of Northwest Quarter-Section	20
N½SWSE	North Half of Southwest Quarter of Southeast Quarter-Section	20
N½SWSE	North Half of Southwest Quarter of Southeast Quarter-Section	20
N½SWSW	North Half of Southwest Quarter of Southwest Quarter-Section	20
N½W½NW	North Half of West Half of Northwest Quarter-Section	40
N½W½SE	North Half of West Half of Southeast Quarter-Section	40
N½W½SW	North Half of West Half of Southwest Quarter-Section	40
NE	Northeast Quarter-Section	160
NEN½	Northeast Quarter of North Half-Section	80
NENE	Northeast Quarter of Northeast Quarter-Section	40
NENENE	Northeast Quarter of Northeast Quarter of Northeast Quarter	10
NENENW	Northeast Quarter of Northeast Quarter of Northwest Quarter	10
NENESE	Northeast Quarter of Northeast Quarter of Southeast Quarter	10
NENESW	Northeast Quarter of Northeast Quarter of Southwest Quarter	10
NENW	Northeast Quarter of Northwest Quarter-Section	40
NENWNE	Northeast Quarter of Northwest Quarter of Northeast Quarter	10
NENWNW	Northeast Quarter of Northwest Quarter of Northwest Quarter	10
NENWSE	Northeast Quarter of Northwest Quarter of Southeast Quarter	10
NENWSW	Northeast Quarter of Northwest Quarter of Southwest Quarter	10
NESE	Northeast Quarter of Southeast Quarter-Section	40
NESENE	Northeast Quarter of Southeast Quarter of Northeast Quarter	10
NESENW	Northeast Quarter of Southeast Quarter of Northwest Quarter	10
NESESE	Northeast Quarter of Southeast Quarter of Southeast Quarter	10
NESESW	Northeast Quarter of Southeast Quarter of Southwest Quarter	10
NESW	Northeast Quarter of Southwest Quarter-Section	40
NESWNE	Northeast Quarter of Southwest Quarter of Northeast Quarter	10
NESWNW	Northeast Quarter of Southwest Quarter of Northwest Quarter	10
NESWSE	Northeast Quarter of Southwest Quarter of Southeast Quarter	10
NESWSW	Northeast Quarter of Southwest Quarter of Southwest Quarter	10
NW	Northwest Quarter-Section	160
NWE½	Northwest Quarter of Eastern Half-Section	80
NWN½	Northwest Quarter of North Half-Section	80
NWNE	Northwest Quarter of Northeast Quarter-Section	40
NWNENE	Northwest Quarter of Northeast Quarter of Northeast Quarter	10
NWNENW	Northwest Quarter of Northeast Quarter of Northwest Quarter	10
NWNESE	Northwest Quarter of Northeast Quarter of Southeast Quarter	10
NWNESW	Northwest Quarter of Northeast Quarter of Southwest Quarter	10
NWNW	Northwest Quarter of Northwest Quarter-Section	40
NWNWNE	Northwest Quarter of Northwest Quarter of Northeast Quarter	10
NWNWNW	Northwest Quarter of Northwest Quarter of Northwest Quarter	10
NWNWSE	Northwest Quarter of Northwest Quarter of Southeast Quarter	10
NWNWSW	Northwest Quarter of Northwest Quarter of Southwest Quarter	10
NWSE	Northwest Quarter of Southeast Quarter-Section	40
NWSENE	Northwest Quarter of Southeast Quarter of Northeast Quarter	10
NWSENW	Northwest Quarter of Southeast Quarter of Northwest Quarter	10
NWSESE	Northwest Quarter of Southeast Quarter of Southeast Quarter	10
NWSESW	Northwest Quarter of Southeast Quarter of Southwest Quarter	10
NWSW	Northwest Quarter of Southwest Quarter-Section	40
NWSWNE	Northwest Quarter of Southwest Quarter of Northeast Quarter	10
NWSWNW	Northwest Quarter of Southwest Quarter of Northwest Quarter	10
NWSWSE	Northwest Quarter of Southwest Quarter of Southeast Quarter	10
NWSWSW	Northwest Quarter of Southwest Quarter of Southwest Quarter	10
S½	South Half-Section	320
S½E½NE	South Half of East Half of Northeast Quarter-Section	40
S½E½NW	South Half of East Half of Northwest Quarter-Section	40
S½E½SE	South Half of East Half of Southeast Quarter-Section	40

Appendix B - Section Parts (Aliquot Parts)

Section Part	Description	Acres
S½E½SW	South Half of East Half of Southwest Quarter-Section	40
S½N½	South Half of North Half-Section	160
S½NE	South Half of Northeast Quarter-Section	80
S½NENE	South Half of Northeast Quarter of Northeast Quarter-Section	20
S½NENW	South Half of Northeast Quarter of Northwest Quarter-Section	20
S½NESE	South Half of Northeast Quarter of Southeast Quarter-Section	20
S½NESW	South Half of Northeast Quarter of Southwest Quarter-Section	20
S½NW	South Half of Northwest Quarter-Section	80
S½NWNE	South Half of Northwest Quarter of Northeast Quarter-Section	20
S½NWNW	South Half of Northwest Quarter of Northwest Quarter-Section	20
S½NWSE	South Half of Northwest Quarter of Southeast Quarter-Section	20
S½NWSW	South Half of Northwest Quarter of Southwest Quarter-Section	20
S½S½	South Half of South Half-Section	160
S½SE	South Half of Southeast Quarter-Section	80
S½SENE	South Half of Southeast Quarter of Northeast Quarter-Section	20
S½SENW	South Half of Southeast Quarter of Northwest Quarter-Section	20
S½SESE	South Half of Southeast Quarter of Southeast Quarter-Section	20
S½SESW	South Half of Southeast Quarter of Southwest Quarter-Section	20
S½SESW	South Half of Southeast Quarter of Southwest Quarter-Section	20
S½SW	South Half of Southwest Quarter-Section	80
S½SWNE	South Half of Southwest Quarter of Northeast Quarter-Section	20
S½SWNW	South Half of Southwest Quarter of Northwest Quarter-Section	20
S½SWSE	South Half of Southwest Quarter of Southeast Quarter-Section	20
S½SWSE	South Half of Southwest Quarter of Southeast Quarter-Section	20
S½SWSW	South Half of Southwest Quarter of Southwest Quarter-Section	20
S½W½NE	South Half of West Half of Northeast Quarter-Section	40
S½W½NW	South Half of West Half of Northwest Quarter-Section	40
S½W½SE	South Half of West Half of Southeast Quarter-Section	40
S½W½SW	South Half of West Half of Southwest Quarter-Section	40
SE	Southeast Quarter Section	160
SEN½	Southeast Quarter of North Half-Section	80
SENE	Southeast Quarter of Northeast Quarter-Section	40
SENENE	Southeast Quarter of Northeast Quarter of Northeast Quarter	10
SENENW	Southeast Quarter of Northeast Quarter of Northwest Quarter	10
SENESE	Southeast Quarter of Northeast Quarter of Southeast Quarter	10
SENESW	Southeast Quarter of Northeast Quarter of Southwest Quarter	10
SENW	Southeast Quarter of Northwest Quarter-Section	40
SENWNE	Southeast Quarter of Northwest Quarter of Northeast Quarter	10
SENWNW	Southeast Quarter of Northwest Quarter of Northwest Quarter	10
SENWSE	Souteast Quarter of Northwest Quarter of Southeast Quarter	10
SENWSW	Southeast Quarter of Northwest Quarter of Southwest Quarter	10
SESE	Southeast Quarter of Southeast Quarter-Section	40
SESENE	SoutheastQuarter of Southeast Quarter of Northeast Quarter	10
SESENW	Southeast Quarter of Southeast Quarter of Northwest Quarter	10
SESESE	Southeast Quarter of Southeast Quarter of Southeast Quarter	10
SESESW	Southeast Quarter of Southeast Quarter of Southwest Quarter	10
SESW	Southeast Quarter of Southwest Quarter-Section	40
SESWNE	Southeast Quarter of Southwest Quarter of Northeast Quarter	10
SESWNW	Southeast Quarter of Southwest Quarter of Northwest Quarter	10
SESWSE	Southeast Quarter of Southwest Quarter of Southeast Quarter	10
SESWSW	Southeast Quarter of Southwest Quarter of Southwest Quarter	10
SW	Southwest Quarter-Section	160
SWNE	Southwest Quarter of Northeast Quarter-Section	40
SWNENE	Southwest Quarter of Northeast Quarter of Northeast Quarter	10
SWNENW	Southwest Quarter of Northeast Quarter of Northwest Quarter	10
SWNESE	Southwest Quarter of Northeast Quarter of Southeast Quarter	10
SWNESW	Southwest Quarter of Northeast Quarter of Southwest Quarter	10
SWNW	Southwest Quarter of Northwest Quarter-Section	40
SWNWNE	Southwest Quarter of Northwest Quarter of Northeast Quarter	10
SWNWNW	Southwest Quarter of Northwest Quarter of Northwest Quarter	10

Family Maps of Lauderdale County, Mississippi

Section Part	Description	Acres
SWNWSE	Southwest Quarter of Northwest Quarter of Southeast Quarter	10
SWNWSW	Southwest Quarter of Northwest Quarter of Southwest Quarter	10
SWSE	Southwest Quarter of Southeast Quarter-Section	40
SWSENE	Southwest Quarter of Southeast Quarter of Northeast Quarter	10
SWSENW	Southwest Quarter of Southeast Quarter of Northwest Quarter	10
SWSESE	Southwest Quarter of Southeast Quarter of Southeast Quarter	10
SWSESW	Southwest Quarter of Southeast Quarter of Southwest Quarter	10
SWSW	Southwest Quarter of Southwest Quarter-Section	40
SWSWNE	Southwest Quarter of Southwest Quarter of Northeast Quarter	10
SWSWNW	Southwest Quarter of Southwest Quarter of Northwest Quarter	10
SWSWSE	Southwest Quarter of Southwest Quarter of Southeast Quarter	10
SWSWSW	Southwest Quarter of Southwest Quarter of Southwest Quarter	10
W½	West Half-Section	320
W½E½	West Half of East Half-Section	160
W½N½	West Half of North Half-Section (same as NW)	160
W½NE	West Half of Northeast Quarter	80
W½NENE	West Half of Northeast Quarter of Northeast Quarter-Section	20
W½NENW	West Half of Northeast Quarter of Northwest Quarter-Section	20
W½NESE	West Half of Northeast Quarter of Southeast Quarter-Section	20
W½NESW	West Half of Northeast Quarter of Southwest Quarter-Section	20
W½NW	West Half of Northwest Quarter-Section	80
W½NWNE	West Half of Northwest Quarter of Northeast Quarter-Section	20
W½NWNW	West Half of Northwest Quarter of Northwest Quarter-Section	20
W½NWSE	West Half of Northwest Quarter of Southeast Quarter-Section	20
W½NWSW	West Half of Northwest Quarter of Southwest Quarter-Section	20
W½S½	West Half of South Half-Section	160
W½SE	West Half of Southeast Quarter-Section	80
W½SENE	West Half of Southeast Quarter of Northeast Quarter-Section	20
W½SENW	West Half of Southeast Quarter of Northwest Quarter-Section	20
W½SESE	West Half of Southeast Quarter of Southeast Quarter-Section	20
W½SESW	West Half of Southeast Quarter of Southwest Quarter-Section	20
W½SW	West Half of Southwest Quarter-Section	80
W½SWNE	West Half of Southwest Quarter of Northeast Quarter-Section	20
W½SWNW	West Half of Southwest Quarter of Northwest Quarter-Section	20
W½SWSE	West Half of Southwest Quarter of Southeast Quarter-Section	20
W½SWSW	West Half of Southwest Quarter of Southwest Quarter-Section	20
W½W½	West Half of West Half-Section	160

Appendix C - Multi-Patentee Groups

The following index presents groups of people who jointly received patents in Lauderdale County, Mississippi. The Group Numbers are used in the Patent Maps and their Indexes so that you may then turn to this Appendix in order to identify all the members of the each buying group.

Group Number 1
ABERNATHY, Martha; EVANS, Martha

Group Number 2
ADAMS, Washington B; DOVE, John

Group Number 3
AGNEW, James; KILLINGSWORTH, Jesse; MYRICK, Henry

Group Number 4
AGNEW, James; WHITE, Elijah; YATES, Isaac

Group Number 5
ALEXANDER, Henry; NUTT, James M

Group Number 6
ALFORD, Julius; TUTT, Gabriel H

Group Number 7
ALLEN, John; BEESON, Jeremiah S

Group Number 8
ALLEN, John; BUSON, Jeremiah S

Group Number 9
ALLEN, John; GRIFFIN, Archibald M

Group Number 10
ANDERSON, James C; SHAMBURGER, Elijah

Group Number 11
ANDERSON, John; SHANBURGER, Elijah

Group Number 12
ANDERSON, Richard; GRANTLAND, Seaton; WALTHALL, Madison

Group Number 13
BAILIFF, John; DOWD, John K

Group Number 14
BALL, John T; DUNCAN, L A

Group Number 15
BALL, John T; DUNCAN, L A; DUNCAN, William C

Group Number 16
BALLARD, Charles; MANN, William

Group Number 17
BAREFOOT, John; TUTT, James B

Group Number 18
BARNETT, Joseph; WILSON, William J

Group Number 19
BEESON, Jeremiah S; KILLINGSWORTH, Jesse

Group Number 20
BELL, Henry; BELL, Mary

Group Number 21
BERRIEN, James W; BOTHWELL, David E

Group Number 22
BLACKWELL, Amanda P; MITCHAM, Amanda P

Group Number 23
BLACKWELL, Elemander; BLACKWELL, Wright

Group Number 24
BOWEN, Horatio; TAYLOR, Swepson

Group Number 25
BOWLING, William F; JORDAN, Henry G

Group Number 26
BOYD, Gordon D; COOPWOOD, Thomas; HENDERSON, John

Group Number 27
BOYD, Gordon D; HENDERSON, John

Group Number 28
BOZEMAN, Lula; JOHNSON, Lula

Group Number 29
BRAGG, Branch K; BROWER, Franklin P

Group Number 30
BRECK, Samuel; CRUFT, Edward; MORROW, Alexander P

Group Number 31
BRIANT, Benjamin; HOLDEN, Ebenezer

Group Number 32
BROWN, John R; COURTNEY, James

Group Number 33
BROWN, John R; HARVARD, Celia

Group Number 34
BUSH, Lewis B; WHITFIELD, Boaz

Group Number 35
BUSTIN, William; HENDRICK, Bernard G

Group Number 36
BYNUM, Benjamin; BYNUM, William; CLINTON, George W

Group Number 37
CAMPBELL, William R; MARSHALL, John R

Group Number 38
CARAWAY, Malona J; CARAWAY, William

Group Number 39
CARR, John P; SMITH, John P; BROCK, John; CLIET, John

Group Number 40
CARTER, Asa R; DEASE, Oliver C

Group Number 41
CARTER, Rachael; COATS, Rachael

Group Number 42
CASTLES, James; MOTH, Lovelace

Group Number 43
CLAYTON, Henry A; MCKINLEY, John

Group Number 44
CLINTON, George W; TOUCHSTONE, Daniel

Group Number 45
COAKER, Peyton; JOHNSON, Jacob

Group Number 46
COLE, Margaret A; COLE, Stephen

Group Number 47
COLE, Stephen M; JACKSON, William H

Group Number 48
COOK, Lewis; HOLDEN, Ebenezer; LANE, Shepherd

Group Number 49
COOPWOOD, Thomas; ESTELL, John

Group Number 50
COOPWOOD, Thomas; FOWLER, Samuel

Group Number 51
COOPWOOD, Thomas; MCCOWN, James

Group Number 52
COOPWOOD, Thomas; TUTT, James B

Group Number 53
CORLEY, Nathan; CRAWFORD, Charity

Group Number 54
COX, Henry E; SWILLEY, James J

Group Number 55
CROSBY, Rowland B; HOUZE, William J

Group Number 56
CURRIE, Jane C; CURRIE, William

Group Number 57
DALE, James; DALE, Samuel

Group Number 58
DAVIS, Edwin B; DAVIS, Thomas

Group Number 59
DAVIS, Edwin B; MONTGOMERY, John D

Group Number 60
DAVIS, Frederick A; NASH, Orsamus L

Group Number 61
DAVIS, James T; DAVIS, Mattie

Group Number 62
DAVIS, Thomas; GRIFFIN, Archibald M

Group Number 63
DEARMAN, John; SWILLEY, James J

Group Number 64
DEASON, Rebecca; DEASON, Sarah

Group Number 65
DOVE, John; LUCAS, Levi

Group Number 66
DUNCAN, L A; DUNCAN, William C

Group Number 67
EASTHAM, Crawford; ODOM, Jacob

Group Number 68
EATHAM, Crawford; ROCHEL, Etherington

Group Number 69
EDWARDS, Rigdon; GRIFFIN, Archibald M

Group Number 70
EVERETT, Samuel; HARRIS, J; DELK, S; AGLETREE, D; RICH, D E; JENNINGS, A

Group Number 71
FAIR, Eliza; MAGRUE, Eliza

Appendix C - Multi-Patentee Groups

Group Number 72
FAIRCHILDS, Jack; THOMPSON, George W

Group Number 73
FENDALL, Adeline; FENDALL, George

Group Number 74
FINLEY, Harrison M; RAGSDALE, Samuel

Group Number 75
FLETCHER, Finley; HENDERSON, William

Group Number 76
FLETCHER, William R; SPINKS, John

Group Number 77
FLEWELLEN, Thomas; GRANT, David B; HARDY, Charles; WHITE, James F

Group Number 78
GADDIS, Dicy; GADDIS, George

Group Number 79
GAINES, William M; HART, James N

Group Number 80
GAINES, William M; MCGRAW, John C

Group Number 81
GARRETT, James T; WARBINGTON, Horatio B

Group Number 82
GILBERT, Dock; GILBERT, Francis

Group Number 83
GRAHAM, Caroline; GRAHAM, Edward

Group Number 84
GRANT, Green W; HERNDON, Thomas H

Group Number 85
GRANT, Greene W; HERNDON, Thomas H

Group Number 86
GRANT, Greene W; HOUZE, William J

Group Number 87
GRANT, Greene W; REYNOLDS, John M

Group Number 88
GRAYSON, Chesterfield; MARSH, Peter

Group Number 89
GREENE, Daniel; HUBBARD, Samuel; LEWIS, Rufus G; TAPPAN, John

Group Number 90
GRIFFIN, Archibald M; HULET, Obadiah

Group Number 91
GRIFFITH, Samuel; WINNINGHAM, James H

Group Number 92
GRIGSBY, Joseph; OWENS, Willis

Group Number 93
GUNN, John; THOMPSON, David B

Group Number 94
GUNN, Margaret; LEE, Robert

Group Number 95
HALTON, William; SPINKS, Rolley

Group Number 96
HAMMOND, Jobe; RUSSELL, William

Group Number 97
HAMMONDS, Job; SCARBOROUGH, Absolam L

Group Number 98
HAMMONDS, Job; SCARBOROUGH, David

Group Number 99
HANDLEY, Mary F; ANDERSON, James M

Group Number 100
HANNA, James; MCCOWN, James

Group Number 101
HANNAH, Andrew; MCCOWN, James

Group Number 102
HARPER, Alexander; HUMPHREYS, William

Group Number 103
HARRISON, Robert; SAPP, Martha

Group Number 104
HATHORN, Nevin C; HUMPHREYS, William

Group Number 105
HATHORN, Samuel B; HUMPHREYS, William

Group Number 106
HENDERSON, Blank; WILLIAMS, Charles

Group Number 107
HENDERSON, William; NICHOLS, Noah

Group Number 108
HENDERSON, William; WILLIAMS, Joel

Group Number 109
HICKS, James M; MCCURDY, Archibald P

Group Number 110
HICKS, John B; TOOL, Eli

Group Number 111
HINES, Dover E; HINES, Narcissa

Group Number 112
HOLDEN, Ebenezer; JONES, James; WILSON, Henry

Group Number 113
HOLDEN, Ebenezer; LEE, Zachariah

Group Number 114
HOLDERNESS, Mckinney; TUTT, James B

Group Number 115
HOLLAM, Mary; MARSH, Peter

Group Number 116
HOLLAND, Mary; LEWIS, Jane; TOOL, Eli

Group Number 117
HOLLAND, Uriah; LEWIS, Stephen; TOOL, Eli

Group Number 118
HOLLIMON, Griffin H; KING, James W

Group Number 119
HOLLINSON, John; SPINKS, Rolley

Group Number 120
HOLLY, Laura; WALKER, Laura

Group Number 121
HOUSTON, Robert F; LANE, Isaac

Group Number 122
HUBBARD, Samuel; LEWIS, Moses; LEWIS, Rufus G; TAPPAN, John

Group Number 123
HUBBARD, Samuel; LEWIS, Rufus G; TAPPAN, John

Group Number 124
HUDSON, Robert; MCCOWN, James

Group Number 125
JACKSON, Emily; JACKSON, William

Group Number 126
JACKSON, Green; JACKSON, Jane

Group Number 127
JEMISON, Henry; WALLACE, James Y

Group Number 128
KILLINGSWORTH, Jesse; MCILWAIN, David

Group Number 129
KILLINGSWORTH, Jesse; MYRICK, Henry

Group Number 130
LACY, Stephen; RAGSDALE, Samuel

Group Number 131
LAMBERT, Nathan; MCLIN, William H

Group Number 132
LANG, William A; SHAW, Dougald C

Group Number 133
LARD, Adam; MARSH, Peter

Group Number 134
LEWIS, Rufus G; WHITSETT, John C

Group Number 135
LEWIS, Stephen; MARSH, Peter

Group Number 136
LLOYD, John E; MORRIS, John H

Group Number 137
MARSH, Peter; NUTT, Jonathan

Group Number 138
MARSH, Peter; RUSSELL, Isaac

Group Number 139
MARSH, Peter; TEMPLETON, George

Group Number 140
MARSH, Peter; WALKER, Ezekiel

Group Number 141
MARSH, Peter; WALKER, James

Group Number 142
MARTIN, Daniel; MITCHELL, Robert

Group Number 143
MARTIN, John P; PHILLIPS, Allinson; SPINKS, Rolley

Group Number 144
MCCOWN, James; MCPHAUL, Daniel

Group Number 145
MCCOWN, James; WHITE, William

Group Number 146
MCCRANEY, Philip; SUGGS, Joshua

Group Number 147
MCCRANEY, Roderick; SPINKS, Rolley

Group Number 148
MCDANIEL, Abram; MCDANIEL, Betsy

Group Number 149
MCDANIEL, William; RAGSDALE, Samuel

Group Number 150
MCGAHA, Tabitha; SPINKS, Rolley

Appendix C - Multi-Patentee Groups

Group Number 151
MCKINNON, William; SPINKS, Rolley

Group Number 152
MCLIN, William H; WHITE, Elijah

Group Number 153
MCSHANN, Ann; MCSHANN, Burnell

Group Number 154
MEEK, Amzi; TRUSSEL, Matthew

Group Number 155
NASH, Ezekiel; NASH, Orsamus L

Group Number 156
NASH, Orsamus L; WALKER, Robert L

Group Number 157
OBRINK, Julia A; OBRINK, William J

Group Number 158
ONEAL, William; WILKINSON, George W

Group Number 159
PARKER, Frank; PARKER, Nancy

Group Number 160
PEARL, Sylvester; TRUSSELL, John; WEIR, Robert

Group Number 161
PIGFORD, Elizabeth; WILLIAMSON, Francis; WILLIAMSON, Wiley W

Group Number 162
REECE, Rebecca; REECE, Squire

Group Number 163
RHODES, Oliver C; SPINKS, Rolley

Group Number 164
RUNNELS, Jesse; SPINKS, Rolley

Group Number 165
RUNNELS, Lewis; SPINKS, Rolley

Group Number 166
RUSSELL, David; RUSSELL, Isaac

Group Number 167
SCARBOROUGH, James R; WILSON, William J

Group Number 168
SHACKELFORD, Emaline; SHACKELFORD, William

Group Number 169
SHAMBURGER, Elijah; TERRAL, Celia

Group Number 170
SPINKS, John; TOOMEY, Pinkney H

Group Number 171
SPINKS, Rolley; TAYLOR, Richard

Group Number 172
SPINKS, Sarah; TAYLOR, William

Group Number 173
STRAIT, John; WELLS, Samuel G

Group Number 174
STRANG, Thomas; ULRIC, John G

Group Number 175
SUMMERS, Lydia; TILLMAN, Henry

Group Number 176
TAYLOR, Preston; TRUSSELL, James M

Group Number 177
TRUSSELL, James; TRUSSELL, John

Group Number 178
TRUSSELL, James; TRUSSELL, William C

Group Number 179
WARREN, Joseph M; WATTS, Haden

Group Number 180
WEBB, Harvey; WEBB, Mary E

Group Number 181
WINNINGHAM, Vandy V; WINNINGHAM, William W

Group Number 182
WITHERSPOON, Ellen; WITHERSPOON, Wat

Group Number 183
WRIGHT, Aleck; WRIGHT, Lucinda

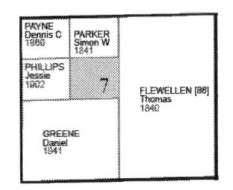

Extra! Extra! (about our Indexes)

We purposefully do not have an all-name index in the back of this volume so that our readers do not miss one of the best uses of this book: finding misspelled names among more specialized indexes.

Without repeating the text of our "How-to" chapter, we have nonetheless tried to assist our more anxious researchers by delivering a short-cut to the two county-wide Surname Indexes, the second of which will lead you to all-name indexes for each Congressional Township mapped in this volume :

> **Surname Index** (whole county, with number of parcels mapped) ....page 18
> **Surname Index** (township by township) ....just following

For your convenience, the "How To Use this Book" Chart on page 2 is repeated on the reverse of this page.

We should be releasing new titles every week for the foreseeable future. We urge you to write, fax, call, or email us any time for a current list of titles. Of course, our web-page will always have the most current information about current and upcoming books.

Arphax Publishing Co.
2210 Research Park Blvd.
Norman, Oklahoma 73069
(800) 681-5298 toll-free
(405) 366-6181 local
(405) 366-8184 fax
info@arphax.com

www.arphax.com

How to Use This Book - A Graphical Summary

Part I
"The Big Picture"

- **Map A** ▸ Counties in the State
- **Map B** ▸ Surrounding Counties
- **Map C** ▸ Congressional Townships (Map Groups) in the County
- **Map D** ▸ Cities & Towns in the County
- **Map E** ▸ Cemeteries in the County
- **Surnames in the County** ▸ Number of Land-Parcels for Each Surname
- **Surname/Township Index** ▸ Directs you to Township Map Groups in Part II

The <u>Surname/Township Index</u> can direct you to any number of **Township Map Groups**

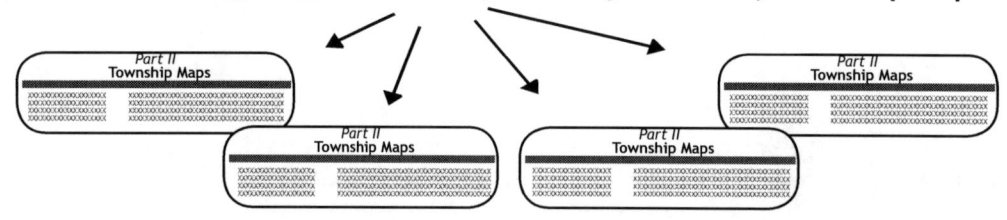

Part II
Township Map Groups
(1 for each Township in the County)

Each Township Map Group contains all four of of the following tools . . .

- **Land Patent Index** ▸ Every-name Index of Patents Mapped in this Township
- **Land Patent Map** ▸ Map of Patents as listed in above Index
- **Road Map** ▸ Map of Roads, City-centers, and Cemeteries in the Township
- **Historical Map** ▸ Map of Railroads, Lakes, Rivers, Creeks, City-Centers, and Cemeteries

Appendices

- **Appendix A** ▸ Congressional Authority enabling Patents within our Maps
- **Appendix B** ▸ Section-Parts / Aliquot Parts (a comprehensive list)
- **Appendix C** ▸ Multi-patentee Groups (Individuals within Buying Groups)

Made in the USA
Charleston, SC
06 February 2016